Design Methodology and Relationships with Science

NATO ASI Series

Advanced Science Institutes Series

A Series presenting the results of activities sponsored by the NATO Science Committee, which aims at the dissemination of advanced scientific and technological knowledge, with a view to strengthening links between scientific communities.

The Series is published by an international board of publishers in conjunction with the NATO Scientific Affairs Division

A Life Sciences **B Physics**	Plenum Publishing Corporation London and New York
C Mathematical and Physical Sciences **D Behavioural and Social Sciences** **E Applied Sciences**	Kluwer Academic Publishers Dordrecht, Boston and London
F Computer and Systems Sciences **G Ecological Sciences** **H Cell Biology** **I Global Environmental Change**	Springer-Verlag Berlin, Heidelberg, New York, London, Paris and Tokyo

NATO-PCO-DATA BASE

The electronic index to the NATO ASI Series provides full bibliographical references (with keywords and/or abstracts) to more than 30000 contributions from international scientists published in all sections of the NATO ASI Series.
Access to the NATO-PCO-DATA BASE is possible in two ways:

– via online FILE 128 (NATO-PCO-DATA BASE) hosted by ESRIN, Via Galileo Galilei, I-00044 Frascati, Italy.

– via CD-ROM "NATO-PCO-DATA BASE" with user-friendly retrieval software in English, French and German (© WTV GmbH and DATAWARE Technologies Inc. 1989).

The CD-ROM can be ordered through any member of the Board of Publishers or through NATO-PCO, Overijse, Belgium.

Series D: Behavioural and Social Sciences - Vol. 71

Design Methodology and Relationships with Science

edited by

M. J. de Vries

Eindhoven University of Technology,
Faculty of Philosophy and Social Sciences,
Eindhoven, The Netherlands

N. Cross

Open University,
Faculty of Technology,
Milton Keynes, U.K.

and

D. P. Grant

Architecture Department,
California Polytechnic University,
San Luis Obispo, California, U.S.A.

Springer-Science+Business Media, B.V.

Proceedings of the NATO Advanced Research Workshop on
Design Methodology and Relationships with Science
Eindhoven, The Netherlands
September 29 – October 2, 1992

ISBN 978-90-481-4252-1 ISBN 978-94-015-8220-9 (eBook)
DOI 10.1007/978-94-015-8220-9

Printed on acid-free paper

TABLE OF CONTENTS

PREFACE

Many business corporations are faced with the challenge of bringing together quite different types of knowledge in design processes: knowledge of different disciplines in the natural and engineering sciences, knowledge of markets and market trends, knowledge of political and juridical affairs. This also means a challenge for design methodology as the academic discipline that studies design processes and methods. The aim of the NATO ARW of which this book is the report was to bring together colleagues from different academic fields to discuss this increasing multidisciplinarity in the relationship between design and sciences.

This multidisciplinarity made the conference a special event. At a certain moment one of the participants exclaimed: "This is not a traditional design methodology conference!" Throughout the conference it was evident that there was a need to develop a common language and understanding to enable the exchange of different perspectives on design and its relationship with science. The contributions that have been included in this book show these different perspectives: the philosophical, the historical, the engineering perspective and the practical designer's experience.

We as the organising committee of this NATO ARW thank all presenters for their contributions to the programme, the rapporteurs for taking notes of the discussions, the NATO Scientific and Environmental Affairs Division and its ARW Programme Director, prof. L. Sertorio, for making this conference possible financially and finally Kluwer for publishing this report. We hope this book will be a valuable resource for many colleagues who are involved in design methodological research and development work.

December 1992

Marc J. de Vries
Nigel Cross
Donald P. Grant

DESIGN METHODOLOGY AND RELATIONSHIPS WITH SCIENCE: INTRODUCTION

M.J. DE VRIES
Eindhoven University of Technology
the Netherlands

1. Relevance of the theme

In many business corporations there is a clear tendency towards increasingly multidisciplinary design activities. A striking example of this is the emerging field of mechatronics, that combines elements of electrical and mechanical engineering and information technology[1]. Mechatronics design is therefore carried out by teams, that consist of experts in different engineering disciplines that have to bring together their individual knowledge to integrate them in the design process.

Multidisciplinarity is, however, not confined to combining different engineering disciplines. More and more business corporation management has become aware of the need to ensure that all further steps in the life of a product are taken account of in the design of this product[2]. Too often it happened that designs were sent back by production departments because the design was not makable. Too often it happened that products did not sell because the desires of the customers had not been asked for in the design phase. Likewise products were difficult to maintain and repair, difficult to assembly, difficult to recycle, because the designers had confined themselves to the technical functioning of the product they were developing. For this reason corporations like DAF and Philips (just to mention two corporations that are located in Eindhoven) are working towards structures that allow influence from e.g. production workers, marketing experts, service engineering, on the design process. Methods like quality function deployment (QFD), value analysis (VA), and failure mode and effect analysis (FMEA) relate to these efforts.

The above trends require new knowledge about the way scientific knowledge should be integrated with other types of knowledge in the increasingly

M. J. de Vries et al. (eds.), Design Methodology and Relationships with Science, 1–14.

multidisciplinary and complex design process. The theme of the NATO Advanced Research Workshop of which this publication is the report fits quite well in these trends. The aim of the workshop is to bring together experts from various fields in design methodology to discuss the frontline research developments with respect to the interrelationship between science and design.

2. Two cultures[3] in design methodology

The theme of the conference suggests at least two groups of experts that can contribute to the meeting: design methodologists and scientists. It seems, however, that these two groups do not communicate as often as could be expected when discussing design issues[4]. This phenomenon can at least partially be explained by looking at the history of design methodology.

Cross identifies four themes in this relative short history, that he labels with four words that typify the activities in those themes: prescription, description, observation and reflection[5].

In the first three themes most contributions were made by architects, mechanical engineers, (cognitive) psychologists and information technologists. Attention focused on the role of flowchart representations for design process (often according to the analysis-synthesis-evaluation assumption) and the extent to which experienced and beginning designers follow the steps in these flowcharts. Computer models were made to represent the thinking modes that were found with designers and from this possibilities for computer assistance to designers were developed. One could think this first 'culture' centred around an international journal like Design Studies. In general the colleagues, who contribute to this journal all are more or less directly involved with design practice.

It is only to the theme of reflection that more philosophical aspects became a major issue of consideration in design methodology. But then already another 'culture' had been established, that focused on aspects of design. Again one can think this second 'culture' centred around an international journal: Technology and Culture. In his analysis of the content of the articles in the first 20 volumes of this journal, Staudinger showed that a considerable amount of articles deal with design issues[6]. Although these articles therefore can be reckoned to be located in the scientific investigation of design and design processes, the title 'design methodology' is never used to characterise these contributions. In this group we find historians, philosophers, scientists, sociologists. Most of these people are not directly involved in design practice.

The first group tends to be more interested in the work of the individual designer or group of designers, while the second group often puts more emphasis on the scientific and social environment in which the design activities take place.

In the NATO Advanced Research Workshop input from both 'cultures' has been asked for. A possible cross fertilisation between the two groups was one of the aims of the conference.

3. Issues in the science-design relationship

3.1. EVOLUTION OF THE SCIENCE-DESIGN RELATIONSHIP

The role of science in design has changed in the course of history[7]. In general in can be stated that originally all scientific knowledge was closely related to direct experience and that gradually different kinds of scientific knowledge have emerged that are more fundamental and abstract and thus at greater distance from everyday life experience. Scientific knowledge becomes more concerned with the invisible (objects at a microlevel) and therefore it is not surprising that it is sometimes stated that science only becomes visible through design[8].

Nowadays we find a strong interdependence between science and design, that confirms the citation we find in the historical account bij Böhme a.o.[9]: 'scientia sine arte nihil est', but at the same time 'ars sine scientia nihil est'.

3.2. SCIENTIFIC KNOWLEDGE IN DESIGN

In earlier years of design methodology is was often assumed that it was possible to develop a model for ideal design processes independent of the nature of the design problem and the kind of knowledge that was used. Nowadays the impression is that this independence is a false assumption. This impression is confirmed when we look at the differences between the various engineering disciplines in their interest in design methodology. It is striking, that most contributions to a journal like Design Studies are from architects, building engineers, and mechanical engineers. Contributions from electrical or chemical engineering are rather exceptional[10]. This bias seems less evident in the Technology and Culture contributions, but as we have stated above, here the focus is not so much on the design process itself as well as on the scientific and/or social environment in which this process takes place. Apparently some engineering disciplines see more relevance in attention for the design process 'as such' than others. Even between the disciplines that have shown interest in the design process 'as such' we see differences in approach, for example between architecture and mechanical engineering[11].

Sarlemijn distinguishes three types of technologies, according to the types of scientific knowledge they use: experience technology, macro-technology and micro-technology, each of which has a different role in design[12]. In particular the micro-technologies are very indirectly related to the design reality and as Layton has pointed out, in such cases there is a need for a translation from the abstract theoretical models to the concrete design reality[13].

Vincenti has proposed a number of categories of design knowledge, based on his studies of design in aeronautics: fundamental design concepts, criteria/specifications, theoretical tools, quantitative data, practical considerations, and design instrumentalities. These various types of knowledge are acquired by different types of knowledge generating activities[14]. In the last part of his survey, Vincenti proposes a variation-selection model for the use of scientific knowledge in design that in fact is not much different from Kroes' suggestion that scientific knowledge is used in a heuristical way[15] and Batty's model of prediction as a bridge between science and design[16].

A special area of investigation is the representation of scientific knowledge with the computer as a tool for the designer (e.g. in CAD-systems, Artificial Intelligence)[17].

3.3. SCIENTIFIC METHODS IN DESIGN

The so-called scientification of technology[18] (and design) does not only refer to the increasing use of scientific knowledge in technology, but at least as much to the increasing influence of scientific methods in technology. Methods like experimental variation and measurement, quantification, mathematical description[19], modelling and abstraction[20] have become increasingy important in technology. This has caused a still ongoing discussion on the potentials and limitations, the desirability and avoidability of the scientific approach to design. The opinions range from a strong believe in the succes of using scientific methods in design to the feeling that a scientific approach to design is incompatible with the dominant role of free creativity in design.

The latter opinion is defended by asking attention for the differences in reasoning patterns between science and design[21]. As March has stated, design has its own type of logic (March uses the term 'productive reasoning' or 'abduction'[22]). Skolimowski has characterised the different natures of science and design as investigation versus creation[23]. Cross et al. describe the 'epistemological chaos' in science and from this conclude that the methodology of science is not a good basis for developing a methodology of design[24]. Cross also stated that designers have their own 'designerly ways of knowing', that are fundamentally different from scientific ways of knowing[25].

Proponents of the first opinion, that defends the use of scientific methods in design, bring forward, that the epistemological chaos is an illusion and that in reality there is a natural evolution from one philosophy of science to the next[26]. In particular Popper's 'conjecture and refutation' approach has been mentioned as one that is not much different from the way designers work[27]. Others have shown that the differences in nature between science and technology are not always as sharp as suggested by Skolimoski and others. In a certain way science can, as technology, be characterised as 'creating': scientific theories are the products of human creativity[28].

Some authors suggest that a more scientific approach to design can be recognised by a more structuredness of the design process[29]. Especially in the German and Swiss tradition (in particular in the area of mechanical design) we find a highly structured and analytical approach[30]. A quite different approach is the praxiological, that Gasparksi has applied to the field of design[31].

The problem for those who have to manage the design process, is to find a balance between structuredness and opportunities for creativity. The balance seems to shift from more openness to more structuredness as we move from the very early stages of the evolution of the product (when it is still in the phase of fundamental research in the industrial research laboratory) to the more concrete phases of pre-development and development (or product creation process, as this phase is often entitled with corporations like Philips).

3.4. DESIGN METHODOLOGY AS A PART OF THE METHODOLOGY OF TECHNOLOGICAL SCIENCES

Finally I want to mention the mere existence of technological (engineering) sciences as a result of the increasing influence of science on technology and design[32].

Design methodology is a scientific discipline and therefore it is not surprising that some authors have tried to derive approaches to the study of design from philosophies of science. For example, Kuhn's concept of paradigmatic revolutions have been used by, Addis and Ropohl to characterise dramatic changes in design philosophies[33]; Hughes and Bijker have applied the concept of social construction to the design of technical artifacts[34]; Usher has taken the Gestalt theories from psychology to develop a description of the design processes of mechanical devices[35]. Although the validity of these uses of theories can be debated, they illustrate the interest to try to gain from what has been reached in other scientific areas.

4. An agenda for future research and development[36]

In his opening presentation, dr. S.L. Marzano from Philips Industrial Corporate Design, set the scene for the discussions on the state-of-the-art of design methodology and the need for future research. He emphasised the need for reflection on the design process that responds to the increasing complexity of the design process. This increase is not only caused by the growing complexity of scientific knowledge, but also because of the emerging awareness, that customer needs and values should more be taken into account. New values, such as self-realisation and environmental concerns should be integrated with the more traditional and scientific skills in the design process. This implies the need for a new qualities and an open-mindedness with designers and the search for new pathways.

Discussions on the state-of-the-art of design methodology revealed that so far more attention has been paid to knowledge and method aspects than to the aspects of human values and needs. In the final session of the conference this was stated by Levy in the terms epistemology, methodology, ontology and teleology (see figure 1), that had been derived from Simon[37].

Among the participants there was serious concern about the lack of success that had been perceived in design methodology so far in implementing 'scientific' design methods. Too often high expectations of simplified models of the design process have been frustrated by the complexity of praxis. This perceived lack of evident success underlined the need to go back to the underlying philosophical assumptions and confront the various approaches (positivist, phenomenological, constructivist, praxiological, etc) with each other. This could lead to a broadening of the body of knowledge for design methodology and thereby the strengthening of design methodology as an academic discipline that both practitioners and other members of the academic community would have confidence in (in the discussions the metaphor of patients having confidence in doctors because of the body of academic knowledge they draw from, was used to illustrate this). Then the four areas of the 'philosophy of design' (epistemology, methodology, ontology and teleology) should to be developed in a more balanced way.

This also implies a shift from studying methods to studying 'beings': both designers (how do they think and behave) and customers (what are their needs and how do these develop). Much can be learned from historical and contemporary case studies. The study of designed objects might lead to new design knowledge when they are seen as extension of human beings, including their values, but so far design methodology has not proceeded very far in such ways of analysing objects as the result of design processes. Another field that was identified in discussions as one that still needs further exploration is the representation of design knowledge and concept generation in computer applications.

When 'integration' can be regarded as the first keyword for future research, 'differentiation' can be seen as the second one. In the past years, doubts about the generality of design concepts and methods have grown considerably. Design processes differ largely between various domains and there is a clear need to investigate what is domain specific and what can be generalised with respect to design concepts and methods. Results of such studies will be useful when considering the possible role of the computer as a tool for designers, which will be different for different types of design problems.

5. Implications for design eduction

In design education the same two keywords, mentioned in the previous section, should play a vital role: integration of various types of skills and knowledge on the one hand and on the other hand the differentiation between various domains of design problems, that ask for different approaches. These skills and knowledge should be learned in an evolutionary way, without trying to jump over phases of learning. A mechanistic approach with rigorous use of simplified models for design processes should be avoided. A more constructivist approach, in which pupils' own concepts are taken as a starting point for teaching, has already proved to be valuable in science education and might be valuable too in design education.

5.1. Design education as a part of the vocation training in traditional engineering disciplines

In the education of engineers, design of course plays a vital role, but surprisingly, explicit attention for design skills is seldom found in syllabi. Only in mechanical engineering and in construction/building engineering students are confronted with subjects in which the design process is dealt with as a separate issue of study. Perhaps this is related to the fact, that design methodology was developed mainly in the context of these disciplines. A broadening of research topics to areas of e.g. electrical engineering, chemical engineering, bio-engineering, and research into the particular characteristics of design processes in these areas would stimulate more explicit attention to design skills in the engineering education programmes of those disciplines.

5.2. Design education in STS programmes

In the past decades, a number of new programmes for future engineers were developed, that were aimed at educating engineers with a broader scope than the traditional engineers (this of course at the same time results in a less indepth technical knowledge and skills repertoire). This broadening includes: knowledge of economic, political, juridical, managerial and ethical aspects of technological developments, skills in communicating, and social research skills. A title that is often used for such programmes is: Science, Technology and Society (STS). In some cases, like in Eindhoven, design methodology is part of such a programme. The results of design methodology researches certainly could be used as a basis for the further development of the design discipline within STS programmes.

5.3. Design education in the context of technology education as a part of general education

Learning about technological developments is more and more seen as a necessary component of the education of all citizens to enable them to live in, cope with, and even have a certain level of control over technology in their world. In most countries worldwide, technology education is part of the curriculum in general education, both at the primary and at the secondary level.

In most cases the subject emerged from a crafts tradition[38]. A recent trend in technology education is the increasing attention to design and stimulating innovative thinking[39]. In most cases, however, the naive use of flowchart representations for the design process can still be found in practical examples of technology education programmes, in which design has been included. More recent design methodological knowledge has not yet influenced these programmes. This should be a concern of both technology educators and design methodologists, because frustration in design activities would certainly not stimulate pupils to opt for careers in which design is a major part of their work. A balanced and realistic view of the practice of design in industrial settings would probably be an important contribution to a positive attitude towards technological professions. And a well prepared future workforce is recognised as a prerogative for the succes of business corporations in technological innovations.

6. Structure of the conference report

The conference report reflects the structure of the programme of the NATO ARW.

Theme 1 is an introductory theme, that shows the emergence of the science-design relationship as an inssue of study in design methodology. In his contribution, Cross pictures the history of design methodology as a discipline, focusing on five themes of study: the development of design methods, the management of design processes, the structure of design problems, the nature of design activities and the philosophy of design methods.

In theme 2 a number of case studies into the science-design relationship are presented. Le Guet Tully discusses the design of locks and the work of Henri Chrétien in the design of mechanical devices. Cases from architecture are dealt with in the paper of Grant. A third area, presented in Spiller's paper is the conceptual design of structures. These cases provide empirical material for a more general methodological discussion in the next theme.

Theme 3 contains contributions, that aim at deriving more overall methodological conclusions from various case studies. Sarlemijn focuses on the different types of scientific knowledges that should be distinguished when discussing the science-design relationship. Eder sees the science-design relationship as one of the components of a 'science of design'. Gasparski explains the praxiological approach as a general methodological approach to design. Bayazit in her contribution offers an epistemological view on design and studies knowledge representation with computers. Eckersley uses a classification by Brown and Chandrasekaran to illustrate the different roles of science in different design problems and shows an example of a computer programme for generating variants. Buchanan focusses on the need to clarify philosophical concepts and assumptions in design research.

In theme 4 the outcomes of design methodological considerations are used in educational issues: how do we educate engineers in such a way that their design skills are improved, and in what way do we educate all young people in such a way that they can appreciate some of the characteristics of design as part of their general education. Blandow discusses the more theoretical aspects of this theme, while McCormick deals with the more practical aspects.

References

Addis, W. (1990), 'The Evolution of Structural Engineering Design Procedures. A History for that Skill Called Design', Transactions of the Newcomen Society 61, 51-64.

Barius, B. (1991), 'Time to market. The holistics of the innovation process from idea to customer', in (eds.), International Conference on Engineering Design ICED '91, pp. 194-197.

Batty, M. (1980), 'Limits to prediction in science and design science', Design Studies 1, 153-159.

Blandow, D. (1992), 'A Modular Conceptual Framework for Technology and Work', Invited Lecture National Conference of the National Association for Science, Technology & Society, Washington, Feb. 1-3, 1991.

Böhme, G., Dale, W. van den and Krohn, W. (1978), 'The 'scientification' of technology', in Krohn, W., Layton, E.T. and Weingart, P. (eds.), The dynamics of science and technology. D. Reidel Publ. Comp., Dordrecht/Boston, pp. 219-250.

Buur, J. and Andreasen, M.M. (1989), 'Design models in mechatronic product development', Design Studies 10, 155-162.

Bijker, W.E., Hughes, T.P. and Pinch, T.J. (eds.) (1987), The Social Construction of Technological Systems. MIT Press, Cambridge, Massachusetts.

Christiaans, H.C.M. and Venselaar, K. (1990), 'Knowledge as a basis for creativity in design', in Hubka, V. and Kostelic, A. (eds.), International Conference on Engineering Design ICED '90, pp. 68-75.

Cross, N., Naughton, J. and Walker, D (1981), 'Design method and scientific method', Design Studies 2, pp. 195-201.

Cross, N. (1982i), 'Designerly ways of knowing', Design Studies 3, 221-227.

Cross, N. (ed.) (1982ii), Developments in Design Methods, John Wiley & Sons, Chichester.

Dixon, J.R. (1989), 'On research methodology towards a scientific theory of engineering design', in Newsome, S.L., Spillers, W.R. and Finger, S. (eds.), Design Theory '88, Springer Verlag, New York/Berlin/Heidelberg, pp. 316-337.

Freeman, C. (1986), 'Succesful industrial innovation', in Roy, R. and Wield, D. (eds.) Product design and technological innovation, Open University, Milton Keynes, pp. 29-33.

Gasparski, W.W. (1989), 'Design Methodology: A Personal Statement', in Durbin, P.T. (ed.), Philosophy of Technology, Kluwer Academic Publishers, Dordrecht, pp. 153-167.

Glynn, S. (1985), 'Science and perception as design', Design Studies 6, pp. 122-126.

Gorokhov, V. (1992), 'Engineering Activity, Designing and Technical Knowledge', History and Technology 8, 107-112.

Hoover, S.P., Rinderle, J.R. and Finger, S. (1991), 'Models and Abstractions in Design, in International Conference on Engineering Design ICED '91, pp. 46-57.

Hubka, V. and Eder, W.E. (1987), 'A scientific approach to engineering design', Design Studies 8, 123-137.

Kroes, P. (1991), 'Technology and science-based heuristics', in Pitt, J.C. and Lugo, E. (eds.), The Technology of Discovery and The Discovery of Technology, The Society for Philosophy and Technology, Blacksburg, VA.

Layton, E.T. (1974), 'Technology as knowledge', Technology & Culture 15, 31-41.

Levy, R. (1985), 'Science, technology and design', Design Studies 6, pp. 66-72.

Lewin, D. (1979), 'On the place of design in engineering', Design Studies 1, 113-117.

March. L. (1976), 'The logic of design and the question of value', in March, L. (ed.), The architecture of form. Cambridge University Press, Cambridge.

McCormick, R. (1991), 'The evolution of current practice in technology education', in: Hacker, M., Gordon, A.T. and Vries, M.J. de (eds.), Integrating advanced technologies into technology education. NATO ASI Series, Vol. F-78. Springer Verlag, Berlin-Heidelberg-New York.

Metcalfe, J.S. and Boden, M. (1992), 'Evolutionary epistemology and the nature of technology strategy', in Coombs, R., Saviotti, P. and Walsh, V. (eds.), Technological Change and Company Strategies, Academic Press Ltd, London.

Newsome, S.L. and Spillers, W.R. (1989), 'Tools for Expert Designers: Supporting Conceptual Design', in Newsome, S.L., Spillers, W.R. and Finger, S. (eds.), Design Theory '88, Springer Verlag, New York/Berlin/Heidelberg, pp. 49-55.

Pahl, G. and Beitz, W. (1984), Engineering design: a systematic approach (transl. K. Wallace). Design Council, London.

Pighini, U. Franscesco, G., Yuan, D.Z., Schettino, A.V. and Rivalta, A. (1983), The determination of optimal dimensions for a city car using methodical design with prior technology analysis', Design Studies 4, pp. 233-243.

Pugh, S. (1990), Total design; integrated methods for succesful product engineering. Addison Wesley, Wokingham.

Roozenburg, N. and Cross, N. (1991), 'Models of the design process - integrating across the disciplines', in International Conference on Engineering Design ICED '91, pp. 186-193.

Ropohl, G. (1991), 'System und Methode. Die neue Philosophie im technischen Handeln', in International Conference on Engineering Design ICED '91, pp. 209-215.
Sarlemijn, A. and Vries, M.J. de (1992), 'The piecemeal rationality of application oriented research. An analysis of the R&D history leading to the invention of the Plumbicon in the Philips Research Laboratories', in Kroes, P.A. and Bakker, M. (eds.), Technological Development and science in the industrial age. New perspectives on the science-technology relationship. Reidel, Dordrecht.
Simon, H. (1969), The Sciences of the Artificial, MIT Press, Cambridge, MA.
Skolimiwski, H. (1966), 'The Structure of Thinking in Technology', Technology & Culture 7, pp. 371-383.
Smith, E.L. and Gregory, S.A. (1983), 'A chemical engineering design view of biotechnology', Design Studies 4, 208-214.
Souder, W.E. (1987), Managing New Product Innovations, D.C. Heath and Comp., Lexington, Massachusetts.
Staudenmaier, J.M. (1985), Technology's Storytellers. Reweaving the Human Fabric, MIT Press, Cambridge, Massachusetts.
Steadman, P. (1979), 'The History and Science of the Artificial', Design Studies 1, pp. 49-58.
Stork, H. (1977), Einführung in die Philosophie der Technik, Wissenschaftliche Buchgesellschaft, Darmstadt.
Tayefeh-Emamverdi, M. (1982), 'Methodical concept development in electro-technology', Design Studies 3, 235-242.
Usher, A.P. (1954), A History of Mechanical Inventions, Harvard University Press, Cambridge, Massachusetts.
Vincenti, W.G. (1990), What engineers know and how they know it: analytical studies from aeronautical history, John Hopkins University Press, Baltimore, Maryland.
Vries, M.J. de (1992), 'Approaches to technology education and the role of advanced technologies: an international orientation', in: A. Gordon (ed.) Advanced Educational Technology in Technology Education. NATO ASI Series. Springer, Berlin-Heidelberg-New York.
Willem, R.A. (1990), 'Design and science', Design Studies 11, 43-47.

Notes

1.Buur 1989.

2.Freeman 1986, Barius 1991, Pugh 1990.

3.A term that has been borrowed from C.P. Snow.

4.Here I disagree with J.R. Dixon (1989), who stated that there can be recognized one 'single goal-directed research community'.

5.Cross (1982ii).

6.Staudenmaier 1985.

7.Böhme a.o. 1978.

8.Willem 1990.

9.Böhme a.o. 1978.

10.Smith and Gregory 1983, and Tayefeh-Emamverdi 1982 are two examples.

11.Roozenburg and Cross 1991 see two types of models for design processes: consensus model versus a type model.

12.In Sarlemijn and De Vries 1992 this characterisation is explained.

13.Layton 1974.

14.Vincenti 1990.

15.Kroes 1991.

16.Batty 1980.

17.Newsome and Spillers 1989.

18.Stork 1977.

19.Böhme a.o. 1978.

20.Hoover a.o. 1991.

21.Steadman 1979.

22. March 1976.

23.Skolimowski 1966.

24.Cross a.o. 1981.

25.Cross 1982i.

26.Levy 1985.

27.Lewin 1979.

28.Glynn 1985.

29.For example Pighini a.o. 1983 mention 'scientific design' as a sometimes mentioned equivalent of 'methodical design'.

30.Pahl and Beitz 1984, Hubka/Eder 1987.

31.Gasparski 1989.

32.Here language is problematic. In German, French and Dutch (just to mention three examples of languages) a distinction can be made between 'Technik' and 'Technologie', 'technique' and 'technologie', 'techniek' and 'technologie'. English does not offer this opportunity to distinguish between the practical activities that result in products, and the scientific and systematic study of these activities. This distinction, however, is necessary for understanding the present state of affairs in engineering. Therefore I chose the term 'technological sciences' as distinguished from 'technology'.

33.Addis 1990, Ropohl 1991. The transferability of Kuhn's revolutionary changes of paradigms to technological developments has been questioned by a.o. Metcalfe and Boden 1992.

34.Bijker and Hughes 1987.

35.Usher 1954.

36.This part has been written with the discussion outcomes as a basis. I want to thank the rapporteurs for taking notes during discussions: S. Kasse, R. Levy (with special thanks for his role in the final session), J. Schlattmann, N. Roozenburg, H.P. Hildre, and J. Heinen.

37.Simon 1969.

38.McCormick 1991 and De Vries 1992.

39.Blandow 1992.

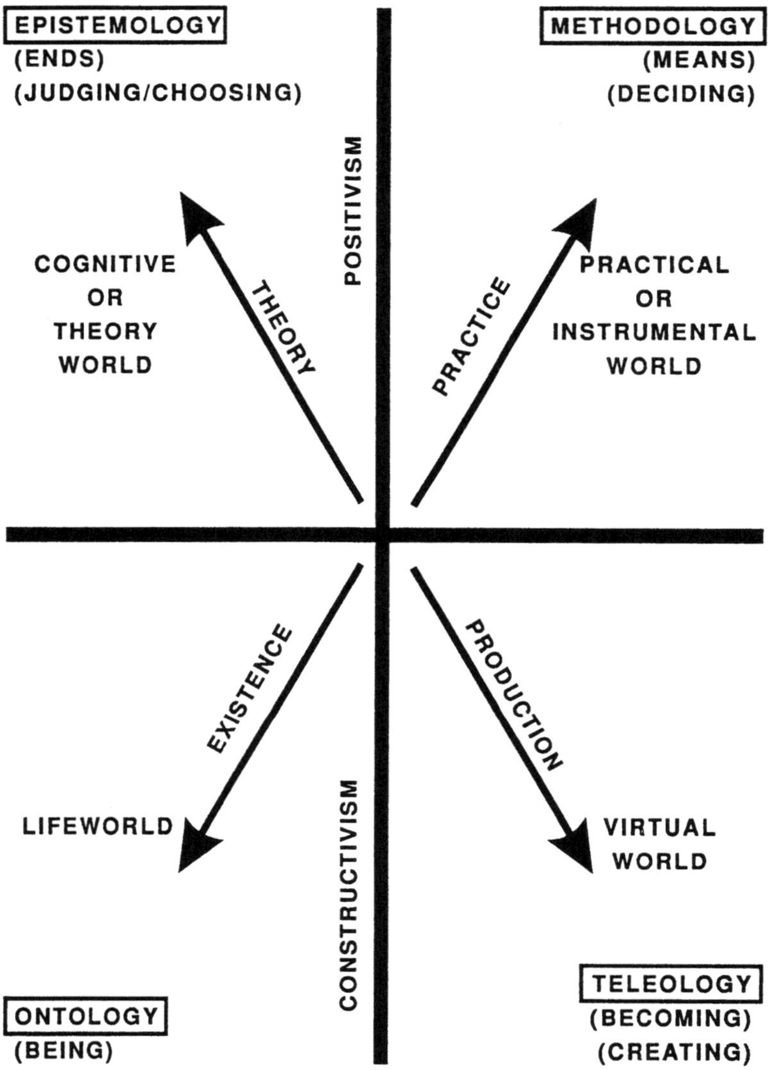

Figure 1. Analytical scheme for state-of-the-art of design methodology

A HISTORY OF DESIGN METHODOLOGY

N. CROSS
The Open University
United Kingdom
and
University of Delft
the Netherlands

ABSTRACT. This Workshop marks the thirtieth anniversary of the event which is normally regarded as the birth of modern design methodology and the design methods movement - the Conference on Design Methods held in London in 1962. The movement almost died in the 1970s, but seems now to have hung on to life and to have re-emerged and grown with some vigour in the last decade. This paper reviews this relatively short history of design methodology, maps out some of the major themes that have sustained it, and tries to establish some agreed understanding for the concepts of scientific design, design science and the science of design.

1. Introduction

This Workshop happens to coincide with the thirtieth anniversary of the *Conference on Design Methods*, held in London in September, 1962 (Jones and Thornley, 1963). This conference is generally regarded as the event which marked the launch of design methodology as a subject or field of enquiry. Of course, the field was based on some earlier work (the earliest reference in Design Methodology literature is probably Zwicky's 'Morphological Method' published in 1948 (Zwicky, 1948)), but the 1962 Conference was the first time that 'design methods' received substantial academic recognition.

So the history of design methodology is still rather a brief one. Some previous 'history' reviews have been by Broadbent (1979) and Cross (1980, 1984). In 1986, The Design Methods Group celebrated its twentieth anniversary with some special reviews, in its journal, *Design Methods and Theories*.

M. J. de Vries et al. (eds.), Design Methodology and Relationships with Science, 15–27.

2. A Brief Overview

The origins of the emergence of new design methods in the 1950s and 60s lay in the application of novel, 'scientific' methods to the novel and pressing problems of the 2nd World War - from which came OR and management decision-making techniques - and in the development of creativity techniques in the 1950s. (The latter was partly, in the USA, in response to the launch of the first satellite, the Soviet Union's 'Sputnik', which seemed to convince American scientists and engineers that they lacked creativity.)

The new 'Design Methods Movement' developed through a series of conferences in the 1960s and 70s - London, 1962 (Jones and Thornley, 1963); Birmingham, 1965 (Gregory, 1966); Portsmouth, 1967 (Broadbent and Ward, 1969); Cambridge, Mass., 1969 (Moore, 1970); London, 1973; New York, 1974 (Spillers, 1974); Berkeley, Cal., 1975, Portsmouth again in 1976 (Evans, Powell et al., 1982) and again in 1980 (Jacques and Powell, 1981) (notably, this latter conference had a similar theme - 'Design:Science:Method' - to that of this Workshop).

The first design methods or methodology books also appeared in this period - Hall (1962), Asimow (1962), Alexander (1964), Archer (1965), Jones (1970), Broadbent (1973) - and the first creativity books - Gordon (1961), Osborn (1963).

However, the 1970s also became notable for the rejection of design methodology by the early pioneers. Christopher Alexander said: 'I've disassociated myself from the field... There is so little in what is called "design methods" that has anything useful to say about how to design buildings that I never even read the literature anymore... I would say forget it, forget the whole thing... If you call it "It's A Good Idea To Do", I like it very much; if you call it "A Method", I like it but I'm beginning to get turned off; if you call it "A Methodology", I just don't want to talk about it.' (Alexander, 1971) And J. Christopher Jones said: 'In the 1970s I reacted against design methods. I dislike the machine language, the behaviourism, the continual attempt to fix the whole of life into a logical framework.' (Jones, 1977)

These were pretty harsh things for the founding fathers to say about their offspring, and were potentially devastating to those who were still nurturing the infant. To put the quotations of Alexander and Jones into context it may be necessary to recall the social/cultural climate of the late-1960s - the campus revolutions, the new liberal humanism and rejection of previous values. But also it had to be acknowledged (and it was) that there had been a lack of success in the application of 'scientific' methods to design. Fundamental issues were also raised by Rittel and Webber (1973), who characterised design and planning problems as 'wicked' problems, fundamentally un-amenable to the techniques of science and engineering, which dealt with 'tame' problems.

Design methodology was temporarily saved, however, by Rittel's (1973) brilliant proposal of 'generations' of methods. He suggested that the developments of the 1960s had been only 'first generation' methods (which naturally, with hindsight, seemed a bit simplistic, but nonetheless had been a necessary beginning) and that a new second generation was beginning to emerge. This suggestion was brilliant because it let the new methodologists escape from their commitment to inadequate 'first generation' methods, and it opened a vista of an endless future of generation upon generation of new methods.

We might wonder what has happened to Rittel's theory of 'generations'. The first generation (of the 1960s) was based on the application of systematic, rational, 'scientific' methods. The second generation (of the early 1970s) moved away from attempts to optimize and from the omnipotence of the designer (especially for 'wicked problems'), towards recognition of satisfactory or appropriate solution-types (Simon (1969) had introduced the notion of 'satisficing') and an 'argumentative', participatory process in which designers are partners with the problem 'owners' (clients, customers, users, the community). However, this approach tends to be more relevant to architecture and planning than engineering and industrial design, and meanwhile these fields were still developing their methodologies in somewhat different directions.

Engineering design methodology developed strongly in the 1980s; for example, through ICED - the series of International Conferences on Engineering Design - and the work of the VDI - Verein Deutscher Ingenieure. These developments were especially strong in Europe and Japan (Hongo and Nakajima, 1991), if not in the USA. (Although there may still have been limited evidence of practical applications and results.) A series of books on engineering design methods and methodology began to appear. Just to mention some English-language ones, these included Hubka (1982), Pahl and Beitz (1984), French (1985), Cross (1989), Pugh (1991). It should also be acknowledged that in the USA there were some important conferences on design theory, and the National Science Foundation initiative on design theory and methods (perhaps in response to German and Japanese progress - like the earlier response to Sputnik?) led to substantial growth in engineering design methodology in the late-1980s. ASME, the American Society of Mechanical Engineers launched a series of conferences on *Design Theory and Methodology*, the most recent being in Miami, Florida (Stauffer, 1991).

So the development of 'second generations' of design methodology in architecture and engineering appeared to diverge from each other in the 1970s and 80s. Roozenburg and Cross (1991) have pointed out that these two fields have tended to diverge especially in their models of the design process, to the detriment of both. Perhaps a third generation of the 1990s might be based on a combination of the previous two; or, as in the model proposed by Cross (1989), on understanding the 'commutative' (Archer ,1979) nature of problem and solution in design. There was also a broader renewal of interest in design methodology in the late 1980s - especially in AI developments, where hope springs again for design automation and/or intelligent electronic design assistants.

A particularly significant development has been the emergence of new journals of design research, theory and methodology. Just to refer, again, to English-language publications, we have had *Design Studies* since 1979, *Design Issues* since 1984, *Research in Engineering Design* since 1989, the *Journal of Engineering Design* since 1990, and the *Journal of Design Management* since 1990.

3. Relationships Between Design Methodology And Science

From the earliest days, design methodologists have sought to make distinctions between design and science, as reflected in the following quotations.

'Scientists try to identify the components of existing structures, designers try to shape the components of new structures.' (Alexander, 1964)

'The scientific method is a pattern of problemsolving behaviour employed in finding out the nature of what exists, whereas the design method is a pattern of behaviour employed in inventing things...which do not yet exist. Science is analytic; design is constructive.' (Gregory, 1966)

'The natural sciences are concerned with how things are...design on the other hand is concerned with how things ought to be.' (Simon, 1969)

Glynn (1985) has pointed out that the above distinctions tend to be based on a positivistic (and possibly simplistic) view of the nature of science, and that scientists too, like designers, create their hypotheses and theories, and use these theories to guide their search for facts. Hillier, Musgrove et al. (1972) also criticized design methodologists for basing their ideas on outmoded concepts of scientific method and epistemology.

Cross, Naughton et al. (1981) went so far as to suggest that the current epistemology of science is in some confusion and therefore is a most unreliable guide for an epistemology of design. This conclusion was challenged by Levy (1985), who suggested that transformations within the epistemology of science should be seen as active growth and development rather than simply chaos, and that it would be naive to try to isolate design and technology from science and society.

However, there may still be a critical distinction to be made: method may be vital to science (where it validates the results) but not to design (where results do not have to be repeatable).

It is also clear that practitioners, whether in science or design, do not have to be methodologists. As Sir Frederick Bartlett pointed out, 'The experimenter must be able to use specific methods rigourously, but he need not be in the least concerned with methodology as a body of general principles. Outstanding "methodologists" have not themselves usually been successful experimenters.'

(Bartlett, 1958.) If 'designer' is substituted for 'experimenter', this observation also holds true in the context of design.

The Design Research Society's *Design:Science:Method* Conference of 1980 gave an opportunity to air many of these considerations. The general feeling from that conference was perhaps that it was time to move on from making simplistic comparisons and distinctions between science and design; that perhaps there was not so much for design to learn from science after all, and that perhaps science rather had something to learn from design. As Archer (1981) wrote in his paper for that conference, 'Design, like science, is a way of looking at the world and imposing structure upon it'. Both science and design, as Glynn (1985) pointed out, are essentially based on acts of perception, and 'it is the epistemology of design that has inherited the task of developing the logic of creativity, hypothesis innovation or invention that has proved so elusive to the philosophers of science.'

More informed views of both science and design now exist than they did in the 1960s. As Levy (1985) wrote, 'Science is no longer perceived in terms of a single fixed methodology focused on a specific view of the world. It is more an expanded rationality for problem-identifying, -structuring and -solving activities.' This makes scientific methodology sound indistinguishable from design methodology. Thus the simple dichotomies expressed in the 1960s are being replaced by a more complex recognition of the web of interdependencies between knowledge, action and reflection.

But in some places, old attitudes die hard. The editorial in Volume 1, Issue 1, of *Research in Engineering Design* was clear about that journal's aim to change design from an art to a science: 'For the field of design to advance from art to science requires research...' (Dixon and Finger, 1989.)

Let us at least try to clarify some of the terminology that is used in discussing concepts such as 'scientific design', 'design science' and 'the science of design'.

3.1. SCIENTIFIC DESIGN

As I said earlier, the origins of design methods lay in 'scientific' methods, similar to decision theory and the methods of Operational Research. The originators of the 'design methods movement' also realised that there had been a change from pre-industrial design to industrial design - and perhaps even to post-industrial design? The reasons advanced for developing new methods were often based on this assumption; modern, industrial design is too complex for intuitive methods.

The first half of this century had also seen the rapid growth of scientific underpinnings in many types of design - e.g. materials science, engineering science, building science, behavioural science. A relatively simple view of the design-science relationship is that, through this reliance of modern design upon scientific knowledge, through the application of scientific knowledge in practical tasks, design 'makes science visible' (Willem, 1990).

So we might agree that **scientific design** refers to modern, industrialised design - as distinct from pre-industrial, craft-oriented design - based on scientific knowledge but utilising a mix of both intuitive and non-intuitive design methods.

3.2. DESIGN SCIENCE

'Design Science' was a term perhaps first used by Gregory (1966), in the context of the 1965 Conference on The Design Method. Others, too, have the development of a 'design science' as their aim; for example, the originators of the ICED conferences, the *Workshop Design Konstruction* (WDK) are 'The International Society for Design Science'. The concern to develop a design science has led to attempts to formulate *the* design method - a single rationalised method, based on formal languages and theories. We have even had presented the concept of 'Creativity As An Exact Science' (Altshuller, 1984).

But a desire to 'scientise' design can be traced back to ideas in the modern movement of design. The designer Theo van Doesburg wrote in the 1920s: 'Our epoch is hostile to every subjective speculation in art, science, technology, etc. The new spirit, which already governs almost all modern life, is opposed to animal spontaneity, to nature's domination, to artistic flummery. In order to construct a new object we need a method, that is to say, an objective system.' (van Doesberg, 1923.) And a little later, the architect Le Corbusier wrote: 'The use of the house consists of a regular sequence of definite functions. The regular sequence of these functions is a traffic phenomenon. To render that traffic exact, economical and rapid is the key effort of modern architectural science.' (Le Corbusier, 1929.)

Hansen (1974), quoted by Hubka and Eder (1987), has stated the aim of design science as being to 'recognize laws of design and its activities, and develop rules'. This would seem to be design science constituted simply as 'systematic design' - the procedures of designing organized in a systematic way. Hubka and Eder regard this as a narrower interpretation of design science than their own: 'Design science comprises a collection (a system) of logically connected knowledge in the area of design, and contains concepts of technical information and of design methodology... Design science addresses the problem of determining and categorizing all regular phenomena of the systems to be designed, and of the design process. Design science is also concerned with deriving from the applied knowledge of the natural sciences appropriate information in a form suitable for the designer's use.'

This definition extends beyond 'scientific design', in including systematic knowledge of design process and methodology as well as scientific/technological underpinnings of design of artefacts. For Hubka and Eder the important constituents of design science are: 1, Applied knowledge from natural and human sciences; 2, Theory of technical systems; 3, Theory of design processes; 4, Design methodology.

Andreasen (1991) points to two important areas of theory in design science that are delineated by Hubka (for mechanical engineering): theory of the design process (general procedures, methods, tools) and theory of machine systems (classification, modelling, etc. of technical systems). This helps to define design science as including both process and product knowledge and theory.

So we might conclude that **design science** refers to an explicitly organised, rational and wholly systematic approach to design; not just the utilisation of scientific knowledge of artefacts, but design in some sense as a scientific activity itself.

3.3. SCIENCE OF DESIGN

There is some confusion between concepts of Design Science and of a Science of Design, since 'Science of Design' seems to imply (or for some people has an aim of) the development of a 'Design Science'. For example, we have praxeology, 'the science of effective action', and in *The Sciences of the Artificial*, Simon (1969) defined 'the science of design' as '... a body of intellectually tough, analytic, partly formalizable, partly empirical, teachable doctrine about the design process.'

This view is controversial. As Grant (1979), wrote: 'Most opinion among design methodologists and among designers holds that the act of designing itself is not and will not ever be a scientific activity; that is, that designing is itself a non-scientific or a-scientific activity.' However, Grant also made it clear that 'the study of designing may be a scientific activity; that is, design as an activity may be the subject of scientific investigation.'

A similar view of 'the science of design' has also been clearly stated by Gasparski (1990): 'The science of design (should be) understood, just like the science of science, as a federation of subdisciplines having design as the subject of their cognitive interests'.

In this latter view, therefore, the science of design is the *study* of design - something similar to what I have elsewhere defined as 'design methodology'; the study of the principles, practices and procedures of design. For me, design methodology 'includes the study of how designers work and think, the establishment of appropriate structures for the design process, the development and application of new design methods, techniques and procedures, and reflection on the nature and extent of design knowledge and its application to design problems'. (Cross, 1984.)

So let us agree here that the **science of design** refers to that body of work which attempts to improve our understanding of design through 'scientific' (i.e., systematic, reliable) methods of investigation.

4. Recent Developments In Design Methodology

I would like to conclude with a brief review of developments in design methodology over the last decade. I will use categories of work similar to those I used in *Developments in Design Methodology* (Cross, 1984), which covered the period 1962-82.

4.1. THE DEVELOPMENT OF DESIGN METHODS

Origination and application of systematic methods

In this category, the last decade has been notable for the development of product quality assurance methods, such as Taguchi methods (Ross, 1988) and Quality Function Deployment (Hauser and Clausing, 1988).

There has also been significant new work in design automation, using expert systems and other artificial intelligence techniques. A new series of conferences on AI and Design has been established, where this work is reported (Gero, 1991).

4.2. THE MANAGEMENT OF DESIGN PROCESS

Models and strategies for executing design projects

We have had a new generation of systematic models of the design process, particularly in engineering design, and particularly from Germany (Hubka (1982), Pahl and Beitz (1984), Verein Deutscher Ingenieure (VDI) (1987)). We have also seen the emergence of 'concurrent' models of product planning and development (Andreasen (1991), Pugh (1991)).

In architecture and planning there has been development of the 'argumentative' process models (McCall (1986), and the paper by Grant in this volume).

4.3. THE STRUCTURE OF DESIGN PROBLEMS

Theoretical analysis of the nature of design problems

There has been significant new work on problem 'types', for example by Schön (1988) and by Oxman (1990). In this category we might also include the new work on formal languages and grammars of design (Stiny (1980), Flemming (1987)).

4.4. THE NATURE OF DESIGN ACTIVITY

Empirical observations of design practice

There have been many more protocol and case studies made in this period. Examples include Schön (1984), Rowe (1987), Davies and Talbot (1987), Wallace and Hales (1987), Stauffer, Ullman et al. (1987), Eckersley (1988), Waldron and Waldron (1988). A conference in Delft on *Research in Design Thinking* brought together several related approaches and recent new work (Cross, Dorst et al., 1992).

4.5. THE PHILOSOPHY OF DESIGN METHOD

Philosophical analysis and reflection on design activity

Some of the comparative discussions of design and science have already been referred to earlier in this paper (Levy (1985), Glynn (1985)). There have been several new studies in the epistemology of design (Buchanan (1989), Zeng and Cheng (1991), Roozenburg (1992)), and we should also include here work in the praxeology of design (Gasparski, 1990).

Some of us have also been theory-building around the concept of 'designerly' ways of thinking and acting (A. Cross, 1984, 1986; Tovey, 1986; N. Cross 1990), although some aspects of this work have been challenged by Coyne and Snodgrass (1991).

5. Conclusion

For some people, design methodology appeared to have died in the 1970s; however, we can now see that it survived, and that there has been some particularly strong and healthy growth in the 1980s, especially in the engineering and product design fields. There is still some confusion and controversy over the use of terms such as design science, but I hope that the discussion here has helped to clarify this.

Design methodology has become a much more mature academic field, but still suffers from a lack of confidence in it by design practitioners and it has had little (acknowledged) practical application.

References

Alexander, C. (1964), Notes on the Synthesis of Form. Cambridge, Ma., Harvard University Press.

Alexander, C. (1971), "The State of the Art in Design Methods." DMG Newsletter 5(3): 3-7.

Altshuller, G. S. (1984), Creativity as an Exact Science. Gordon and Breach.

Andreasen, M. M. (1991), "Design Methodology." Journal of Engineering Design 2(4): 321-335.

Archer, L. B. (1965), Systematic Method for Designers. London, The Design Council.

Archer, L. B. (1979), "Whatever Became of Design Methodology?" Design Studies 1(1): 17-18.

Archer, L. B. (1981), A View of the Nature of Design Research. In R. Jacques and J. Powell (ed.) Design:Science:Method. Guildford, UK, Westbury House.

Asimow, M. (1962), Introduction to Design. Englewood Cliffs, NJ., Prentice-Hall.

Bartlett, F. C. (1958), Thinking: an experimental and social study. London, UK, Allen and Unwin.

Broadbent, G. (1973), Design in Architecture. Chichester, UK, John Wiley & Sons Ltd.

Broadbent, G. (1979), "The Development of Design Methods." Design Methods and Theories 13(1): 41-45.

Broadbent, G. and A. Ward, (ed.) (1969), Design Methods in Architecture. London, UK, Lund Humphries.

Buchanan, R. (1989), Declaration by Design: rhetoric, argument and demonstration in design practice. In V. Margolin (ed.) Design Discourse. Chicago, Ill., University of Chicago Press.

Coyne, R. and A. Snodgrass (1991), "Is Designing Mysterious? Challenging the dual knowledge thesis." Design Studies 12(3): 124-131.

Cross, A. (1984), "Towards an Understanding of the Intrinsic Values of Design Education." Design Studies 5(1): 31-39.

Cross, A. (1986), "Design Intelligence: the use of codes and language systems in design." Design Studies 7(1): 14-19.

Cross, N. (1980), The Recent History of Post-industrial Design Methods. In N. Hamilton (ed.) Design and Industry. London, The Design Council.

Cross, N., (ed.) (1984), Developments in Design Methodology. Chichester, UK, John Wiley & Sons Ltd.

Cross, N. (1989), Engineering Design Methods. Chichester, UK, John Wiley & Sons Ltd.

Cross, N. (1990), "The Nature and Nurture of Design Ability." Design Studies 11(3): 127-140.

Cross, N., K. Dorst, N. Roozenburg, (ed.) (1992), Research in Design Thinking. Delft, Netherlands, Delft University Press.

Cross, N., J. Naughton, et al. (1981), Design Method and Scientific Method. In R. Jacques and J. Powell (ed.) Design:Science:Method. Guildford, UK, Westbury House.

Davies, R. and R. J. Talbot (1987), "Experiencing Ideas: identity, insight and the imago." Design Studies 8(1): 17-25.

Dixon, J. R. and S. Finger (1989), "Editorial." Research in Engineering Design 1(1): 1.
Eckersley, M. (1988), "The Form of Design Processes: a protocol analysis study." Design Studies 9(2): 86-94.
Evans, B., J. Powell, R. Talbot, (ed.) (1982), Changing Design. Chichester, UK, John Wiley & Sons Ltd.
Flemming, U. (1987), The Role of Shape Grammars in the Analysis and Creation of Designs. In Y. E. Kalay (ed.) Computability of Design. New York, John Wiley & Sons Ltd.
French, M. J. (1985), Conceptual Design for Engineers. London, UK, The Design Council.
Gasparski, W. (1990), "Editorial: Contributions to Design Science." Design Methods and Theories 24(2): 1186-1194.
Gasparski, W. (1990), "On the General Theory (Praxeology) of Design." Design Methods and Theories 24(2): 1195-1215.
Gero, J., (ed.) (1991), Artificial Intelligence in Design '91. Oxford, UK, Butterworth-Heinemann.
Glynn, S. (1985), "Science and Perception as Design." Design Studies 6(3): 122-133.
Gordon, W. J. J. (1961), Synectics. New York, Harper & Row.
Grant, D. P. (1979), "Design Methodology and Design Methods." Design Methods and Theories 13(1): 46-47.
Gregory, S. A., (ed.) (1966), The Design Method. London, Butterworth Press.
Gregory, S. A. (1966), Design Science. In S. A. Gregory (ed.) The Design Method. London, Butterworth.
Hall, A. D. (1962), A Methodology for Systems Engineering. Princeton, NJ, Van Nostrand.
Hansen, F. (1974), Konstruktionswissenschaft. Munich, Germany, Carl Hanser.
Hauser, J. R. and D. Clausing (1988), "The House of Quality." Harvard Business Review (May/June): 63-73.
Hillier, B., J. Musgrove, et al. (1972), Knowledge and Design. In W. J. Mitchell (ed.) Environmental Design: Research and Practice. Los Angeles, Ca., University of California.
Hongo, K. and N. Nakajima (1991), "Relevant Features of the Decade 1981-1991 for Theories of Design in Japan." Design Studies 12(4): 209-214.
Hubka, V. (1982), Principles of Engineering Design. Guildford, UK, Butterworth Scientific Press.
Hubka, V. and W. E. Eder (1987), "A Scientific Approach to Engineering Design." Design Studies 8(3): 123-137.
Jacques, R. and J. Powell, (ed.) (1981), Design:Science:Method. Guildford, UK, Westbury House.
Jones, J. C. (1970), Design Methods. Chichester, UK, John Wiley & Sons Ltd.
Jones, J. C. (1977), "How My Thoughts About Design methods Have Changed During the Years." Design Methods and Theories 11(1): 48-62.

Jones, J. C. and D. G. Thornley, (ed.) (1963), Conference on Design Methods. Oxford, UK, Pergamon Press.
Le Corbusier (1929), CIAM 2nd Congress, Frankfurt.
Levy, R. (1985), "Science, Technology and Design." Design Studies 6(2): 66-72.
McCall, R. (1986), "Issue-Serve Systems: a descriptive theory of design." Design Methods and Theories 20(3): 443-458.
Moore, G. T., (ed.) (1970), Emerging Methods in Environmental Design and Planning. Cambridge, Ma., MIT Press.
Osborn, A. F. (1963), Applied Imagination - Principles and Procedures of Creative Thinking. New York, Scribener's Sons.
Oxman, R. (1990), "Prior Knowledge in Design." Design Studies 11(1): 17-28.
Pahl, G. and W. Beitz (1984), Engineering Design. London, The Design Council.
Pugh, S. (1991), Total Design: Integrated Methods for Successful Product Engineering. Wokingham, UK, Addison-Wesley.
Rittel, H. (1973), "The State of the Art in Design Methods." Design Research and Methods (Design Methods and Theories) 7(2): 143-147.
Rittel, H. and M. Webber (1973), "Dilemmas in a General Theory of Planning." Policy Sciences 4: 155-169.
Roozenburg, N. (1992), On the Logic of Innovative Design. In N. Cross, C. Dorst and N. Roozenburg (ed.) Research in Design Thinking. Delft, NL, Delft University Press.
Roozenburg, N. and N. Cross (1991), "Models of the Design Process: integrating across the disciplines." Design Studies 12(4): 215-220.
Ross, P. J. (1988), Taguchi Techniques for Quality Engineering. New York, McGraw-Hill.
Rowe, P. (1987), Design Thinking. Cambridge, Ma., MIT Press.
Schön, D. (1984), "Problems, Frames and Perspectives on Designing." Design Studies 5(3): 132-136.
Schön, D. A. (1988), "Designing: rules, types and worlds." Design Studies 9(3): 181-190.
Simon, H. A. (1969), The Sciences of the Artificial. Cambridge, Ma., MIT Press.
Spillers, W. R., (ed.) (1974), Basic Questions of Design Theory. Amsterdam/New York, North-Holland/Elsevier.
Stauffer, L., D. Ullman, et al. (1987), Protocol Analysis of Mechanical Engineering Design. In W. E. Eder (ed.) Proceedings of International Conference on Engineering Design, Boston. New York, American Society of Mechanical Engineers.
Stauffer, L. A., (ed.) (1991), Design Theory and Methodology - DTM '91. New York, American Society of Mechanical Engineers.
Stiny, G. (1980), "Introduction to Shape and Shape Grammars." Environment and Planning B 7: 343-351.
Tovey, M. (1986), "Thinking Styles and Modelling Systems." Design Studies 7(1): 20-30.

van Doesberg, T. (1923), "Towards a Collective Construction." De Stijl : Quoted by Naylor, G. (1968) The Bauhaus, Studio Vista, London.

Verein Deutscher Ingenieure (VDI) (1987), Systematic Approach to the Design of Technical Systems and Products: Guideline VDI 2221. Berlin, Germany, Beuth Verlag.

Waldron, M. B. and K. J. Waldron (1988), "A Time Sequence Study of a Complex Mechanical System Design." Design Studies 9(2): 95-106.

Wallace, K. and C. Hales (1987), Detailed Analysis of an Engineering Design Project. In W. E. Eder (ed.) Proceedings of the International Conference on Engineering Design, Boston. New York, American Society of Mechanical Engineers.

Willem, R. A. (1990), "Design and Science." Design Studies 11(1): 43-47.

Zeng, Y. and G. D. Cheng (1991), "On the Logic of Design." Design Studies 12(3): 137-141.

Zwicky, F. (1948), The Morphological Method of Analysis and Construction. In Studies and Essays. New York, Interscience.

SCIENCE AND THE DESIGN OF MECHANICAL AND OPTICAL DEVICES: A FEW CASE STUDIES

F. LE GUET TULLY
Observatoire de la Côte d'Azur,
France

ABSTRACT. By examining a few case studies dealing with the process of designing various very common devices, I shall try and analyse how the results of science, and/or the use of a scientific method, have affected the design process. The choice of historical and contemporary cases is clearly enormous. I chose to study one device designed by several designers and several devices designed by one designer. In the first part I follow the evolution over the centuries in the design of the lock, a very ancient and common device. In the second part I shall consider a few widely used devices which were conceived and designed by Henri Chrétien, a 20th century French astronomer and inventor. From these historical and contemporary case studies, I shall try to draw some conclusions about the clear but intricate interaction between science and the process of design.

Foreword

I am no expert in design, nor in design methodology nor in design science. I am an astronomer who became interested in the history of astronomy. I happen to be responsible for the archives of a French astronomer who was born towards the end of last century, Henri Chrétien (1879-1956).

While studying his life and work I naturally became interested in the history of technology because Chrétien invented and designed a precision instrument called the Ritchey-Chrétien telescope, of which the Hubble Space Telescope is an example, and two very common precision devices which are still in use today - night reflectors and the optical device used in CinemaScope.

Chrétien's work therefore provides what I believe are interesting contemporary case studies that illustrate aspects of the complex relationship between scientific knowledge and method and the design process.

What about historical case studies that illustrate the same relationship ? When Dr de Vries first announced the subject of this workshop, I tried to think of some and soon realised how vast the choice was. Which historical period to

M. J. de Vries et al. (eds.), Design Methodology and Relationships with Science, 29–61.

choose: Classical Antiquity, the Middle Ages, or one more recent either prior to or after the 17th century scientific revolution ? Which cultural area to choose: the Mediterranean World, Islamic and related cultures, India, the Far East, Precolumbian America, Africa ? Which field to choose: transport, architecture, energy and power, medicine, horology, printing, textiles, machine tools, military or agricultural technology, etc. ?

Confronted with the difficulty in answering these questions, I decided to try and study a very common device which has existed since the dawn of time: the lock.

1. Introduction

Not being an expert in design, I have what is probably a very naive approach to the subject and apologize for this, since it is only thanks to the invitation of Dr de Vries that I realised, what must be clear to all of you, how much the process of design is, and always has been, present everywhere in daily life.

Ever since man polished stones, cooked his food, cultivated wheat, bred sheep or built a wooden shelter, design has been a characteristic of human activity.

Of course until societies reached a stage where work began to be specialised, design was the responsability of each individual. Today in our western societies design has to be in the hands of a few trained specialists.

Not only has every object we use to be designed, but every component that goes into its fabrication, even the most apparently insignificant, has had to be thought of and designed. Examples range from the kitchen knife to the television set, from the buttons on our clothes to the telescope launched by a satellite, from the lampshade to the nuclear power plant.

Although it is clear that more scientific knowledge and method have gone into a jet airliner than into the design of a hammer, in other cases it is harder to evaluate the contribution of science in the design process.

However this evaluation should be of interest for a better understanding and analysis of what makes, or has made, the success or the failure of a design.

One cannot deny it is difficult to decide precisely whether some specific knowledge is scientific or not, or to draw a line exactly between scientific and non-scientific method.

Philosophers of science debate fiercely about what is science. I am thinking for example of the book by Alan Chalmers entitled 'What is this Thing Called Science ?'[1] in which the author discusses the work of Popper, Lakatos, Kuhn and Feyerabend. From the philosopher's point of view it is extremely difficult if

---

[1] Chalmers, A. (1982) 'What is this Thing Called Science ? An Assessment of the Nature and Status of Science and its Method.', University of Queensland Press, St Lucia, second edition.

not impossible to give a universal and timeless definition of science and of its method.

From the layperson's point of view, science and its method are believed to possess specific characteristics which give them special power. 'It is scientific', or 'it is scientifically proven' means it is useless to discuss it, or to have a different opinion.

Although it may not be the right place for such a debate, I would just like to pinpoint two facts. First the consensus amongst scientists about what is scientific and what is not varies with time. Until the Renaissance, astronomy and astrology, which both predicted events, were considered as scientific. After the Copernican Revolution, astronomy, mechanics, optics, chemistry etc... developed new methods of investigation which gave birth to modern science. On the other hand astrology did not and is still based on the same analogical principles. Consequently the consensus has changed and today's scientists no longer accept astrology as a science.

Secondly, and this may be more relevant to the problems we wish to consider here, a given branch of knowledge may have been reserved to a few specialists at one time and shared with all later on. For example you do not need to be a scientist today to know about free fall, differential equations or the human genome. Today these subjects are taught at high school. A few centuries ago, you had to be a scientist or mathematician to understand the first two, and a few decades ago the last one was understood only by experts.

These are just a few semantic remarks about 'scientific knowledge' and 'scientific method' before coming back to the subject of my talk and examining the case studies already mentioned.

In the first part I shall briefly treat the evolution over the centuries in the design of this very ancient and common device, the lock, while in the second part I shall study the intellectual processes and the general context which led the French astronomer Henri Chrétien to think up two widely used inventions: night reflectors and an optical system for making wide screen cinema.

2. A Historical Case Study: the Evolution of Locks

Fastening devices must be nearly as old as mankind itself, or at least as old as private property.

If the design of tools and weapons for hunting arose from human beings' need to feed themselves in order to survive, the design of locks arose from another seemingly essential need, the protection of private property or precious objects.

What is a lock ? According to the Encyclopedia Britannica it is 'a mechanical device for securely fastening a door or lid, usually including a sliding bolt which may be shot by a removable key'.

2.1. EGYPTIAN LOCKS

Locks have been in use for at least 6000 years. Those of the Egyptian locksmith (fig 1) are probably the oldest to be preserved. One such lock was found in the ruins of Nineveh, and was apparently used to secure the door of a room in the Palace of King Sargon in Khorsabad, near Mossoul. Isaiah (XX.22) was probably referring to a similar lock when he says 'and the key of the house of David will I lay upon his shoulder'.

The lock and the key, which are entirely of wood, are described as follows in the Encyclopedia Britannica : 'The vertical piece of wood, the staple, is fixed to the door post and contains movable pins in the upper part, in this case six in number; the cross piece is the bolt. The pins in the staple fall into corresponding holes in the bolt and so prevent the latter moving until they are raised to the level of the top of the bolt by pins fixed on the key. Only the key made specially to fit the lock could operate it and quite a reasonable degree of security was achieved. This lock, the earliest of which the construction is known, is the prototype of the modern cylinder lock.'

What other sort of locking device was there at that time? Probably latches and simple bolts, the first ones working by simple vertical lifting and lowering of a bar into a groove or a hole, the latter operating by the simple horizontal motion of a bar sliding into a socket or striking plate.

Compared to both of these, the Egyptian lock is very elaborate. But one should note that the precise purpose which these three devices fulfil is not the same. Latches and simple bolts are used by somebody inside a room to prevent unauthorised people from coming in[2], while Egyptian locks were meant to stop unauthorised people from getting at something.

As will be seen later this difference is important. A simple means of locking up is sufficient as long as simple purposes have to be fulfilled: latches and simple bolts provide an efficient answer for people who wish to lock themselves in during day to day circumstances. But when it comes to protecting some valuable object in the absence of its owner, or a sacred place in the absence of the priest, these straightforward ways of locking were not satisfactory and a more elaborate device had to be developed.

In fact the final aim of this elaborate locking system is to dispense with a guardian; instead of having a human being present to ensure safety, the king, priest, or whoever owns something that needs protecting, holds a key. In the case of latches or simple bolts, human beings are not to be replaced, they are the ones who lock themselves in.

---

[2] There exists a modification of the latch - a lever going through the door and operated by pressure of the thumb - which allows a door to be opened from the outside. But in this case safety is not ensured since unauthorized people can open the door and come in. Thus this modified latch only helps to hold a door closed, providing protection from cold or wind or stopping animals coming in.

Since there is nothing more straightforward than lifting a metal bar or shooting a bolt, one can say that the design of latches and simple bolts contains no scientific knowledge nor are they the result of a particularly scientific way of thinking.

Although the conception of the Egyptian lock did not require specific scientific knowledge (but how does one define scientific knowledge a few thousand years B.C. ?), its design was the result of a more complex and elaborate thought process.

First of all, to a simple device made of a staple attached to a door with some sort of striking plate fixed to the doorframe and a bar moving vertically or horizontally, the Egyptians added a portable part, the key, which carries the authorization for opening the lock.

Secondly they introduced vertical movable pins in the staple in order to stop the horizontal motion of the bolt if the right key was not inserted in the lock.

It is obviously difficult to qualify this invention as relying on scientific method or knowledge. However there are a few ingenious features which show how non-trivial the Egyptian lock was:

- the lock's function is split into two parts (a) the fastening of the door or chest is achieved by a bolt, and (b) the authorization for opening the lock is contained in the design of the key;
- movable parts are introduced in order to stop the motion of the bolt unless they are exactly aligned at the right height with a special key;
- the specificity of the 'key - lock' system is obtained by varying the spatial distribution and cross-section of the pins.

I should add that in the Egyptian lock the key was introduced sideways and had to be strong enough to pull the bolt into the open position. Since from its design it appears that the whole device was fixed on and accessible from the outside, the system could not be opened from the inside and there was still neither a handle, nor a keyhole.

As we shall see later on, the Egyptian lock is the predecessor of the modern cylinder lock.

2.2. FROM ANCIENT GREECE TO 18TH CENTURY LOCKS

By an inexplicable oblivion, the principle of movable pins invented by Egyptian locksmiths disappeared from the Mediterranean world[3] and only reappeared at the end of the 18th century.

[3] The Egyptian lock did not disappear completely since until recently similar ones were made in Africa, in the Faroe Islands or in the Tanimbar Islands (west of New Guinea), G. Berthier (1992), private communication.

It seems that the wooden sliding bolt was invented by the Hittites[4] around 800 BC. The Laconians and the Romans developed it further.

2.2.1. Laconian and Roman locks. In Laconian locks the key lifted movable blocks and released the lock, while the bolt was slid back by pulling a cord. Although the fastening of these locks was ensured by movable parts, they were not as secure as the Egyptian lock since their design was much less elaborate, consisting of only three blocks side by side instead of three or more pins with different spatial distribution and shapes.

During the Roman Empire the fastening of the locks, which were made of bronze or iron, was ensured by the spur of a leafspring that held the bolt (fig 2). The design of the key is even simpler than in the Laconian case, so simple in fact that Roman locks were very easy to pick by means of a simple hook.

While the Egyptian device is attached to the outside and the key introduced sideways, the Laconian and Roman locks are fixed on the inside and the key is introduced in the bolt from the outside, perpendicularly and not parallel to it, through a hole made in the door or in the top of the chest. Since these locking devices are hidden inside, the only part visible from outside is therefore the hole for the key and the slot which allows the key to move the bolt. It should be noted that the keyhole is on one side of the device only and that, like the Egyptian lock, it can only be opened from the outside.

So, as far as security goes, the sole advantage of the Laconian and Roman locks over those of the Egyptians is that the whole device is hidden inside and therefore not easily accessible by the wrong people.

As in the Egyptian lock, the role of the key is twofold: to free the bolt, by lifting the blocks or the spur, and to move the bolt along in order to put it in the open position.

What can we conclude concerning the design of Roman locks ? Although their purpose does not seem to be any different from that of Egyptian locks, the Roman design is less ingenious. The only real improvement in the efficiency of the lock comes from the use of metal instead of wood. The designer therefore has made use of new technological knowledge - metalwork and the existence of springs - but has forgotten or left out the more complex and clever ideas of movable pins and of keys that have to be cut exactly.

Why in the same cultural area do we find such a decline in the design of a fairly common object ? I have no answer to that question and believe if explanations are to be found, they can only come from a careful study and comparison of both civilisations.

---

[4] Information from the Deutsche Museum in Munich, Germany.

2.2.2. Locks from the Middle Ages until the industrial revolution. From the 12th century onwards, Roman sliding keys were gradually replaced by devices such as keys, knobs or handles which had to be turned.

To start with, a half turn of the key allowed the spring to be pushed, so freeing the bolt. This system only worked from outside and the key could not be removed from the lock unless the door was locked again (fig 3).

Then locks appeared with keys that made a complete turn. In these the bolt is fixed in the closed or open position by spurs which are held in place by a spring. A half turn of the key allows one to free the bolt and a complete turn moves the bolt into the open or closed position. Safety is ensured by fixed wards which stop the wrong key from turning.

Another design makes use of a bolt with a chamfered edge, so that the bolt is automatically engaged in the strike plate when one closes the door. The door is opened from outside with a key and from inside by pulling a knob.

In the case of handles, the chamfered bolt is moved into the open position because it is pushed by turning the handle.

The ordinary locks that we still use today in France are a juxtaposition of these two systems. They contain two bolts, one being operated by the key, while the other, which is chamfered, is operated by the handle.

During the last few centuries a large variety of locks have been made according to this system. Their main characteristics are that

- the bolt is held or freed with a tumbler;
- they make use of a key which operates from either outside or inside, thus implying the existence of a keyhole, a weak point as far as security goes;
- the turning of the key frees and moves the bolt;
- security is ensured by fixed wards inside the lock and by a key cut according to the shape and layout of these wards.

2.2.3. Why was there so little change in the design of locks? When compared to earlier devices, this later lock does not provide better security because the design is such that the cutting of the key does not have to be exact. A key that is cut more deeply than need be is not stopped by the wards and therefore a 'minimum' key, such as a simple hook, opens the lock as well as the right key (fig 4).

In fact designers and craftsmen spent a lot of time improving the appearance of locks and keys. Each of these was a unique work of art and as centuries passed changes only came from improvements in the art of the locksmith and in the art of metal working. No attempt was made to alter the design in order to improve the lock's efficiency.

In fact who really needed a secure means of locking things up in those bygone days ? Castles, churches and cities had doors or gates adorned with beautiful locks, but one can infer from the design that their role was more to show off the power of the king, the prince, the Church or the head of a city, rather than to stop unwanted people gaining access. From the pieces exhibited in museums today it seems that the symbol of the key was far more important than the actual function of the lock: the lock only showed that access was restricted to a few who therefore had power over those who did not have access, the protection actually being ensured by real human guardians. Since these magnificent locks were also certainly the best that existed, one is entitled to believe that efficiency was not the main objective of their designers.

Regarding locks for chests, only wealthy people possessed precious items that justified the expenditure to acquire a unique locking device. As in the case of doors or gates, chests containing valuables were probably not left unattended by their owners. Consequently, here again, a rather rudimentary locking system was not a real drawback.

People living in the countryside, or lower class people in cities, surely did not have much to protect and in any case could not have afforded handmade locks[5].

In conclusion one can say that during this period, for various historical, political and social reasons, locks were not very common and the artistic appearance surpassed the technological qualities in the designer's mind. Because locksmiths were aiming at beauty rather than efficiency, the technological design of locks stood still and, although as early as the end of the 16th century scientific and technological knowledge increased in many fields, one finds no spin-off in the design of locks.

2.3. The return of locks with movable parts.

2.3.1. Introduction of multiple lever locks. This stagnant situation was to end with the industrial revolution which first took place in England. It is from this country that the first changes came: between 1774[6] and 1849 over eighty locks were patented[7].

In a famous patent dated May 27 1774, Robert Barron re-invented the principle of movable parts. It seems that, contrary to some famous later

---

[5] Popular stories for children sometimes provide information about the use and shape of locking devices. In 'Little Red Riding Hood', although the front door of the grandmother's house is locked, the wolf can open it very simply from outside by pulling a small piece of wood (la chevillette) in order to remove a small round bolt (la bobinette). The bears in 'Goldilocks' do not seem to have a lock on their front door since Goldilocks can get in in their absence. On the contrary, in 'The Three Little Pigs' the wolf cannot get in through the door from outside, which seems to imply that there was an efficient, if not elaborate locking system inside.

[6] In France the system of patenting was introduced in 1791.

[7] Berthier G. (1992), private communication.

inventors, he did not know about the Egyptian lock and therefore genuinely rediscovered it.

Barron became aware that the single acting tumbler lock gave practically no security because of two weak points in its design:

- it was easy to make false keys by taking prints of the fixed wards and
- so long as the tumbler was lifted enough for the talon[8] on the bolt to pass underneath, the bolt was free to move. This meant that the opening of the lock did not therefore depend on the exact distance the tumbler was shifted but only on a minimum distance.

Barron's invention consisted in introducing notches on the bolt as well as a double acting tumbler. The levers and their spurs penetrated into the notches and the lock could only be opened by a key which lifted both tumblers to a precise height. Failing this the bolt was kept in a blocked position either because a spur was too high or too low.

It was the first time since the ancient Egyptians that movable parts in the locking device had to be lifted simultaneously to the same height and that their displacement had to be exact.

The main characteristics of a modern lock are already present in Barron's design: several tumblers instead of one, simultaneous and exact lifting of the tumblers and a unique position for the movable parts to allow the bolt to pass.

It contained however a limitation. First of all it was not a very convenient design because a lot of space was needed in the talon of the bolt in order to make the corresponding notches. Although Barron himself proposed it in a patent dated 1778, it was not really possible to add more levers to increase the security because each extra one needed a bulkier talon.

In this patent Barron also proposed an ingenious improvement of his system. Instead of making notches in the bolt and spurs on the levers, he suggested doing the reverse, i.e. have the blocking spur on the bolt and notches in the levers. It overcame the space problem which arises when the notches are in the bolt. This excellent design was widely adopted and is still used in many cheap locks today (fig 5).

2.3.2. Towards cylinder locks. A few years after Barron had designed his multiple lever lock, Joseph Bramah, an English engineer and fruitful inventor[9], also took

---

[8] The part of the lock that the key presses on when it is turned.

[9] Joseph Bramah (1748-1814), the son of a Yorkshire farmer, worked as a cabinet-maker in London, where he subsequently started his own business. His first patent for some improvements in the mechanism of water closets was taken out in 1778. In 1784 he patented the lock known by his name, and in 1795 he invented the hydraulic press. In 1806 he devised for the Bank of England a numerical printing machine specially adapted for bank notes. Other of his inventions included the beer engine for drawing beer, machinery for aerating water, planing machines and improvements in steam engines and boilers and in papermaking machinery. In 1785 he suggested the possibility of screw propulsion for ships, and in 1802 the hydraulic transmission of power. He constructed waterworks at Norwich in 1790 and 1793. He died in London on December 9, 1814 (from the Encyclopedia Britannica which refers to a book by J.W. Roe, 'English and American Tool Builders', New Haven, 1916).

out patents for locking devices. In 1784 he invented the lock known by his name[10].

Although recognizing advantages inherent in Barron's design, Bramah pointed out that Barron's lock could easily be picked by a key covered with wax. The reason is that the shape of the key is determined by the position and height of the levers, so that when they are are held in the lower position by the springs, their bottom ends are not all at the same level and this can be printed in wax. This defect in the design therefore enables one to cut the right key after successive trials.

In order to overcome this problem, Bramah designed a device where sliding parts replace the movable levers. The principle of the design is very similar to that of an Egyptian lock: it is by pushing the sliding parts that they are brought into alignment and free the bolt .

He soon gave to his design the shape of a cylinder, which could only be made after he had designed a new machine-tool. More than two centuries later, his device is still in use in England and France[11].

2.3.3. How much science was there in these new designs? In conclusion, it appears that the two important changes brought about in the design of locks at the end of the 18th century took place in England after the patenting system was introduced and during the early years of industrialisation.

The new designs, including those based on the Egyptian one, were therefore brought about by new needs, themselves the result of industrialisation. The growth of the need for locks is linked to the appearance of an urban middle class so that more people possessed items to protect, and also probably to the subsequent increase in the number of thefts. Consequently new problems arose, such as the production of cheap locks and the need for a very large number of different keys.

What then was the relationship between the newly designed locks and science? There does not seem to be a direct one; the incentive for new designs came from clearly formulated needs, while the new designs were invented by talented people who were well aware of the needs and technical possibilities of the time. The fact that, of the eighteen patents taken out by Bramah, only two dealt with locks while the other sixteen were in entirely different fields, shows that the man had an exceptionally open mind. Scientific knowledge seems to have been less important than curiosity, cleverness and general education.

---

10 In France it is known as a "serrure à pompe".

11 Berthier G. (1992), private communication.

2.4. 19TH CENTURY INDUSTRIAL LOCKS

2.4.1. The English lever lock. In 1818, a few decades after Barron had designed his multiple lever device, Jeremiah Chubb took out his first patent for locks.

At that time the design of locks was based on Barron's idea, the multiple lever lock, with usually at least six levers, the correct key raising them so that the talon on the moving bolt could pass through the gate. Different combinations, almost infinite in number, were obtained by raising or lowering the position of the horizontal slot, or gate, through which the talon of the bolt passes.

Chubb's improvement of Barron's device consisted in an additional detector lever which comes into operation if any key other than the correct one is used in an attempt to open the lock. The detector lever is lifted too high, where it remains until the correct key is turned in the reverse direction, allowing the lock to be used again. In this way the owner of the correct key is informed when some unauthorized person has tampered with the lock.

Chubb soon founded a society, with his brother and nephew, which enabled him to produce locks in large quantities. While Barron's name was quickly forgotten, Chubb's became famous for the making of multiple lever locks. He kept improving their design until the middle of the 19th century.

As a result, the modern English lever lock came into general use and several societies for the making of these in the North of France were founded during the second half of the 19th century.

2.4.2. Yale's cylinder lock. The other outstanding invention of the 19th century is that of Linus Yale, a North American, who in 1848 conceived the idea of adapting the Egyptian lock to modern requirements and brought about a revolutionary change by separating the key mechanism from the lock itself, thereby making it possible for a very small key to be used, as it did not have to pass through the door (fig 6).

The key operates in a cylinder which consists of an outer barrel fixed to the door and a cylindrical plug which is rotated by the key and has a tongue at the far end which projects into the lock. The upper pins, five in number, in the fixed outer barrel, fall into corresponding holes in the plug, which contains five similar pins which are raised to the level of the circumference of the plug by the correct key. The top pins are kept in the right position by means of phosphor bronze springs. If a key is inserted which raises the pins in the plugs to the required height, the cylinder and key are free to turn, the key is able to turn and the projecting tongue moves the bolt. If the wrong key is inserted, the pins are not raised to the right height and the plug cannot be turnedn

A further degree of security is provided by the irregular shape of the key. The notches of the key may be cut to eight different depths. Hence since there are five notches, the number of different keys possible is eight to the power of five, i.e. 32,768. The notches are cut automatically by a milling machine to the required depth, and the pins in the plug are made afterwards to correspond.

The number of different keys is further increased by milling grooves of various shapes in the keyway inside the plug.

Today Yale's design is still widely used, especially in western countries, for locks on front or street doors and all over the world for padlocks.

2.4.3. Was there any science in the 19th century locks? Chubb improved and developed the English lever lock, while Yale designed a cylinder lock whose security depends on moving parts and in which the key mechanism is separated from the lock itself. Both devices are still widely used today.

Is there a relationship between the successful design of these two locks and science ?

The answer is that the new designs were first induced by new social needs, the need for producing large quantities of different locks at low prices and the need for an improved efficiency in protecting houses and goods.

After these needs were clearly formulated, the designers provided two ingenious ways of solving the problem. Although both designs do not seem to contain much contemporary scientific knowledge, they could not have been produced without making use of some sort of scientific method in order to increase the number of combinations and without modern technological knowledge about working metal, milling machines, machine tools, etc...

However the most important contributions to the new designs came from the ingenuity of the designers, who adapted the Egyptian lock to modern requirements in one case, and thought of a clever arrangement between the functions assigned to the bolt and to the levers in the other.

2.5. Modern locks and science.

2.5.1. From old safety problems to new ones. As everyone knows, scientific knowledge increased at an impressive rate from the end of the 19th century onwards.

Before examining how electricity and later on electronics and computer technology were brought into the design of locks, it must be noted that most locks produced today are still cylinder and English lever ones.

Where does this lack of evolution in the design of common locks come from ? Most likely from the fact that the items developed in the 19th century still provide a satisfactory answer, at a reasonable cost, to ordinary safety problems.

However, while the day-to-day security problems of 20th century citizens continue to be solved by 19th century solutions, the appearance of new safety problems provided the impetus for new designs which make use of new technological resources. For example the growth of safe deposits, the progressive replacement of human beings by machines for economical reasons, the development of restricted access for large numbers of people to strategic areas, factories, laboratories etc., posed new safety problems which gave birth to new highly technological locking devices.

2.5.2. Electricity and a further division of the lock's function. As early as the beginning of this century electric motors and electromagnets were used for moving the bolt in and out.

Since the invention of the key-lock system, the final opening of a door had always been the result of a human hand turning a key or a handle after the bolt had been freed, i.e. the freeing of the bolt was mechanically coupled to the part in which the key was introduced. When electricity became widely available lock designers were able to treat independently the mechanical action of moving the bolt and the authorisation for opening the lock.

This important step in the evolution of locking devices opened such a wide range of new possibilities that the trend towards separating the two functions - opening and authorising - has been carried on throughout the 20th century.

2.5.3. Science and evolution of opening authorisation. Recently three different ways of recognizing the authorised person have been developed. One of them is similar to the recognition by a human guard: instead of being recognised as a whole by a human brain, the authorised person is recognised either by fingerprints, the back of the eye or even personal smell. In these cases use is made of results in pluridisciplinary research[12].

Another way of authorising the opening of a lock is based on the use of an item which carries coded information: either a key, as before, or a magnetic card, a memory chip card, an electromagnetic badge, etc.... The choice depends on the situation. For example in large hotels where an important number of people are authorised for a short time, the most convenient system is the magnetic card because of its low cost. In big firms, a good choice is the electromagnetic badge. Apart from the key, all the devices used here rely on 'high technology' developed in the last few decades.

At this point it must be noted that since the invention of keys, millennia ago, keyholes have always been a weak point in locks. It is only with the introduction of 20th century new technology that the disappearance of this long inherent weakness became possible.

The third way relies on the use of keyboards. In this case the coded information is not carried on an object but is known by the authorised person. One speaks of virtual codes.

In the three cases, it is also necessary to provide solutions allowing the authorisation to be withdrawn. With the guardian or the biometric recognition system, the straightforward way to stop a person coming in is to provide the information to the guardian or to the system. In the two other cases, it is necessary to interact with the lock itself, i.e. with the device which reads and interprets the code.

---

[12] including biology, optics, electronics and computer science.

There are situations where this can be a serious drawback. For instance the increasing number of thefts in large hotels can be fought efficiently by changing the code of the bedroom lock each time a client leaves, but it turns out to be very costly to have a member of the hotel staff modify the code.

Recently an alternative procedure for a safe and economical invalidation of a code has been developed by a French inventor[13]. First it avoids the human intervention of a staff member because the invalidation of the previous code is done by the client him/herself the first time the magnetic card is used. Secondly it ensures great safety because each new code is produced by a mathematical algorithm and, as a result, even if the holder of the magnetic card were able to read the information about the old and the new codes, he/she would not be able to find out what the next code will be.

2.5.4. Science and evolution in the motion of the bolt. Whatever the locking device is, some mechanical energy is necessary in order to move the bolt in and out. Since the dawn of civilisations the force exerted by a human hand on the key has fulfilled this purpose. With the introduction of electricity and electromagnets at the beginning of the 20th century, the use of human energy was reduced to the pushing of a button. This solution has been widely used for the outside doors of buildings, prison gates, and more generally in all situations where doors or gates are opened from a distance.

However this attractive solution has drawbacks which put severe limitations on further developments of electric motors and electromagnets in locking devices. Since the amount of electrical energy necessary to replace the force exerted by the hand on a key or a handle is quite large, it is necessary to use a rather high current in the connecting cable. The voltage drop in the cable being proportional to the intensity of the current and to its length - according to Ohm's law - above some length the system becomes inefficient. Consequently the distance between the controls and the mechanical part under control is limited.

Several solutions have been proposed to circumvent this problem. I wish to mention the elegant one designed by Lewiner[14] because it appears as a good example of how a clearly formulated problem can work as an incentive for the production of a new design which at the same time is very ingenious and makes use of the resources of available technology.

Lewiner's main idea is to employ the physical force exerted by the human hand without loosing the benefit of a remote control for the opening. An

---

[13] French patents nr. 2 568 032 and 2 568 040 by Jacques Lewiner and Claude Hennion (1984). During the last fifteen years J. Lewiner and his co-inventors have patented a large number of inventions related to locking devices and safety problems. Many of the new designs are based on results of pluridisciplinary research done at the Ecole Supérieure de Physique et de Chimie Industrielles de Paris. In 1991, on the occasion of the bicentenary of the French patenting system, J. Lewiner was cited as the most prolific private French inventor.

[14] French patent nr. 2 568 032 by J. Lewiner (1984).

additional advantage of his design is that the distant part of the wiring of the remote control is installed in the doorframe and not in the door itself in order to avoid the well-known difficulties of wiring movable parts.This goal is achieved by adding to the normal hand operated chamfered bolt an auxiliary one operated from a distance. When at rest the auxiliary bolt is held in the engaged position by a spring which pushes it into the striking plate. The remote control acts upon the position of this auxiliary bolt by means of a movable plug located in the striking plate. When the plug is in a recessed position at the bottom of the striking plate, the auxiliary bolt, pushed by the spring, occupies the space that is left empty. When the plug is in the other position and therefore occupies the space inside the striking plate, the auxiliary bolt is forced inside the lock in a nonengaged position. A sophisticated mechanical system controls the action obtained by the rotation of the handle. If the auxiliary bolt occupies the empty space inside the striking plate, the rotation of the handle allows both bolts to be moved into the lock and the door opens. If the plug occupies even partially the empty space inside the striking plate, the rotation of the handle is disconnected from the motion of the main bolt and the door does not open (fig 7 & 8).

2.5.5. Emergence of new functions in locking devices. To start with, the growth of safe deposits has meant the design of special locks which require each locker to be opened by two authorized persons together. More recently the need for increasing safety in banks , jewellery shops or other places where precious or valuable objects are kept, has given rise to two door systems, one of which being always closed. In order to achieve the necessary interdependence of the opening and closing of the doors, the design of these systems relies on electronics and computer science.

Practical problems arise with the wiring of such systems and the trend today is towards systems which can be easily installed by someone who is not an electronics engineer, and easily operated when turning from the normal mode - one door open and one closed - to an emergency one - two doors open, in case of danger for example, or two doors closed.

Another function emerging from the new needs related to safety problems is the monitoring of the interconnection in systems consisting of an increasing number of elements. It seems that further development is to be found in network systems.

2.6. Tentative conclusion

On the one hand, locks and keys evolved extremely slowly until the industrial revolution and, as far as common devices for daily purposes go, they have not evolved at all since the 19th century. On the other hand, during the last decades designers have applied the latest results of interdisciplinary research to new locking devices and systems.

This apparent paradox seems to be due to the fact that the locking devices needed now in everyday situations are similar to those that were needed centuries or even millennia ago, while new needs have recently arisen from economical activities with important financial implications.

In some cases the issues at stake have had such important financial consequences for the customers that the societies making safety devices were ready to buy patents and pay highly skilled designers to offer satisfactory answers to new security problems. This in turn worked as an incentive for the production of new designs. It seems likely that without this strong incentive, most of the new complex devices and systems which rely on sophisticated technology would not have been thought of. Can we therefore assert that the more complex the safety problem is, the more dependent the locking device's design must be on technology?

In the beginning locks were invented to replace custodians. When improvements were made to locks in the past, those interested in forcing the locks found ways to overcome the new obstacles. Thus designers and inventors added other improvements to their already complex devices. And so on, in an apparently endless game constantly sustained on both sides by human imagination and creativity.

One is therefore led to conclude that, if financial considerations were ignored, the most complex and safest locking device would be the human custodian. Will this assertion remain true with the development of interconnecting machines used in networks ?

3. A Contemporary Case Study: Devices Designed by Henri Chrétien

After having examined an historical case study covering several millennia we shall now turn to contemporary case studies and investigate the optical devices designed in the twenties by a French inventor, Henri Chrétien (1879-1956).

3.1. A FEW WORDS ABOUT HENRI CHRÉTIEN

Henri Chrétien, an astronomer of modest origin, was born in Paris in 1879. At the age of thirteen, after obtaining his 'certificat d'études primaires', he became apprenticed to a printing company and entered the private school set up by its famous founder Napoléon Chaix.

Chaix had foreseen the growing need for short-lived documentation linked to the development of the newly created railway companies and, because of the lack of well-trained typographers, had created his own school.

During the five years that he spent there, the young Chrétien acquired an important part of his future technical skills as well as his taste for science. It is said that he realised he was able to understand mathematics while working on the proofs of the 'Journal des Mathématiques Elémentaires', one of the regular publications of the Chaix company, and thus decided to take his baccalaureat without going to a grammar school.

It was during these years that he met Camille Flammarion, the worldfamous populariser of astronomy, and developed his interest in astronomy. After leaving the printing school, Chrétien studied mathematics and physics at the Sorbonne. Through Flammarion he met professional astronomers and was soon asked to give regular popular lectures in astronomy.

It seems that his industrial and technical training left him with an equally marked interest in technology, because instead of finishing his degree at the Sorbonne, he entered the newly created Ecole Supérieure d'Electricité and became an electrical engineer.

At the same time he was also practising astronomy and soon became so highly valued by professional astronomers that in 1905 he was offered a position at Nice Observatory in order to create a department for developing the new field in astronomy called astrophysics.

Having obtained his degree as an electrical engineer, he returned to the Sorbonne and after graduating moved to Nice with his wife and baby.

Now an expert in applied optics he designed and built astronomical instruments, one of which was destined to a late but brilliant future: the aplanetic Ritchey-Chrétien telescope on which most telescopes made since the 1960s are based.

3.2. TOWARDS NIGHT REFLECTORS

Unlike most astronomers, he was not sent to the front when the First World War broke out because of his poor health and small size. He therefore arranged to be sent to the newly created 'Technical Aeronautical Service', where he was asked to work on improvements to the triple mirror for secret signalling.

At the time of the First World War, there existed military catoptric[15] devices which made use of mirrors only, and a more sophisticated catadioptric reflecting system, the Fizeau device which had been designed for a scientific experiment and made use of mirrors and lenses, i.e. of reflection and refraction.

The first system had the disadvantage of being bulky and made out of glass (fig 9). Therefore it was heavy and difficult to adjust.

Fizeau's device (fig 10), consisting of a refracting telescope and a plane mirror installed in the focal plane, was perfectly suited for measuring the speed of light, but could not be used for secret signalling because the reflected beam was only visible on or very near the axis of the incident beam. In any case it was also very bulky.

---

[15] Catoptric means that the optical system which sends the light back (cata) has one optical surface, namely a mirror, while catadioptric means that it has two optical surfaces (lenses).

However Chrétien, a shrewd astronomer, soon realised that the efficiency of secret signalling could be greatly improved by making use of the Fizeau catadioptric reflecting device he had had the opportunity to study in Nice[16]. Some of the drastic changes he introduced in order to make it appropriate for night signalling were scientific, others were ingenious ones.

First of all he modified the optical system, replacing the plane mirror by a spherical one whose radius was equal to the focal length. Doing so he increased the intensity of the reflected beam and allowed it to be seen even though the axis of the device is not strictly parallel to the incident beam.

Secondly, he thought of using an astronomical refractor with a very short focal length in order to make the reflecting device sufficiently small. This meant using the instrument of a professional astronomer in a non astronomical situation.

However, reducing the size of the refractor implied reducing the luminosity of the reflecting device.

In order to overcome this problem, Chrétien had the bright idea of making a multiple device consisting of a number of small reflectors set side by side according to any convenient arrangement (fig 11).

He designed his 'multiple autocollimating device for optical signalling' and in 1917 applied to have it patented. In spite of the superiority of the modified Fizeau combination over the other existing secret reflecting devices, the French authorities never issued the patent because of the war.

In 1923 Chrétien applied for a new patent, which only differed from the previous one in so far as it dealt with optical signalling in general, and introduced new applications such as night-time advertising.

While the previous device could be considered as a precision instrument, the new one was much simpler. Contrary to secret signalling, optical signalling in general is not based on a very small divergence of the reflected beam. Since increasing the divergence produced by an optical system is much easier than reducing it, the second device could be made of much cheaper components.

Chrétien created the generic name 'cataphote' for his invention and suggested many possible applications, including night advertising and reflective ball shaped lenses, which later were used in extra-luminous screens (fig 12). However he did not foresee the application for which his design was best suited at that time. He sold his patent to an engineer who successfully exploited Chrétien's invention by applying it to night-time reflectors which have made driving in the dark much safer.

---

[16] Perrotin and Prim (1908), 'Détermination de la vitesse de la lumière', Annales de l'Observatoire de Nice, vol. 11, Gauthier-Villars, Paris.

3.3. The design of reflectors: from an expressed military need to a new civilian invention

The origin of the design of night reflectors goes back to a war situation during which a scientist is confronted by a clearly stated problem, namely military secret signalling. For this he makes use of his scientific knowledge and method and, aided by his inventive character, designs a new device which answers the military needs.

Next, one finds a scientist whose invention has not been applied, but who is determined to go further.

From what I know, Chrétien had a Belgian friend, Eugène Frey, who was interested in luminous stage settings. It is probably because of Frey that he thought of applying his first design to night advertising and made the corresponding modifications.

Once the new design was scientifically achieved, Chrétien looked for various applications, but in a theoretical rather than practical way.

It seems he could apply his scientific knowledge and shrewdness to new designs if he was aware of a specific problem, but he could not think of other applications for the device he had designed.

In this case there is clearly a strong relationship between the design of the device and the scientific knowledge of the designer. Moreover it seems very unlikely that night reflectors would have been invented by somebody who was not an astronomer and who did not know about Fizeau's experiment.

It is also interesting to recall that the original problem of secret night signalling was linked to the fact that glass was the only material available for making mirrors that were to be used as reflecting devices. The advantage of Chrétien's device, which was also made of glass, was its small weight, size and price. Since then new materials much lighter than glass have become available. As a result the design of today's reflectors is no longer the one patented by Chrétien. It is much cheaper to mould small reflecting surfaces than to make small lenses.

As a conclusion this case study emphasizes the broad interactions at work between science, technology, the military art and material techniques before the device could be thought of and designed. It also shows the crucial role played by the training, the professional experience and the inventiveness of the designer. However, since this case study does not deal simply with the design of an object, but also with the invention of a new device, it is not clear whether the present intermingling of various features, including science and technology, is representative of the act of design in general.

3.4. The design of another optical device: the Hypergonar lens for wide screen cinema

3.4.1. Cinerama versus Cinemascope. During the last three years of his life Chrétien became known worldwide as the French scientist who invented Cinemascope, 'an anamorphic process of wide-screen film projection in which an image of approximately twice the usual width is squeezed into a 35 mm frame and then screened by a projector having complementary lenses' as the Collins English dictionary describes it.

The successful launching of Cinemascope in 1953 relied on a very efficient advertising campaign which used the mythical image of the old inventor, working alone in the secret of his laboratory. At that time Chrétien was seventy-four, and he still had the great sense of humour he had shown all his life. He played his part very well and not only was Cinemascope brilliantly launched, but it survived many of its rivals, including Cinerama which had been launched about a year earlier.

If nearly forty years later Cinemascope is still a widely used process, while Cinerama has more or less disappeared, it cannot be due to the advertising campaign. There are surely other reasons.

Both Cinerama and Cinemascope were attempts to fight the disastrous effect that the development of television was having on the film industry after World War II. But why did one process survive and not the other ?

Cinerama[17], which was invented by Fred Waller, aimed at reproducing the normal field of vision on a curved screen. It used three cameras equipped with 27 mm lenses, which is approximately the focal length of the human eye, mounted as one unit and set at 48-degree angles to each other.

The three pictures were shot simultaneously on three separate rolls of film. A single rotating shutter in front of the three lenses assured simultaneous exposure on each of the films. The cameras were standard 35mm studio cameras with only slight modifications (larger frame height, increased speed of the film). In the theatre, the three projectors were located in separate booths; they were also standard 35mm with the matching modifications. The three machines were electronically interlocked with motors that automatically kept the three images in perfect synchronisation on the screen.

In order to put the audience even more in the picture Cinerama was equipped with a sound system to match visual realism: six fully directional soundtracks on a separate film which was synchronised with the three picture panels[18].

There were two serious drawbacks to this complex wide-screen process. First of all only a small number of cinemas were authorised to show the films because very elaborate equipment was needed. Secondly, for each performance, five operators were necessary - three projectionists and two control engineers, one

---

[17] Carr, R. E. and Hayes,R.M. (1988) 'Wide Screen Movies A History and Filmography of Wide Gauge Filmmaking', Mc Farland & Company, Inc. Publishers, Jefferson, North Carolina.

[18] The system was designed by Hazard Reeves, one of the entrepreneurs behind Cinerama and a noted electronics engineer.

for the sound and one for the picture - in order to avoid problems arising from the synchronisation of the three picture panels and to ensure that the sound reproduction speed remained constant. Consequently, although Cinerama offered sights and sounds that no television could reproduce, it was never economical.

It should be noted that when it was launched in 1952 Cinerama did make use of new technological developments, but it did not involve much science.

In fact there was no more science in Cinerama than there was in the Triptych process the French film maker Abel Gance (1889-1981) developed around 1925 in order to make his film 'Napoleon'.

According to Kevin Brownlow, the British film historian and archivist, 'besides being the most inventive director in France (Gance) was also the most ambitious'.

Gance said that when writing the scenario of 'Napoleon', he had the feeling that in certain scenes he was 'lacking space', and thought then of 'stretching the screen'[19]. He wrote in one of his carnets: 'I vaguely thought that if I put one camera on the right, one in front and one on the left, I would have an enormous panorama'.

He ordered the equipment from his friend and camera pioneer André Debrie[20] who eventually made the apparatus, named Triptych. It consisted of a 'pyramid', as Debrie called it, with one camera mounted above the other, linked to a motor by flexible shafts[21]. Of course there were parallax problems and,from the beginning, 'Debrie doubted the device would be adopted by enough theatres to make it commercial'[22].

He was right and years later, when Kevin Brownlow asked Abel Gance what he thought of Cinerama, his answer was: 'It's exactly the same as my idea. They haven't even solved the problem of the joins between the screens !'

Although provocative, this statement was not completely untrue.

So what is so different with Cinemascope? Why did it cause filmmaking to change so 'drastically and permanently'[23]?

---

[19] Brownlow, K. (1983), 'Napoleon, Abel Gance's classic film', Jonathan Cape Ltd, London, p 131-132

[20] "Debrie had become an inventor of motion picture equipment at a surprisingly early age. His father, Joseph Debrie, was a manufacturer of laboratory apparatus. In 1908, an Englishman going on safari ordered some of his equipment and then wanted a camera as well. Joseph asked André, then aged seventeen, if he could do it. André, having no preconceived ideas about moving picture cameras, produced the Parvo, a camera which was so well designed it could truthfully be advertised as the smallest, lightest and strongest camera on the market" (from Kevin Brownlow, p 53).

[21] Kevin Brownlow, p 135.

[22] Kevin Brownlow, p 134

[23] Carr & Hayes, p 57.

The answer to that question is that Triptych and Cinerama, although they relied on the normal use of several standard studio cameras, required very complex and expensive equipment in the cinemas. On the other hand Cinemascope used only one standard camera equipped with a special anamorphic lens. And, apart from a wide screen, no special equipment was required in the cinema itself, except one of these anamorphic lenses which could easily be fitted to a normal projector.

The difference in the diffusion of both inventions thus comes from their respective economical aspects. Already in 1927, when he patented it, Chrétien knew that his lens was more practical, more economical. But at that time, the film industry was not interested.

There is also a more basic difference between the two wide screen processes, which is the amount of science each of them relies upon.

3.4.2. The Hypergonar anamorphic lens. As we have seen, Gance was not a scientist and his invention came from a 'feeling' coupled to a very simple idea: 'join' three existing cameras.

How much science is there in the anamorphic lens which Chrétien invented and named Hypergonar from the Greek for 'wide' and 'angle'.

Anamorphosis is the optical effect of distorting an image in one direction, while leaving it unchanged in the other direction. The image seems abnormally compressed or extended. Distorting mirrors, which are simply cylindrical reflecting surfaces provide the simplest example of anamorphosis. But such devices do not give images of sufficient quality to be used in photography or cinematography. However mirrors can be replaced by cylindrical lenses.

According to the anamorphic theory of the German physicist Ernst Abbe (1840-1905), the necessary condition for obtaining a real image with anamorphic lenses is the use of cylindrical lenses whose axes are not parallel. On the other hand, if the cylindrical lenses of the system have crossed axes, then the anamorphic image is real but it suffers serious astigmatism.

This means that all the rays coming from the same point do not concentrate exactly onto the same point of the image: the real image produced by anamorphic lenses is blurred.

Abbe showed that one cannot obtain an image which is simultaneously real and stigmatic. And since any optical device obviously demands stigmatic images, Abbe concluded that anamorphic lenses were of no use for such devices.

When Chrétien started his own study of anamorphosis, he agreed that good stigmatic images were of course essential, but he pointed out that forming a real image was not necessary. His scientific contribution to the question was to reconsider one of the points which had stymied Abbe.

His view was that cylindrical lenses with parallel axes must be used in order to overcome astigmatic aberrations. From there on he worked out a way of forming a real image from the virtual one produced by the system. It turns out that this can be done very simply using an ordinary converging lens (fig 13).

As I said before, Chrétien had a profound knowledge of applied optics and his deep understanding of technical problems helped him to think up a very ingenious and elegant device.

When he became famous and was interviewed, he always denied he had invented anything at all and kept referring to Abbe. In a way he was right. He did not add anything to Abbe's theory, but he had a new idea which, although very simple since it uses an ordinary lens in order to make a real image out of a virtual one, solved the problem Abbe had not been able to overcome.

The subtlety of his system comes from the combination of two cylindrical lenses, one divergent and the other convergent, whose distance apart is calculated precisely so that the virtual image is formed in the plane of the object itself. As a consequence, not only is this optical system a-focal (and does not interfere with the ordinary lens of the movie camera as far as focusing is concerned), but it is extremely compact, and therefore practical.

3.5. Science, technology and design in the case of Henri Chrétien

Chrétien's case is complex. Although he was not trained as a designer, he designed several new optical devices of which three are still in use today: an aplanetic telescope, night reflectors and the Hypergonar lens for wide screen cinema. He can therefore be considered as a creative designer.

From the two cases we have studied it seems clear that Chrétien's scientific, technical and practical training had an influence on his achievements: he provides an example of the successful intermingling of science, technology and design.

He certainly made use of his wide knowledge in applied optics and mathematics when designing these new devices. But how did he actually come to think of them ? In France it was very uncommon for someone with a university degree and holding an official position, astronomer or university teacher, to do anything else than pure and 'clean' science[24] Although Chrétien reached the summit of his academic career by becoming a Professor at the Sorbonne, he never denied his former training as an apprentice nor his engineering background. Consequently, and contrary to many of his colleagues, he never had any objection towards applied and 'messy' science.

This is probably why he retained a freedom of spirit which enabled him to apply his scientific knowledge and experience to design problems.

However, in my opinion, this characteristic would not have been sufficient to make him become such a fruitful designer. As I said earlier an important additional clue to his successful achievements is his broad natural curiosity which ranged from observational astronomy, to cinema through military optics, traffic

---

[24] It still is, although there seems to be a slight tendancy towards a higher valorisation of applied science.

signalling, ship engines or handicapped people and enabled him to be aware of many unsolved design problems.

4. Reasoning Versus Luck in Creative Design: Shall We Let Galileo Conclude ?

As we have seen on several occasions, creative design can occur either when tackling a clearly enunciated problem or serendipitously.

Galileo, who was well aware of these two possibilities because of the dispute about the invention and design of the telescope, provides us with an interesting opinion.

Let us recall the context of the telescope's invention. According to Winkler and van Helden[25] the optical device which later gave birth to the telescope 'emerged from the anonymous craft condition', after 'a hazy period between the speculations of the Renaissance magi and the first application for a patent on the device in the Netherlands in early October 1608'.

The first applicant, Hans Lipperhey, 'a spectacle-maker from Middleburg, a humble, very religious and God-fearing man, presented to His Excellency[26] certain glasses by means of which one can detect and see distinctly things three or four leagues removed from us as if we were seeing them from a hundred paces.'

In Sidereus nuncius[27], Galileo wrote: 'About ten months ago a rumor came to our ears that a spyglass had been made by a certain Dutchman by means of which visible objects, although far removed from the eye of the observer, were distinctly seen as though nearby. About this truly wonderful effect some accounts were spread about, to which some gave credence while others denied them.'

He then describes how he improved the instrument: 'And first I prepared a lead tube in whose ends I fit two glasses, both plane on one side while the other side of one was sperically convex and of the other concave. Then applying my eye to the concave glass, I saw objects satisfactorily large and close. Indeed, they appeared three times closer and nine times larger than when observed with natural vision only. Afterwards I made another more perfect one for myself that showed objects more than sixty times larger. Finally, sparing no labor or expense, I progressed so far that I constructed for myself an instrument so excellent that

---

[25] Winkler, M. G. and Van Helden, A. (1992), 'Representing the Heavens, Galileo and visuel astronomy', ISIS, 83, pp. 195-217.

[26] Prince Maurice.

[27] Galileo, G. (1610), 'Sidereus nuncius', Venice

things seen through it appear about a thousand times larger and more than thirty times closer than when observed with the natural faculty only.'

But as Winkler and Van Helden remark, Galileo did not mention that the eight-powered spyglass was presented to the Venetian Senate by a mathematician and that it was a gift for everybody and for nobody in particular.

Galileo clearly expresses his own view of the invention of the telescope: 'To discover the solution of a known and designated problem is a labor of much greater ingenuity than to solve a problem which has not been thought of and defined, for luck may play a large role in the latter while the former is entirely a work of reasoning. Indeed, we know that the Hollander who was first to invent the telescope was a simple maker of ordinary spectacles who in casually handling pieces of glass of various sorts happened to look through two at once, one convex and the other concave, and placed at different distances from the eye. In this way, he observed the resulting effect, and thus discovered the instrument. But I, incited by the news ..., discovered the same by means of reasoning.'

Galileo's opinion is unambiguous.

ACKNOWLEDGEMENTS: I wish to thank Mr G. Berthier, former President of the Société Fontaine, 190 rue de Rivoli, 75001 Paris, France, for the precious information he provided and for allowing me to reproduce some figures from his article 'Historique des Sûretés'. I am also grateful to J.A. Tully for his helpful suggestions and critical reading of the manuscript.

References

ABOUT LOCKS:

(1) Berthier, G. (1992 private communication) 'Historique des Sûretés', unpublished.

(2) Lecoq, R. (1973) 'La Serrurerie Ancienne, Techniques et Oeuvres', Librairie Gedalge, Paris.

ABOUT THE HYPERGONAR LENS:

(1) Chrétien, H. (1951) 'La cin_matographie panoramique par le procédé Hypergonar', Bulletin de l'A.F.I.T.E.C, nr. 11, 1952.

ABOUT HENRI CHRÉTIEN:

(1) Le Guet Tully, F. (1991) 'A case study: Henri Chrétien, French scientist and engineer', in B. Gremen (ed), The Interaction between Science and Technology, Wageningen University.

(2) Le Guet Tully, F. (1991) 'On the history of the catadioptric reflector', Proceedings of the XIXth International Congress of ICOHTEC, Vienna.

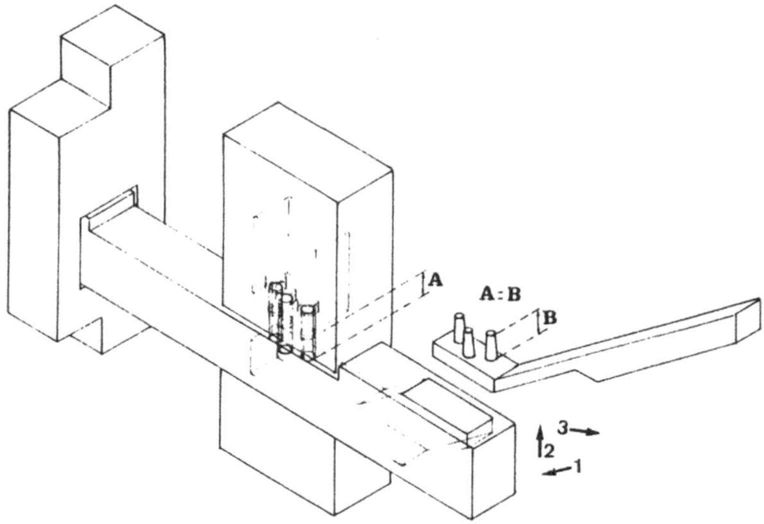

Figure 1

Principle of the Egyptian lock.

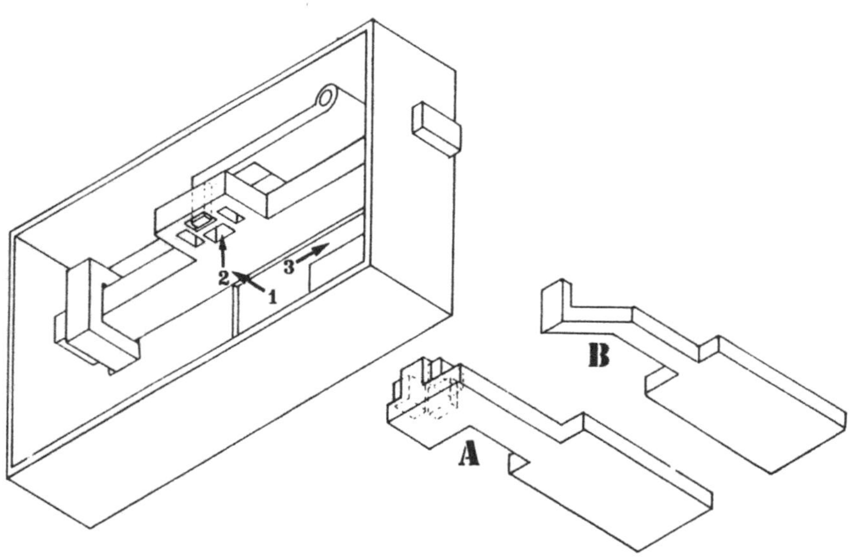

Figure 2

Principle of the Roman lock.

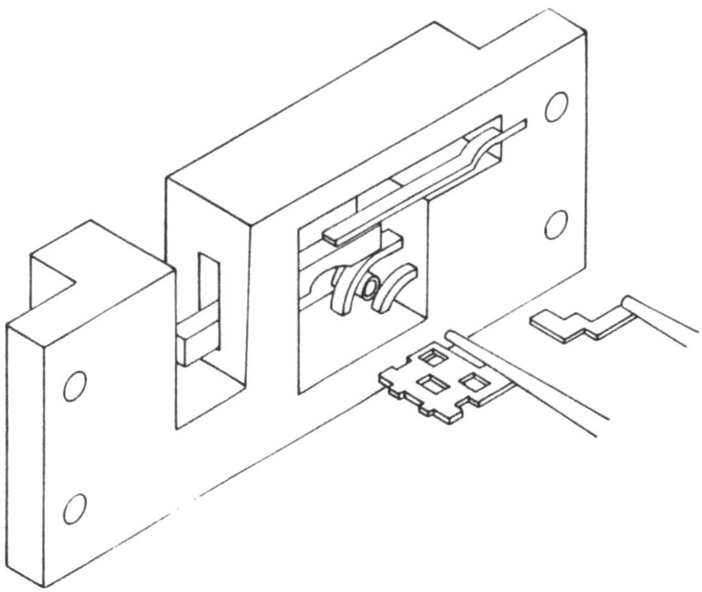

Figure 3

Lock with key that makes a half turn.

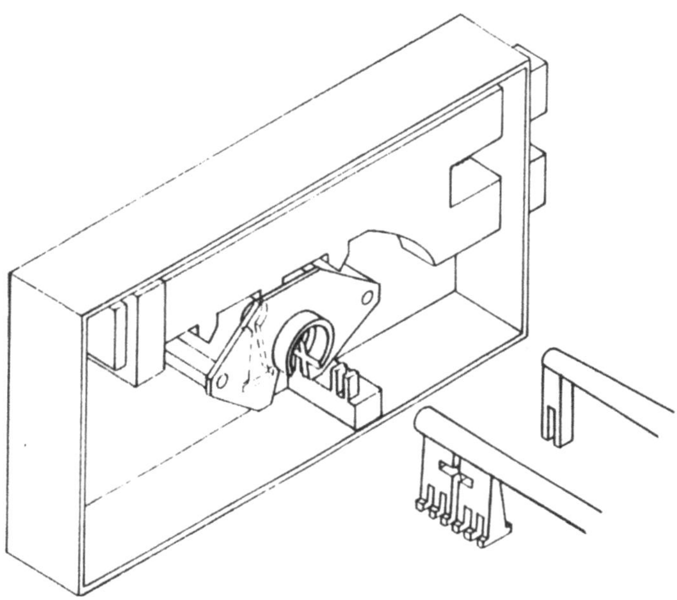

Figure 4

Lock with key that makes a complete turn.

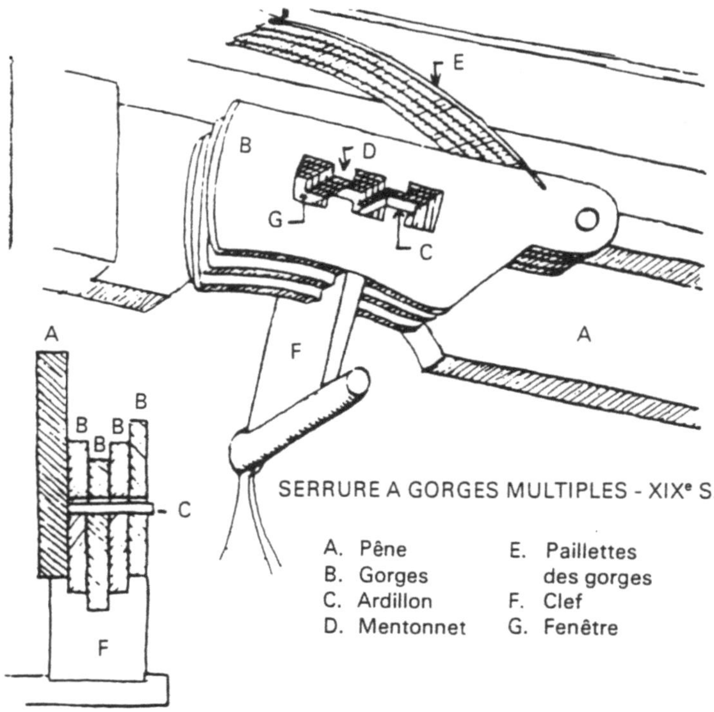

Figure 5

Multiple lever lock.

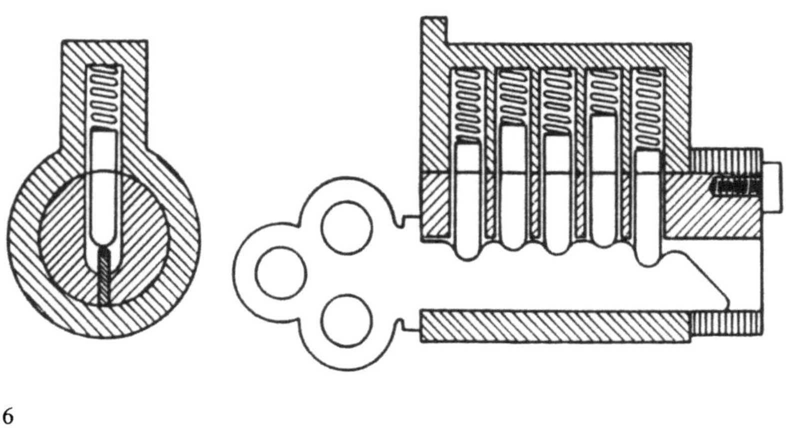

Figure 6

Yale's cylinder lock.

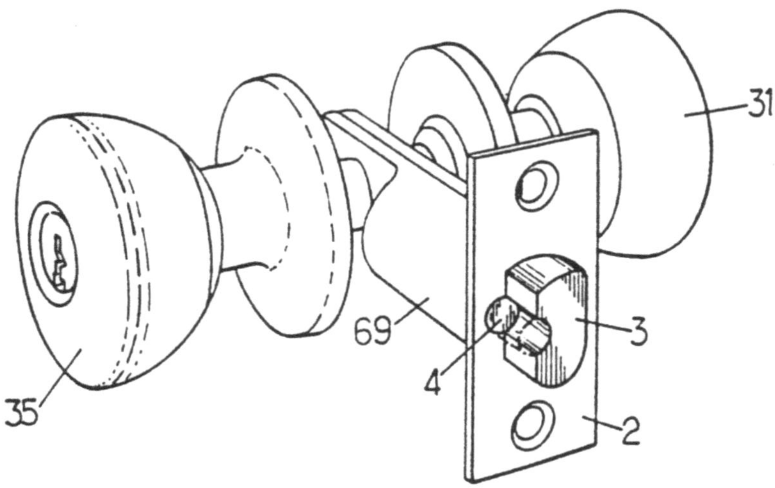

Figure 7

Lewiner's two bolt lock, outside view (model with tubular shape).

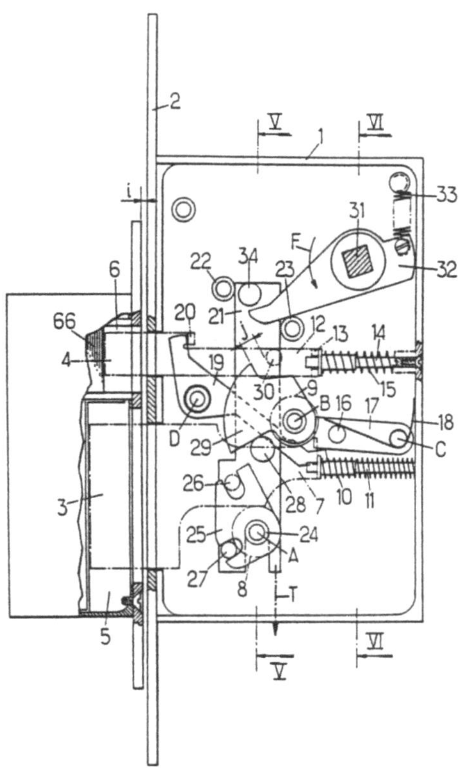

Figure 8

Lewiner's two bolt lock, one of the inside views.

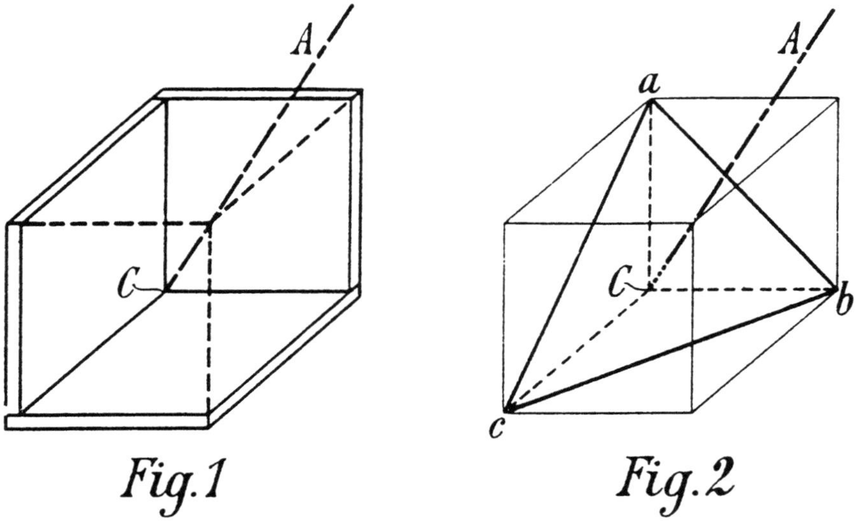

Figure 9

The triple mirror and tetrahedral prism: glass reflecting devices patented in France by Société Carl Zeiss in 1905.

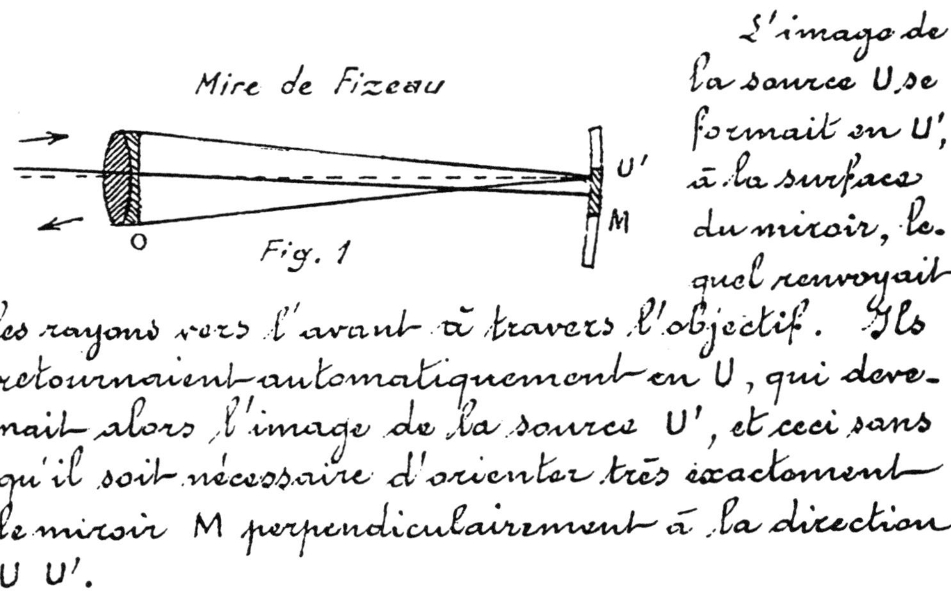

Figure 10

Principle of Fizeau's optical combination for measuring the speed of light.

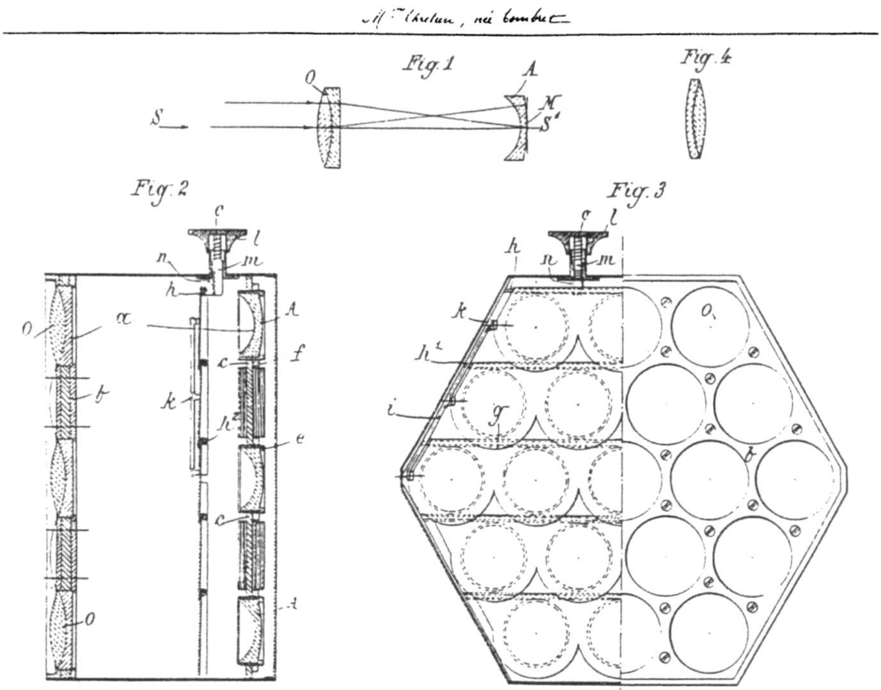

Figure 11

Optical principle of Chrétien's catadioptric reflector for secret signalling and arrangement of reflectors for the multiple device.

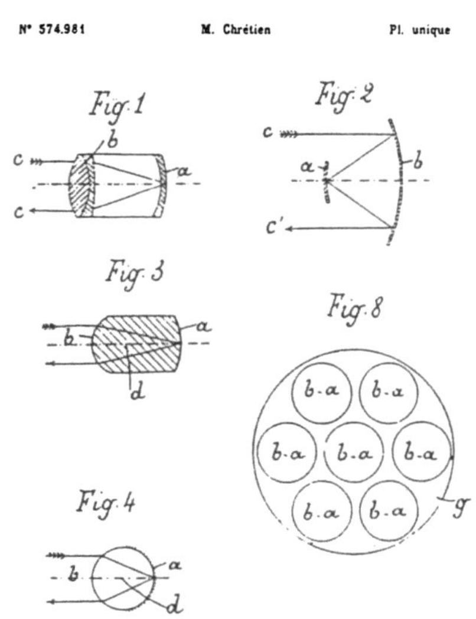

Figure 12

Optical principle of the catadioptric reflector for night signalling; notice the very short focal length.

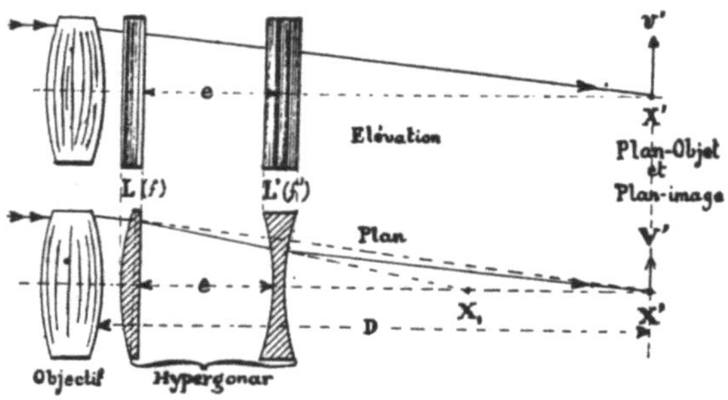

Figure 13

Optical principle of the Hypergonar lens.

CREDITS FOR THE ILLUSTRATIONS:
Figures 1, 2, 3, 4 & 6: G. Berthier, 'Historique des Sûretés', unpublished. Figure 5: R. Lecoq, 'La Serrurerie ancienne', 1973. Figures 7 &8: French patent n° 2 575 510 by J. Lewiner, 1984. Figure 9: French patent nr. 360 159 by Society Carl Zeiss, 1905. Figure 10: H. Chrétien, 'Calcul des Combinaisons Optiques', Paris, 1927. Figures 11 & 12: French patents by H. Chrétien, The Chrétien archives, Nice. Figure 13: Bulletin de l'A.F.I.T.E.C., nr. 11, 1952.

HOUSING LOCATION FOR LOW INCOME RESIDENTS: AN ARCHITECTURAL CASE STUDY OF SIMULATING CONFLICTS OF INTEREST AND GENERATING COMPROMISE PROPOSALS

D.P. GRANT,
California Polytechnic State University
USA

ABSTRACT. Design and science are compared, and the essential difference between the two is defined in terms of subjective values. In science, subjective values must be excluded from deliberations, while in design subjective values are the core of the decision process. However, some characteristics of scientific thought are useful in the design process, and design methodology is in part the application of science-like order and discipline to the process of design. Models of design in architecture, applications of design methodology in architecture, and overlay models for site selection and their application to selecting sites for low-income housing are described, using a modern digital computer version of the overlay method. The process of site selection is by its nature dependent on the subjective values of the decision maker. If there are multiple points of view in a multi-client problem situation, each point of view can be modeled by a separate iteration of the process, with value judgments differing as appropriate to the different points of view. Differences in output for multiple points of view are descriptive of conflicts of interest. Compromise proposals can be developed by proposing mutual changes in judgments to two or more parties with conflicting values. The method allows sensitivity analysis to insure that a proposed trade-off by two or more parties does in fact bring about a compromise agreement as to which sites are suitable, so that only effective compromise proposals are directed to the participants.

1. Design Methodology In Architecture And Its Relationship To science

As this workshop is directed towards the relationships of science with design methodology, I will endeavor to present an example of design methodology in architecture in a way that sheds light on these relationships. First, I will trace the development of design methodology in architecture and contrast it with the methods of science. Second, I will describe the development of models of designing and specify how the case study of low-income housing location relates to various models of designing. Third, I will describe some specific design methods used in architecture during recent decades, and relate the housing

M. J. de Vries et al. (eds.), Design Methodology and Relationships with Science, 63–101.

location model reported here to these other models; and fourth, I will provide a description of the low-income housing location model and some observations on attempts to implement it in the field.

Design methodology in architecture is largely a post-1945 phenomenon. J. Christopher Jones recalls publishing an article on design methodology related to product design in 1948, and that may be the earliest appearance of published work in an area closely related to architecture. Engineering design methodology subsequently appeared in books by Asimow (1962), Alger and Hays (1964), Harrisberger (1966) and others. Many of the methods that are used in architecture today appeared in the literature of other disciplines, such as Operations Research (Churchman, Ackoff and Arnoff, 1957), The science of decision-making (Kaufmann, 1968), Psychology (Osgood, Tannenbaum and Suci, 1957) and Engineering Economics (Grant and Ireson, 1970). Other sources were found in Information Science, Systems Engineering, Logic, Praxeology, and in various forms of mathematics such as Graph Theory, Set Theory, Matrix Algebra and Boolean Algebra.

Architectural design methodology, borrowing from the above sources and others, appeared in the work of Christopher Alexander (1964), Bruce Archer (1963), Geoffrey Broadbent (1973), and in the unpublished work of Horst W.J. Rittel, especially in his teaching at the University of California at Berkeley (1963-1990) and at the University of Stuttgart (1970s-1990). J. Christopher Jones's book DESIGN METHODS: THE SEEDS OF HUMAN FUTURES (1970) has been a book on methods widely read among architects even though it is broader in scope than architecture, and relates more to industrial design models with repetitive production of identical objects than to architecture with it focus on unique, one-off designed objects.

Science provided and continues to provide a model for thought and action that both attracts and repels architects. The prestige and proven effectiveness of scientific discipline in other fields was and is an attraction, in the hope that successes can be achieved in architecture similar to successes achieved in other fields when scientific rigors of thought and action were brought to bear. Specific attractions were orderliness, the uses of quantification and mathematical models of thought, and retraceability of decision processes for purposes of communication, justification, delegation, and growth and learning through reviews of past decisions. The repulsion factor that science has for architects centers on the necessary attempt at the exclusion of values and subjective perceptions in science, whereas in architecture subjective values and perceptions are the very core of thought and action. Science attempts to add to a body of reliable statements not distorted by subjective factors, while design tries to change the present into a future that conforms more closely to subjective perceptions and values. The scientist constantly attempts to seek out and exclude any subjective influences in the posing of hypotheses and the testing of them, while the designer must constantly attempt to guide the decision process towards an ever closer conformity between a value-based image of the future and the image that the designer will create as a result of his or her design decisions. The fundamental question that a scientist must ask is whether or not deliberations have successfully excluded subjective factors, including personal, social and cultural values, while the most fundamental question that a designer must ask is, whose values and which of those values are the legitimate basis for making design decisions. In this sense the misguided ideal of a "scientific design" is a contradiction in terms. Design is not unscientific in a pejorative sense, it is a-scientific. The decision tasks of the designer are qualitatively different from the decision tasks of the scientist. The late Professor Horst Rittel was fond of turning the slogan of the organic architects, "Form Follows Function," into a more accurate observation: "Form Follows Fiction." The implication is that design is the activity of writing a scenario for the future, with appropriate costumes, sets and script, that will conform to a value-based image of the future.

It should be noted that design is a perfectly legitimate object for scientific study, as are design history and design education, and that scientific study is a desirable thing in order to learn about the natures of these things. Perhaps the knowledge so gained can be used to propose ways to improve the effectiveness and efficiency of the process of designing. However, while design, design education and even design history may be fruitfully studied in a scientific way, the activity of design is an a-scientific activity, and cannot be carried on by scientific method or with scientific discipline, especially in the area of excluding personal, social or cultural values. Methods for designing must be appropriate to the nature of the activity of design, that is to say, must directly and openly shape the decision process in conformity with the appropriate subjective values that must be served. The organic school of architecture holds, for example, that no design process or product is ever repeatable, as each design situation is unique, and that part of that uniqueness lies in the values and personalities both of the clients and of the designer, and of the relationships among them, a concept that would rightly be a scandal in science.

Some of the attitudes and even some of the methods of science have proved to be stimulating and fruitful for design thought, but at core the two activities, design and science, are fundamentally different and incompatible, and the reason for this is the diametrically opposed attitudes towards subjective values that are appropriate to each.

An example of scientific thought that proved most stimulating to design thought was the concept of paradigmatic revolution described in Thomas Kuhn's THE STRUCTURE OF SCIENTIFIC REVOLUTIONS (1970). The ruling paradigms in architecture have shifted among many contrasting poles of thought in a way similar to the shifts of scientific thought that Kuhn describes. Design thought has shifted back and forth between rationalist and empiricist paradigms, as has science. However, design thought has also shifted back and forth between other poles that may not have a parallel in science: between the dominance of form versus the dominance of content, between the dominance of theory and ideology versus the dominance of focus on the building as a visual object, and between a preference for the emotion of recognition versus a preference for the emotion of surprise, to offer three examples. In the design of the Temple of Athena known as the Parthenon, Iktinus and Kallikrates focussed their efforts on designing an image on the retina of the viewer, distorting the objective reality of the building to achieve the desired perception of the building. In the work of later designers like Brunelleschi, Alberti and Palladio, the focus was on the object itself, and on its objectively measurable dimensions and proportions, rather than on the image of the object in the perceptual system of the viewer. A similar contrast can be seen in the work of Frank Lloyd Wright, who focussed on the perceptual experience of the viewer, and that of Ludwig Mies van der Rohe and Le Corbusier, who focussed on the objective dimensions and proportions of the building itself. Thus, while some aspects of design thought in architecture can be viewed as having interesting parallels with scientific thought, others are so different as to be contradictory and incompatible, and would violate the most fundmental assumptions in science if applied there. It is not thinkable, for example, for a scientist to distort the objects of study so that they will conform to a predetermined image on the retina of the viewer, but in architecture, precisely that kind of distortion was undertaken in the design of one of the most admired monuments from architectural history, the Parthenon. Iktinus and Kallikrates might have profited from having available some data from the modern sciences of optics and of perception, but they would have put such scientifically derived data to a very a-scientific sort of use, in order to create a building distorted to achieve the value-based image of what the building should be perceived to be. To put it mildly, the accumulation of reliable and "true" statements, the overriding goal of science, is not by any stretch of the imagination the goal of architectural thought and action.

One characteristic of scientific thought that both attracts and repels is the matter of quantification. Decision-making in architecture over the 4,800 years since Im-ho-tep has focussed more on qualitative matters than on quantitative matters. To a certain degree it is probably accurate to say that, at least in architecture, the more susceptible a matter is to being quantified, the less significant it is in human terms; and that the more important a matter is in human terms, the less likely it is to be quantifiable. People to whom this point of view is attractive are fond of pointing out that the master builders of the Gothic cathedrals, while they could doubtlessly add and subtract, probably couldn't multiply and divide. Such people also probably find that they are often using the term "number crunching" in a derogatory sense in the closing years of the 20th century. Quantification is now well established in technical and economic aspects of architec- tural decision making, and few argue with this. The question that rouses intense disagreement today is that of whether human values can or should be expressed as numeric symbols. In the case study presented here, human values are represented in numeric form, and that is a controversial aspect of the work reported. An example of quantification in architectural design from the ancient world is the precise measurement used to achieve entasis in the

Parthenon and other classical buildings in which entasis was a feature. Examples from other times include the numeric theories of beauty in use by architects of the Italian Renaissance, including the Pythagorean Lambda, the Golden Ratio, and the arithmetic, geometric and harmonic proportions of Palladio. The system of scale and proportion known as Le Modulor, developed and applied by Le Corbusier and his followers in the 20th century, is a further example. All of these fall within the category of applying quantified models to questions of aesthetics. The non-numeric mathematics of set theory and graph theory have had much influence on architectural thought in recent decades, and indeed the most powerful method yet developed for architectural design is the application of planar graphs and their duals to architectural design in plan, section and elevation. Even so, quantification still both attracts and repels architects in these closing years of the century, and the repulsion felt towards quantification has been the basis upon which many architects have rejected and criticized the whole idea of a methodology of design. A commonly held belief that architectural design must be wholistic in its approach whereas science is analytical is a related factor that has led many architects to be suspicious of both the relationships of science and design, and of the whole concept of a body of knowledge known as design methodology.

Design methodology, then, has emerged in architecture mostly in the years since World War II. It was first accepted in design education and design practice in the forms in which it was presented by J. Christopher Jones, Christopher Alexander, Horst W.J. Rittel, Geoffrey Broadbent, and others. One feature of design methodology that still both attracts and repels architects is the concept of quantifying the decision process, especially in the area of value judgments and intangibles. Engineering design methodology, decision theory, operations research and other fields have focussed largely on quantification and mathematical models, while to many architects this is still a highly arguable and unattractive prospect. The methodology of British and Australian building science and architectural science has also focussed on quantified analysis and manipulation of tangibles, while the design methodology of architecture has focussed more on values and judgments in the realm of intangibles. A curious paradox is that British architectural design methodology has tended more towards quantification and tangibles, while American and German architectural design methodology has tended more towards dealing with intangibles and qualitative distinctions. Given other aspects of British, American and German cultures, one would have expected the opposite to be the case.

Overviews of architectural design methodology include Rittel (1970; 1971), Jones (1970), Grant (1972a), Broadbent (1973), Cross (1984) and Heath (1984). The primary periodical on the topic of design methodology is DESIGN METHODS: THEORIES, RESEARCH, EDUCATION AND PRACTICE, published by the Design Methods Institute, P.O.Box 3, San Luis Obispo, California, 93406 USA, which is the successor to the periodicals DESIGN METHODS NEWSLETTER (1966-1971); and DMG-DRS JOURNAL: DESIGN

RESEARCH AND METHODS (1971-1975). This series of titles is now in its 26th year of continuous publication. Another periodical which publishes substantial amounts of material on design methodology along with more general material on design is DESIGN STUDIES, co-sponsored by the Design Research Society in Britain, Nigel Cross, Faculty of Technology, Open University, Walton Hall, Milton Keynes, MK7 6AA, United Kingdom.

2. Models Of Designing

2.1. A FIRST MODEL OF DESIGNING

An early step in the development of a design methodology was of course the development of a linear model of design. Various versions of this model contained from three to nine steps, the three crucial steps being ANALYSIS, SYNTHESIS and EVALUATION. The choice of terms was unfortunate. The terms SYNTHESIS and ANALYSIS seem to have been borrowed from chemistry, perhaps in the hope that the prestige and successes of science would be brought over along with the vocabulary. A better choice of three terms for the most important three steps would have been TRANSLATION, IDEA PRODUCTION, and SELECTION. Jones (1970) employed the terms DIVERGENCE and CONVERGENCE for SYNTHESIS and EVALUATION, and these terms would have been better, too. However, the literature of design and of methodology generally uses the original three terms - ANALYSIS, SYNTHESIS and EVALUATION - and so those are the terms that will be used here.

```
          ANALYSIS     SYNTHESIS    EVALUATION
--->|--->|----------->|----------->|----------->|--->|--->
          Translation  Idea Produc-  Selection
          Definition     tion        Grading
          Formulation  Divergence    Convergence
          Transforma-  Creativity    Choice
            tion       Conjecture    Refutation
```

Figure 2.1. A First Model of Designing

2.2. A SECOND MODEL OF DESIGNING

The 1960s witnessed a general rejection of the idea that important human experiences could be described by linear models, and so a second model of design was taken up, borrowed from the field of Cybernetics. In this model, ANALYSIS, SYNTHESIS and EVALUATION are three spokes radiating out from a center, and the process of design is seen as a spiral process crossing these three spokes many times. In the center of the model, where the three spokes intersected, is a cloud of fog labelled "Solution Country." The criterion for having reached completion in "solution country" is having exhausted time, money and/or patience. This model was used by Horst Rittel to illustrate the concept that design problems are "wicked" problems, with no clearcut criteria for when a solution is complete and other troublesome characteristics. The underlying idea for this model is that a designer constantly does all three things -

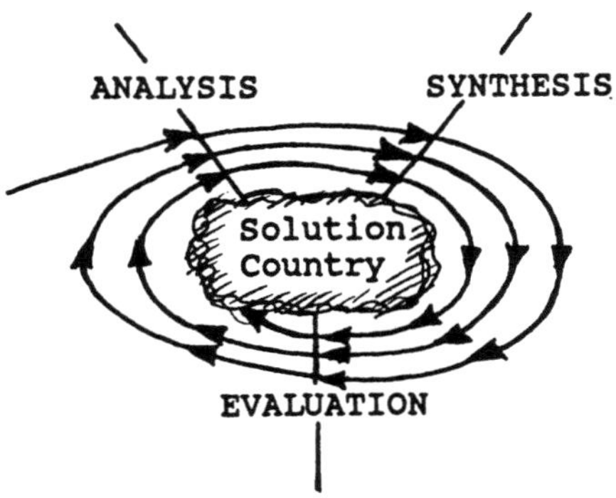

Figure 2.2. A Second Model of Designing

analyzes, synthesizes, and evaluates - over and over, repeating the three-activity sequence dozens or hundreds or thousands of times, each time redefining or reformulating the problem in the light of what was learned in earlier iterations. This model of design employs the cybernetic concept of an isomorphism, or repeated pattern, as the model for portraying what the designer does.

2.3. A THIRD MODEL OF DESIGNING

Like the first, linear model, the second, isomorphic model seemed to imply that ANALYSIS causes SYNTHESIS, and some designers do not believe that that is what they experience in their work. Geoffrey Broadbent theorized that the design process was more properly modeled as a "Y" or a three-pointed star in which the most important sources of ideas in the SYNTHESIS stage came from outside sources, and not from intellectual analysis of the program itself. He posed four such sources of ideas for designers: the Pragmatic or trial-and-error source; the Iconic source wherein the designer followed past precedents; the Canonic source in which rules and principles govern the design process; and the fourth and most profound, the Metaphoric or Analogic source, wherein ideas were drawn from different entities altogether for application to the concept of an architectural design. This third model seems to be the most fruitful model to date. It carries with it the suggestion that the development of a design methodology should be the development of procedures specific to ANALYZING (defining formulating, or translating) a problem; to SYNTHESIZING (producing ideas) in the realms of the Pragmatic, Iconic, Canonic and Metaphoric; and to EVALUATION (selection, convergence, or variety reduction). See Figure 2.3 for an illustration of this model.

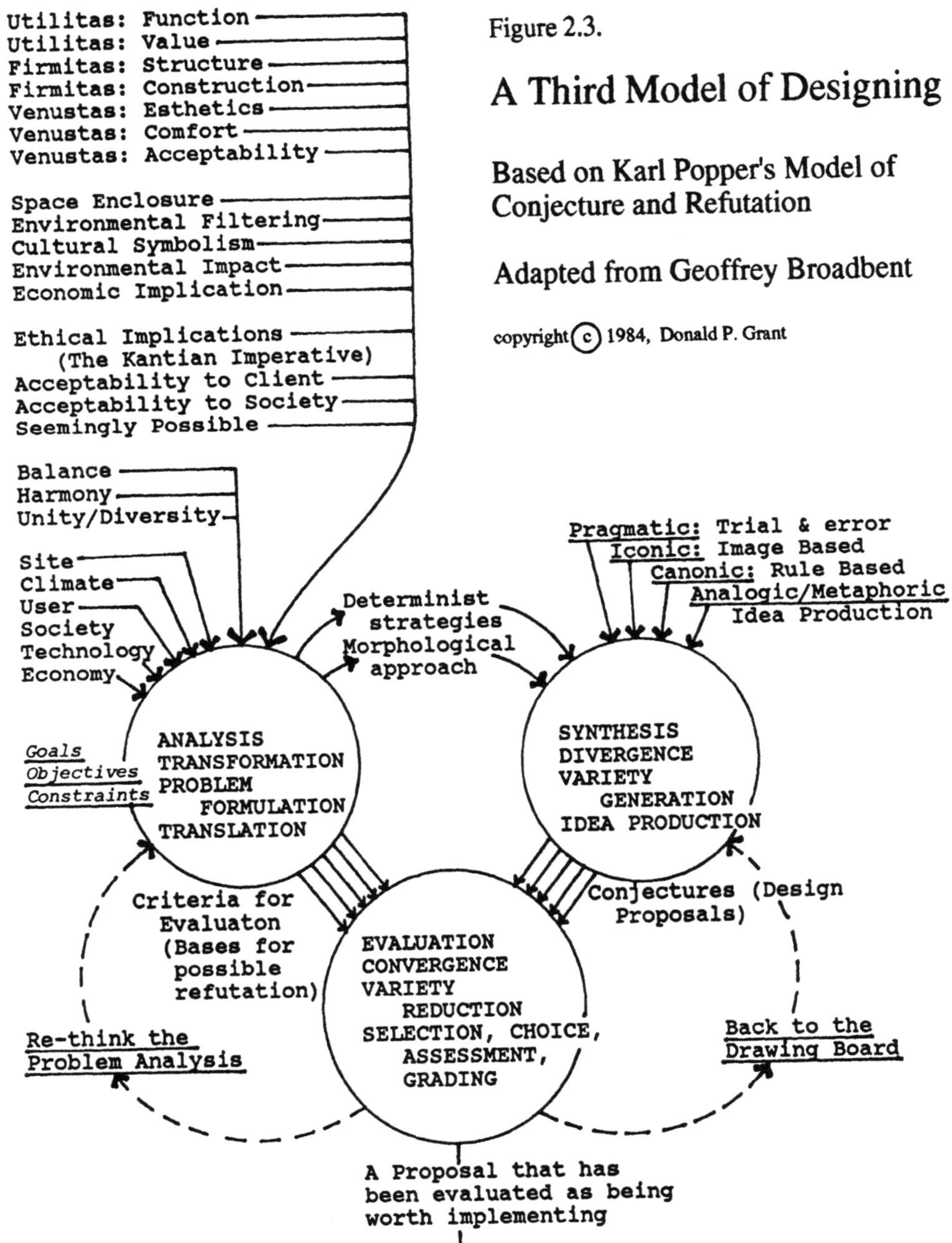

Figure 2.3.

A Third Model of Designing

Based on Karl Popper's Model of Conjecture and Refutation

Adapted from Geoffrey Broadbent

2.4. REFINEMENTS TO THE THIRD MODEL OF DESIGNING

The third model also suggests some added paths. There should be a "back-to-the-drawing-board" path, whereby the designer returns from the EVALUATION process to the SYNTHESIS process to produce more ideas to be evaluated. There should also be a "rethink-the-whole-problem" path, whereby the designer returns from the EVALUATION process to the ANALYSIS process, in order to rethink/re-analyze/ redefine/reformulate/retranslate the problem itself. Experience suggests that a designer should expect to follow these paths of redefinition and redesign more often than not, and should expect only occasionally to exit from the model on a path labelled "implementation." The idea also suggests itself that there should be at least a tentative path from ANALYSIS to SYNTHESIS. Designs produced in a highly deterministic process, such as economic determinism, might indeed be instances in which ANALYSIS causes SYNTHESIS, and designs produced by means of Fritz Zwicky's morphological approach might also be viewed as a variation of this theme, although in both cases it seems that these approaches are part of the CANONIC or rule based approach to architectural design. In the case study to be presented here, involving conflicting interests over the location of low-income housing, an approach to SYNTHESIS or the production of design proposals will be seen in which ANALYSIS does imply SYNTHESIS through a CANONIC or rule based process.

So there are three successive models of design, the last of which seems to have the most to offer. The first model was discredited because it seemed to imply that human experience could be modelled by a linear model, and many people did not agree with that supposition. In popular culture during the 1960s, that idea was rejected by Marshall McLuhan and his followers, and in scientific/technological culture it was rejected on favor of the cybernetic model of a repeated process or isomorphism. The second model was in turn rejected because it implied that intellectual ANALYSIS caused creative SYNTHESIS, and some designers disputed the truth of this notion as not corresponding to their experiences when designing. The third model suggests the addition of significant paths in the process of designing: the path from EVALUATION back to SYNTHESIS, the path from EVALUATION back to ANALYSIS, a secondary path of causation from ANALYSIS to SYNTHESIS, and paths from outside the model into the processes of ANALYSIS and SYNTHESIS, the most significant parhaps being the concepts that ideas came from without, along paths labelled PRAGMATIC, ICONIC, CANONIC and METAPHORIC/ANALOGIC. The method used for the low-income housing case study reported here is largely a third-model approach, in which proposals are developed in a canonic or rule-based approach.

2.5. MODELS OF DESIGNING: DMUC, DMUR OR DMUU?

Another way of modelling the design process is in terms of where it falls in the area of decison-making. Processes of quantification and prediction in design to date have generally fallen within the realm of DMUC, or Decision Making Under Certainty, wherein it is assumed that the predicted consequence of a decision has a probability of occurence of 1.0. That assumption represents of fairly naive state of mind, and it should be expected that design thought will move towards DMUR, Decision-Making Under Risk, and subsequently to DMUU, Decision-making Under Uncertainty, as design thought gains in sophistication. Gaming and simulation seem to be fruitful future directions for design thought, and indeed, have already been discussed for some years as directions in which design methodology might develop. Probabilistic decision trees in the manner of Fritz Zwicky's morphological trees with probability added, seem to offer immediately useful methods for some aspects of design, just as they are emerging in construction cost estimating and in the appraisal of real property values. The case study of low-income housing location reported here is based largely on a DMUC or certainty-based method, although it is easily seen how movement towards a probabilistic or DMUR approach might be implemented, and it will even be suggested that, given adequate resources, the problem might be undertaken as a large-scale DMUU simulation.

2.6. MODELS OF DESIGNING: BLACK BOX, GLASS BOX, GRASSHOPPERS AND FOG

Yet a third way of modelling the design process is on a spectrum from "black box" to "glass box," as suggested by Christopher Jones. The "black box" theory of design implies that something goes on inside the designer's mind, but we cannot see what it is that is happening. The "glass box" theory is that we can see what is happening inside the human mind as design takes place. An extreme version of the "black box" view is that we not only cannot see into the designer's mind, but also that we shouldn't look, because inspection may cause the creativity to disappear. Frank Lloyd Wright refused to take part in a psychological study of architectural designers because be believed that attempts to observe creativity might destroy the creative processes so observed, implying sort of a "grasshopper box" or "locust box." Creativity in this view might be described as a lot of grasshoppers or locusts careening around inside a closed box in a sort of Brownian motion that occasionally spins off creative ideas. Take off the lid to observe the motion, and all the locusts or grasshoppers would escape, and the creative process would be destroyed. That would seem to be the extreme version of the "black box" theory: we cannot see or know what is going on inside the designer, and to try to look inside would be disastrous. The extreme version of the "glass box" is the idea that not only is all knowable about creativity, but indeed that all is known. Not many people hold that view. A more realistic view might be that design idea production is like a glass box with a lot of fog inside, so that while in theory we might be able to see and know what goes on, in fact we do not yet see or know, but perhaps someday we might, if the fog clears. The method applied to low income housing location reported here attempts to be a glass-box-with-fog-method, in which we try to model value judgments in an open and explicit way, with limited success.

2.7. MODELS OF DESIGNING: GENERAL SYSTEMS APPROACHES VERSUS METHODS SPECIFIC TO INDIVIDUAL DISCIPLINES AND PROFESSIONS

A fourth and final classification for methods is into the categories of being general to many disciplines versus being specific to the subject matter of one discipline or profession. One tendency in postwar methodological thought has been to seek general patterns or systems that occur in widely different fields. I will refer to this tendency as the general systems approach, with obvious reference to the work of Bertalanffy (1968) and the Society for General Systems Research. The other tendency has been to develop models from the work specific to one profession or discipline Most of the methods used in architectural design methodology are derived from work in other disciplines, and so it seems fair to say that architectural design methodology has been the fortunate recipient of gifts from other disciplines in the general systems mode. For this reason it seems worthwhile in the future to maximize interaction and communication among design methodologists in architecture and methodologists in other fields, although there will always be annoyances arising from this interaction. For example, it is difficult for architects to work with engineers in design because their two attitudes towards costs and benefits are so different. Given a program or brief and a budget, an architect is conditioned by his or her education and professional subculture to seek ways in which to maximize the benefits derived from the given body or resources. The engineer, on the other hand, is conditioned by education and professional subculture to fix the benefits desired and to attempt to minimize the budget necessary to achieve them. For this reason, the methods favored by the two professions tend to be in conflict rather than in harmony. For another example, when attempting to work with social scientists, the goal of the architect is to learn what the situation is like and then to develop ways to intervene in the situation and change it, while the goal of the social scientist is to learn what the situation is, and then to analyze data and pose hypotheses, but never to intervene or take responsibility or act to change the situation. That makes for very different methods and for dissatisfaction of each discipline with the other. The profession of City and Regional Planning has moved in the course of its development from physical urban design, involving making plans for action, towards endless analysis and modeling that seldom evolves into plans for action to achieve a desired, value-based future state. For that reason, very different methods characterize the work of planners than characterize the work of architects, and again, considerable dissatisfaction is encountered when the two disciplines try to work together. William Alonso touched upon this problem in his paper titled "Beyond the Interdisciplinary Approach" (1972), and Churchman speculated on the topic of interdisciplinary collaboration as well (1969). As an overall observation, a methodology that evolves in the "general systems" frame of mind is probably going to be most valuable in the further development of an architectural design methodology, but the approach has its sources of discontent. In the method reported here for low-income housing location, one source of the model is the traditional map overlay technique that has been used for land use planning, roadway location and meeting scheduling for a century-and-a-half. Another is the method for ranking and weighting objectives developed by Churchman, Ackoff and Arnoff in Operations Research (1957, Ch, 6, pp. 136-154). Yet another is the data organization used in various forms by the Harvard GRID system, the U.S. Census GRIDS system, and the cellular mapping method used by the ORL-ETHZ Institute in Zürich (Ackerknecht, 1972) to organize data in a geographic format for the whole of Switzerland up to the tree line. Indeed, there is a growing discipline of geograffically-based information systems for environmental, social, economic and planning data that has emerged in the 1980s and 1990s. Also borrowed from another field is the generalized, geograffically organized spread sheet approach by which the overlay method, traditionally a labor-intensive graphic technique, is evolved into a digital method for use on modern micro-computers and mini-computers. Thus, the method used in the case study reported here is decidedly a multi-disciplinary, general systems approach to methodology, as are most of the other architectural design methods that will be mentioned briefly here.

3. Methods In Architectural Design

3.1. THE USE OF METHODOLOGY IN EDUCATION AND IN THE PROFESSIONAL PRACTICE.

An overall observation is that methodology in architecture is a more active component of education than it is of professional practice. Design methodology has been taught in many schools of design in Britain and on the continent for twenty to thirty years now, and design methodology is known as a field of study to most people even if they do not choose to use design methods. In the United States, on the other hand, most architecs and architectural educators probably do not even know the field exists, or if they do know, have only a sketchy idea that it is something that Horst Rittel taught at Berkely and that his students attempt to teach at other schools with varying degrees of succes. In the course of professor Rittel's teaching in Berkeley, from 1963 until his death in 1990, many hundreds of students took his classic introductory course, and some students also went on to take a series of graduate seminars with professor Rittel on varieous topics within the field of design methodology as he defined it. One of the Ph.D.

options in architecture at Berkeley has been Architectural Design and Design Methods, and indeed that is the educational backgroud of the author of this paper. Teaching at a larger "mass production" type school at California Polytechnic, I have had well over a thousand students take my introductory course, and a few take follow-up courses as well. It has been my experience that design methodology is a very valuable educational tool, and has greatly helped those students who learned its lessons to grow and develop as designers. Given the large numbers of students who have studied design methodology in architecture at Berkeley and at San Luis Obispo, there is of course inevitable carryover into professional practice, but even so, I believe that it is fair to say that design methodology has had its main impact in education, rather than in practice. However, the same characteristics that make methodology valuable in education also recommend it in practice. These characteristics include the encouragement of thoroughness, the objectification of decision processes so that they can be retraced, communicated, argued, justified, defended, and delegated, and the characteristic of making it possible to trace and reveal and discuss the value judgements underlying decisions. The movement towards accountability and frequent legal liability suits in the United States makes in seem likely that the orderly peocedures known as design methodology will become inceasingly attractive to the professional in practive with the oassage of time, so it seems teasonable to expect that design methodology will be used more and more in progessional practice in the future.

3.2. AN EARLY METHOD: HIERARCHICAL DECOMPOSITION, OR CLUSTERING

Let us look now at specific methods used in architecture over the past 25 years. The first method to be widely used was the method of hierarchical decomposition or clustering, widely exposed in Christopher Alexander's book NOTES ON THE SYNTHESIS OF FORM. This method deals more with possible ways of composing programs or briefs than it does with the design of architectural forms. Alexander has in the years since the publication of NOTES ... renounced this method, indeed has denounced the whole idea of design methodology, and gone on to produce several widely-read books on a different approach to designing. The nature of hierarchical decompusition or clustering was tailor made for implementation on the digital computer, and many articles and papers on this topic characterize the early literature of both design methodology and of computer-aided design. In terms of the models of design, this is a method of ANALYSIS. Many readers of Alexander's NOTES ON THE SYNTHESIS OF FORM expressed some dissatisfaction with the fact that it did not carry over into SYNTHESIS as effectively as they hoped it would. The method has been little seen in recent years, and was probably more a useful step in the growth and development of design methodology than it was a succesful method in its own right. Positive outcomes of the method were the suggestion that matrix organisation could be useful as a means of analyzing a design program or brief, and the message that relatively modern and non-numeric forms of mathematics might have something to offer to design thought. Later methods of circulation-based plan synthesis and evaluation like CRAFT and CORELAP, and the powerful method of planar graphs and their duals for synthesizing and evaluating building plans or forms, probably take their origin from early efforts in the area of hierarchical decomposition and clustering. The method is useful in a general sense in helping a designer to increase familiarity with the problem at hand, and although it is easily implemented on a digital computer, its main value may be derived from simple pencil-and-paper explorations in seeking clustered relationships among program components. The method is legitimate and is still used by some, although it was never as thorough and complete a method for plan development as some of its early adherents hoped that it would be.

3.3. ANOTHER EARLY FAMILY OF METHODS: TRAFIC AND CIRCULATION-BASED FLOOR PLAN SYNTHESIS AND EVALUATION AS IN CRAFT, CORELAP AND RELATED METHODS

The circulation and traffic based methods, including CRAFT, CORELAP and dozens of independently invented similar methods, were borrowed by architectural design methodologists from the work of industrial engineers. In general, this family analizes program or brief data into a matrix, and then manipulates the data in the matrix before converting it into a directed graph for use in generating and/or evaluating floor plan layouts. In industrial engineering, the early uses of this approach were in the design of factory floor plan layout for efficiency of movement, and in a very short time this kind of method was picked up by architects for use in such complex problems as hospital floor plan design. Similar methods had been used for some time by traffic planners and urban planners, known as accessability models and gravity models for plan analisys, synthesis and evaluation. Traffic and circulation based models were over-used at first. They were used to generate floor plans, when in fact it is hard to conceive of very many floor plans that could justifiably be laid out exclusively on the basis of traffic and circulation. In general, it is probably seldom if ever justified to use these models for plan synthesis, but it is a very legitimate method for evaluating floor plans against those design objectives that deal with traffic and circulation. These methods, like hierarchical decomposition and clustering, were widely learned and used for a while because their appearance coincided with the growing availability of digital computers for use in design and planning.

3.4. DESIGN BY OBJECTIVES

The group of methods that make up the "Design by Objectives" family are also known by various other names, including as an "Alpha-Beta Model for Design Evaluation". Procedures included are defining goals, objectives and constraints; developing criterion or efficiency functions for specific objectives and constraints;

developing an Aggregation Function for combining multiple, partial measures of predicted performance into a single, overall measure of predicted performance; and the controversial area of ranking and weighting objectives and constraints. The "Design by Objectives" frame of mind is truly a general systems approach, and similar approaches have emerged in many fields, including management science, librarianship, information science, criminology and others. Many other methods, including the traffic/circulation methods and the method of planar graphs and their duals can be considered to be special case applications of "Design by Objectives". The method used in the low-income housing location case study reported here is a very direct application of the "Design by Objectives" method, made specific to a geographically-based information system and to modern spread sheet programs. The key to the use of this method is the following steps:

3.4.1. Explicit statement of goals, objectives and constraints in the form. "A should be the case" or "B should not be the case".

3.4.2. Explicit statement of aggregation function. Often this is a weighted average function, but can also be any of several other forms.

3.4.3. Ranking and weighting the relative importance of objectives and constraints (that is, deliberating Alpha-values). This is a controversial aspect of this method, as reported in Grant (1974b).

3.4.4. Explicit statement of the criterion function or efficiency function for each objective and each constraint.

3.4.5. Organizing of data for the application of the method.

3.4.6. Deliberation of the appropriate level of generality for considering the output of the method: a ranking? a performance indicator on a scale of values? a simple accept/reject decision?

In terms of the model of design, "Design by Objectives" is properly viewed as being primarily in the areas of ANALYSIS and EVALUATION, but may be suggestive to some users in the area of SYNTHESIS.

The form of the "Design by Objectives" method that underlies the housing case study reported here is in the DMUC decision framework, with the simple assumption that each consequence predicted has a probability of occurance of 1.0. The next stage in developing the method will probably be the passage over to a DMUR format, with predicted probabilities for alternative outcomes, and then possibly to the level of large scale simulations in a DMUU format without specific probability estimates.

3.5. AN OVERVIEW OF ARCHITECTURAL DESIGN METHODS

Most applicable methods fall into catagories made up of two or three of the basic activities of ANALYSIS, SYNTHESIS, and EVALUATION. Evaluation systems require at least ANALYSIS, SYNTHESIS, and EVALUATION.

The first family of methods mentioned, that is, hierarchical decomposition or clustering methods, is primarily a method of ANALYSIS only.

A method from the second group of methods, traffic and circulation, can be applied, either as an Evaluation method, including ANALYSIS and EVALUATION components, or as a Comprehensive method, adding in SYNTHESIS. Using these methods for SYNTHESIS is probably not appropriate in most situations, and often provides an example of using a method for purposes for which it is not appropriate.

The "Design by Objectives" methods are primarily evaluation methods, with ANALYSIS and EVALUATION components. While the list of goals, objectives and constraints might in fact be usefully suggestive in SYNTHESIZING a variety of design ideas, this is a secondary use of the approach. Additional evaluation methods will be described in the following section.

3.6. EVALUATION SYSTEMS: ANALYSIS + EVALUATION

Evaluation systems might be usefully divided into those that use money for their measures of performance, and those that do not use money for this purpuse.

3.6.1. Evaluation systems in which money is the measure of performance are well documented in a literature of their own, and include procedures that range from simple budgetary limits to more complex methods such as Return-on-Investment Analysis and Benefit/Cost Analysis. Offshoots of these methods include life-Cycle Costing and such management methods as PERT-COST and CPM. Inherent problems limit the objectivity of these methods, as well as their ability to yield consistent results. Some of these problems are problems of inflation and deflation in the buying power of money over time, lost opportunity costs, subjective utility, life-phase subjectivity of value, measuring the value of intangibles in monetary units, and the question of moving into probabilistic DMUR and non-probabilistic DMUU formats. Architectural case studies in this area are frequently encountered in the areas of Return-on-Investment and Benefit/Cost analyses. The trend recently has been toward Life-cycle Cost Evaluation, in either of the above frameworks, ROI or B/C.

3.6.2. Evaluation systems with measures of performance other than money. Many people find the monetary framework for evaluation inappropriate for considering important human values, indeed, even hostile to the consideration of important human values. One of the reasons that Churchman, Ackoff and Arnoff developed their system for ranking and weighting goals, objectives and constraints was in order to have an alternative evaluation framework in wich it was not necessary to express human values in monetary terms. In general, non-monetary evaluation systems can be divided into those that use numbers as symbols for intangibles, and systems that use other things than numbers as symbols for intangibles.

3.6.2.1. Evaluation using non-monetary numbers as measures of performance. All kinds of rating scales using arbitrary scales of numbers as measures of performance have been developed. The 4.0 scale used by American universities as a grading system is one example. The entire family of methods described as the "Design by Objectives" family is another example. In the latter methods, numbers called Alpha-values are used to communicate judgments about relative importance, and numbers called Beta-values are used to communicate judgments as to quality, fitness, suitability or desirability. In the housing case study reported here, a third class of numbers called Gamma values is used to communicate judgments about relative power in the decision process. All of these uses of numbers are to some degree problematic. The actual judgments are not numeric in themselves, but are only preferential. Numbers are used as a preference judgments of various types, and this is an unacceptable use of numbers in the opinions of some people. This controversial point applies very directly to the housing location study reported here, in that the method reported does use numeric Alpha-, Beta-, and Gamma-values in an attempt to portray human preference judgments.

3.6.2.2. Non-numeric symbols for use in evaluation. The primary method in architectural design methods that falls within this category is the method of Image Profiling in the Manner of the Semantic Differential. The Semantic Differential method was developed in Psychology by Osgood, Tannenbaum and Suci (1957) as a means of analyzing judgmental data about abstract entities to which numbers could not be easily applied. The approach makes use of a scale of paired, opposing terms, and a profile traced between these pairs to characterize some abstract thing. In design, the application is primarily to matters such as image, mood, symbolism and feeling. Numbers do intrude at a certain point, as measures of how much one profile deviates from another, but this represents only a minor intervention of numbers into the process; the basic preference judgments that make up the core of the method are semantic, not numeric, distinctions.

3.6.3. Application of evaluation systems. Since design has become an area of very public controversy in the United States, evaluation systems are frequently called for as frameworks for argument. Sometimes these are posed in non-monetary terms, in which the end decision is one of which alternative course of action to follow, or whether to take (or allow) no action at all. Environmental Impact Reports (EIR's) are probably the most frequently encountered examples of this.

Evaluation, made up of ANALYSIS and EVALUATION (Selection) processes, has always been a necessary part of making a conscious decision. In architectural design methodology, what is different is that these processes are made explicit. Benefits are recordability and memory, thoroughness, and communicability for purposes of delegation, justification and defense, and long-term professional learning by reviewing past decisions. With the advent of frequent liability suits the matter of justifying and defending a past decision to prove that there was no negligence is perhaps the most timely current reason for interest in Evaluation systems.

3.6.4. Limitations of Evaluation Systems: Valuing Intangibles. Although considerable sophistication has been achieved in various kinds of Evaluation systems over the past three decades, there still remains the ultimate problem of valuing intangibles. How does one value a human life, a lost view, a lost cultural monument or landmark, or the gain or loss of aesthetic pleasure by one or more persons? When construction was begun on Eero Saarinen's Arch in St. Louis, Missouri, it was estimated that 13 lives would be lost in the course of the construction project. Is the value of the completed arch greater than the value of the 13 lives, or less? In fact, no lives were lost in the construction, which surprised everyone involved in the project, but the question of the relative values of intangibles remains forever an unresolved question, whether the measures of performance used are monetary or non-monetary. This dilemma of relative values is encountered very directly in the housing location study reported here.

3.7. IDEA PRODUCTION SYSTEMS: ANALYSIS + SYNTHESIS

Creative idea production became a topic of much discussion and research in the United States of America after the first Sputnik went into orbit. There was great anxiety that American science and technology were no longer in the forefront of creative thought, so research projects were funded in order to try to discover the nature of the creative process. The primary thrust was in the sciences and technologies, but some investigations lapped over into architecture, art, music and other fields. For several years the topic "Creativity" was so obnoxious a fad that one grew tired of hearing of it. One useful outcome of all this activity was the posing of a model of the creative process:

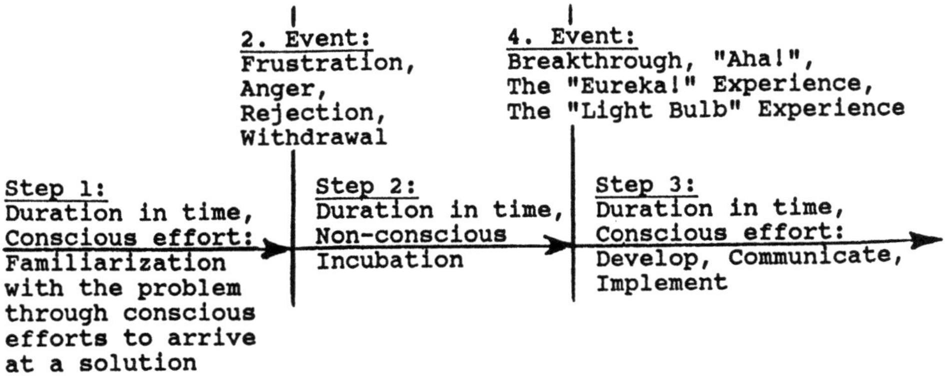

The creative process was seen as three activities with duration over time, two of them periods of conscious effort and one of them the non-conscious period of incubation. Two instantaneous events divided these three tasks from each other: withdrawal from the problem in anger and frustration, and the instant of breathrough, the light bulb experience, the Aha! or Eureka! experience. Researchers also listed some personality characteristics found among effectively creative people: tolerance for ambiguity, resistance to premature closure, tendency to doodle with no set goal in mind, and so on. Creativity workshops and growth groups proliferated, most of them very flaky and fatuous. However, some usable design methods also emerged, and it is with these that design methodology concerns itself.

In the ancient world, the creative person, whether a poet or musician or sculptor or architect, was considered to be a conduit through with the Gods or Muses spoke. The artist was not an originator, but a channel. This same notion re-emerged in the Romantic movement in England and elsewhere, but with the difference that the artist was a conduit for Nature, rather than a conduit for Gods or Muses. An opposite view of art intervened between the Ancient and the Romantic concepts of the artist-as-conduit, and this was academic art. The artist studied the best that had been done before, and cultivated knowledge and taste, and then continued the tradition in his or her own work, possibly adding something to tradition in the process. To be an artist in the academic tradition one did not try to stay "open" for the free flow of impulses from the Gods, Muses or Nature, but instead learned the canons of taste and conformed to them. During the 20th century, a new concept of creativity emerged, that is, creativity as a combinatoric phenomenon. Creativity was redefined as the bringing together of existing entities in new combinations, whether by free associaton as in brainstorming and synectics, or through a set of disciplined exercises as in the several morphological methods developed by Astrophysicist Fritz Zwicky.

In the modern free-association methods, brainstorming and synectics, ANALYSIS is provided by the group coordinator or leader in the form of stimulus statements. SYNTHESIS comes from the participants in the form of their accumulated free-association responses, and EVALUATION takes place in a subsequent review after the sessions.

In the combinatoric methods, including the morphological methods, specific steps are followed in organizing data so that hitherto unassociated ideas can be forced into combinations that might be new. Combinations might be randomly derived, or have some probability framework introduced into the process. The method of the morphological box seems to suggest future developments in the area of simulation (DMUU) by its very nature. A recent use of combinatorics can be seen in the marketing methods of Japanese housing producers. Potential buyers sit in front of a computer screen and combine parts from the housing manufacturer's catalogue in various ways in a search for the house appropriate to them. The method closely resembles Zwicky's methods of the morphological box and the morphological tree. The housing location case study reported here most closely resembles the morphological methods as means for idea development.

3.8. Comprehensive methods: ANALYSIS + SYNTHESIS + EVALUATION

The comprehensive methods in use in architectural design methodology include the Overlay Method that is the base for the Housing Location Study reported here, as well as the Method of Planar Graphs and their Duals. Several reductionist strategies for architectural form generation can also be seen to be comprehensive methods, albeit narrow in scope.

3.8.1. The Overlay Methods. The traditional overlay method for space planning and the modern digitized version of the overlay method qualify as comprehensive methods, since the processes included in the methods touch upon all three key aspects of the design process: ANALYSIS + SYNTHESIS + EVALUATION.

The overlay methods are also useful in illustrating the role of three different kinds of activity in the process: data handling, value judgment, and pattern recognition.

The overlay methods include many individual methods already described: defining goals, objectives and constraints; deliberating Alpha-values of relative importance; deliberating Beta-values of relative desirability; developing criterion functions and aggregation functions; and organizing data for the application of value judgments.

When the deliberation of the relative power in the decision process of different persons or groups is included in the process, the area of maximum controversy in design is faced squarely: whose values should be included in the decision process, and with what weight relative to the values of other participants in the process. It is in situations of conflict like this where the objectification characteristic of design methodology seems an attractive characteristic. It is in this extension of the overlay models that the Housing Location Study is posed: the exploitation of the objectification characteristics of the method to simulate conflicts of interest and to generate compromise proposals for multiple clients whose values are in conflict.

3.8.2. The Use of Planar Graphs and their Duals to Develop Architectural Forms. These methods, derived from that part of Topology called Graph Theory, are probably the most powerful methods developed to date for use by Architects in the course of designing the physical plan and form of buildings. Indeed, the late Professor Horst Rittel once said that he thought that when an overall theory of architecture was created, it would probably be in this area, which he called the Method of Cell Configurations (Rittel, Arch.130 Lectures, 1969, 1971). Grant (1975, 1983; 1979) developed step-by-step procedures for the use of this method, and other work centered on this method can be seen in Levin (1964), March and Steadman (1971), Grason (1970), Yu Da (1992) and others. In a general sense, the method is a development of the traditional method called "Bubble Diagramming," but it is so much more powerful as a design tool that there is scarcely any comparison in the usefulness of the two methods. The method has been taught in various forms at several schools of architecture, and having been found by students to be so effective, is used by many in their professional work. Like the overlay methods, this method so clearly distinguishes and sets forth the value judgments that must be made in designing that it is also a good vehicle for illustrating the design process itself.

Basically, the method requires the designer to make value judgments about which spaces or rooms to design, and how to relate to each other and to external environmental characteristics, and expresses these value judgments in a matrix. The matrix is converted into a graph, usually non-planar, and then this graph is planarized to yield a graphic version of program relationships. The dual of the planar graph is constructed, and this new planar graph is a topological model for a large family of floor plans that satisfy the original program expressed in the matrix. Since the working of the method is in the form of matrices, and graphs that correspond to the matrices, the method is easily digitized for computer application. A danger of using the technique is the tendency to work so hard for mathematical rigor and objectivity that the subject design problem is reduced to uselessness in terms of designing a building. This tendency can be seen in the work of March and Steadman (1971) and Steadman (1983). When such reductionist applications are used, the results are probaby inferior to the results of traditional, intuitive design processes.

The method of planar graphs and their duals is a useful vehicle for theory development. For example, it clearly demonstrates that buildings with elaborate programmatic requirements for relationships with the external environment - views, breezes, solar access, and so on - will almost always develop as complex, concave polygons; and that conversely, a simple convex polygon is incapable of meeting elaborate requirements for environmental relations.

The method is clear and comprehensible enough that it can be used as a medium for collaborating with clients to generate floor plans, and according to informal feedback from former students who are now in professional practice, is reliable, effective and efficient as a working tool in the process of building design.

A challenge for methodological development in site selection and master planning will be to combine the method of planar graphs and their duals with the overlay method. The overlay method is most suitable for Locational, "Nature- of-the-spot" objectives and their parameter maps, while the dual graph approach is most suitable for objectives and their parameter maps dealing with relational, adjacency types of parameters. The challenge is to develop an approach in which locational parameters are dealt with using overlay maps, and then relational parameters are dealt with using planar graphs and their duals on the evaluated overlay surface. The method should probably be developed as an adaptive, sequential process in which several iterations are performed alternating between overlay and dual graph processes, in the decision making approach called by Bellman "Dynamic Programming" (Bellman, Cooke and Lockett, 1971). The concept of "fuzzy sets," raised by Zadeh (1973) in the late 1960s and adopted by Japanese management theorists in the early 1990s, may also be a concept of some promise, and also poses one format in which DMUR, probabilistic thought might be brought into the site selection and master planning process.

3.8.3. Reductionist Strategies for Generating Building Form. The most familiar versions of reductionist strategies are processes by which building form is an almost automatic function of economic variables; that is to say, an aggregation of figures about market rents, vacancy rates construction costs, land cost and other economic factor generates the form of a building in a highly determinist manner. Skidmore, Owings and Merrill's early Building Optimization Program (BOP) seems to have been an effective example of this approach to design.

In general, architects seem to hold buildings whose form is derived purely from economic and other determinist factors in some contempt. However, it should be kept in mind that such admired forms as kayaks, canoes, tipis, igloos, and specialized house forms to deal with specific climates were also the products of highly determinist processes, with little added for the sake of self-conscious aesthetic effect. In this frame of mind it is interesting to look at such highly determinist design products as the standard kitchen plan of the McDonald's fast food chain, and to speculate as to whether this deterministically produced design has an elegance of its own, derived from a fitness for its purpose, just as the kayak and canoe are widely admired for the same reason.

Design methodology is feared by some architects, out of concern that there might be some reductionist and/or determinist tendency concealed within a method. The negative argument is that if the method determines architectural form from program requirements, the role of human creativity is minimized or destroyed altogether. The positive argument is the argument of the organic school of thought in architecture, that is, that beauty is fitness, and the most beautiful building is the one that takes its form most directly from the nature of its required functions. The slogan that we take from American sculptor Horatio Greenough, and associate with the architectural thought of Louis Sullivan and Frank Lloyd Wright, is "Form Follows Function." It is a good thought, but the slogan has been so often used and misused that it has taken on a triteness that undercuts the perception of its basic wisdom.

Reductionist, determinist strategies of generating building form can produce trivial results that cause one to question the appropriateness of calling them "architecture." However, reductionist, determinist processes in generating form, whether of buildings or of tools or other artifacts, may also be the path by which the greatest beauty is achieved. There is a dilemma here that must be faced by every working designer.

4. Simulating Conflicts Of Interest And Generating Compromise Proposals For Low-Income Housing Using A Digitized Overlay Method

4.1. THE PLANNING OF SPACE IN ARCHITECTURE

The planning of space in architecture usually takes one of two forms: the selection of specific locations for architectural elements, or the master planning of an entire space by assigning a use to every point in the space. An example of selecting locations is the selection of one or more sites for building low-income housing from among many possible sites, like an existing city. An example of master planning is the assignment of a zoning classification to every point in a city, with no points or spaces left out.

This case study is of the selection of a few sites for low-income housing from among a large number of potential sites, for example, seeking ten sites of 5 hectares each from among a city of several thousand hectares.

4.2. MULTIPLE CLIENTS WITH CONFLICTING VALUES

Fundamental to this case study is the assumption that the design decision process has many clients. Borrowing from Churchman (1969), "Client" is defined here as a person whose interests are affected by a decision. In an absurdly extreme sense, everyone on earth is a client, but only a few of those people have their interests affected with enough immediacy that they have an ethical right to be included as a participant in the decision process. Deciding where to draw the boundary as to whose values are to be represented is perhaps the most political and most rightly controversial of the ethical questions that a designer must face, especially in areas like land use and site selection.

In low-income housing decisions, there are always several important clients whose points of view should be considered. First, there are the potential residents themselves, and the factors that are important to them. Then there are less immediate clients. The managers of low-income housing have concerns like maintenance, the density and size of projects and the cost of land for projects. The local planning officials have interests based on how the location of low-income housing affects the overall nature of the community and how it relates to things like transportation. Police and fire officials have their own interests, as do school districts and welfare agencies. The city government has its own unique set of interests, part of which requires the balancing of one constituent group's interests against those of other constituents. Middle and upper economic class residents often come forth to express the opinion that, where ever low-income housing is to be located, it should be far away from them, and preferably not in their school districts, either. Conflict seems integral to the whole idea of low-income housing location. Systematic, thorough, repeatable and transparent methods for simulating conflicts of interest and generating compromise proposals are important contributions that design methodology can make to this area of decision making. One possible approach is to use the concept of indifference curves in utility fields (Grant, 1992a, 1992b, 1992c). Other sources that might prove stimulating to work in this area are found in the work of Coser (1956) on conflict, that of Alinsky (1972) and Franklin (1961) on community organizing, that of Rubin and Brown (1975) on bargaining and negotiation, and in the IBIS, or Issue-Based Information System developed by Rittel and Kunz (Grant, 1977a; Grant, 1982).

4.3. OVERLAY MODELS FOR PLANNING SPACE

Typically, location decisions involve many goals, objectives and constraints. The overlay method requires the selection of a parameter corresponding to each goal, objective or constraint, and then the construction of a map of the data relevant for each parameter. In the traditional method, the maps are used in their original graphic form. In the digitized approach, a field of symbols substitutes for the map, usually organized into grid cells. The overlay method consists of making appropriate value judgments about relative importance (Alpha-values) and relative desirability (Beta-values), and then applying these value judgments to the data maps or grids to arrive at output judgments about the overall desirability of each point in the space mapped.

The kinds of maps included are typically things like zoning, land cost, distance to schools, distance to shopping, distance to nuisances like noise sources or dangerous traffic, soil types, quality of school district and other such considerations. Each of these parameters is mapped separately. See Figure 4.3.1. for a list of suggested parameters for housing location.

4.3.1. The Traditional Graphic Overlay Method and its Disadvantages. In the traditional version of the overlay method, the first judgment that is required is about which parameter-maps to include in the process. The process of doing this is:

4.3.1.1. Deliberate goals, objectives and constraints.

4.3.1.2. Decide what parameter to map for each of the above.

4.3.1.3. Deliberate Criterion Functions for each of the above, and choose the graphic scale by which Beta-value judgments will be shaded on each map. The usual Beta-value scale uses black for most undesirable and clear (transparent) for most desirable, and intermediate grays for intermediate degrees of desirability.

4.3.1.4. Map the parameter data for each of the above.

4.3.1.5. Shade each map with the appropriate Beta-values, derived from the Criterion Function for each map.

4.3.1.6 Typical Parameters for Site Selection

4.3.1.6.1. <u>Locational, "Nature-of-the-Spot" Parameters</u>

a. Existing density of settlement.
b. Socio-economic characteristics of the population.
c. Existence of low-income housing units at present.
d. Income level of present residents.
e. Elementary school district. This may be mapped more than one time, for example, for level of achievement, for degree of crowding, for presence of special programs, for budget-per-student, for quality of physical facility, for frequency of shootings, stabbings and drug-related incidents at each school.
f. Present zoning or use classification.
g. Probability of future zoning changes - may be a predicted future zone designation, or a probability of change, or a combination of these two.
h. Hours of fog per year.
i. Rainfall data.
j. Wind data.
k. Availability of desirable scenic views
l. Soil conditions. May be too expensive to map all sites, and obtained only for a few highly ranked sites that merit the expense of soil tests.
m. Cost of land. May be too expensive, changeable to map.
n. Existence and condition of existing plant life. May be too expensive, changeable to map.

o. Views from a given site. May be too expensive to map.
p. Predicted paths of landslides, rockslides.
q. Condition of housing and buildings.
r. Condition of streets and utilities.
s. Percentage of unbuilt lots or undeveloped ground.
t. Propensity to flooding.
u. Feeling, mood, image, symbolism, subjective responses to a site and its surroundings. May be unmappable.

4.3.1.6.2. Relational, adjacency or proximity parameters

a. Nearness to public transportation stops or stations.
b. Nearness to the center of the town.
c. Nearness to jobs open to low-income people.
d. Nearness to medical facilities of several kinds.
e. Number of streets to cross to reach school, weighted for traffic speed and density.
f. Distance to elementary school.
g. Nearness to various kinds of recreation facilities.
h. Nearness to food and other shopping.
i. Some measure of contribution to sprawl.
j. Distance to busy streets.
k. Nearness to air pollution, hazards and noise sources.

4.3.1.7. Having done this for all maps, stack the maps on a light table and light from beneath.

4.3.1.8. The areas showing through lightest are the most desirable areas, those darkest the least desirable, and intermediate shades between the lightest and the darkest are of intermediate desirability.

4.3.1.9. Seek areas of appropriate size and shape that show through very lightly. These are the most desirable sites given the value judgments made in the course of the process.

This method is simple and is easy to understand. Clients and others are able to perceive how the process works, and if they participate in making the necessary value judgments they usually feel that this is a fair way of aggregating such judgments and have confidence in the fairness of the outcome.

There are several disadvantages to this traditional, graphic version of the overlay method.

The first disadvantage is that the aggregation function implies by default that all parameter-maps are of equal importance, that is, that the Alpha-value judgment indicating relative importance for each is 1.00. The only way to increase the relative importance of one parameter-map would be to include more than one copy of its shaded Beta-value map in the stack on the light table, a problem that compounds both the second and third disadvantages. The second disadvantage is that it is difficult to discriminate different degrees of desirability when the maps are stacked on the light table. The cumulative sum of several layers of very light gray might be black, or at least very dark, even though the individual judgments were not in the undesirable dark gray range. Even clear mylar or other apparently transparent sheets stop some light, so the tendency is for all points to regress towards black, implying undesirability in the output without having had individual judgments of undesirability as input. The effects of this problem are to limit the number of parameter-maps that can be included in the process; to discourage increasing relative importance for any maps by including multiple copies of the shaded Beta-value parameter-map; and to make misreadings probable by giving an output judgment of undesirable even though the input judgments might all have been in the desirable range.

The third disadvantage relates to the economics of re-iteration. Once the entire process has been completed, it is often desirable to repeat it. One reason for repetition might be to try to model the problem from a second and different point of view. Another reason for reiteration might be to try out a different set of value judgments from the initial point of view, to inspect the effects of changes in value judgments. An example of this would be to inquire as to how sensitive the output would be to my placing greater or lesser importance on one or more parameter-maps. The traditional graphic version is very labor-intensive, and therefore very expensive. It is expensive to do the first iteration, and almost equally expensive to do each subsequent iteration. Thus the method in its traditional form is not friendly to many repeated iterations trying out different judgments. All of the disadvantages of the traditional, graphic version of the overlay method are handily overcome if the parameters can be mapped as numeric or alphabetic symbols instead of as graphic symbols, and if judgments are entered as numeric symbols instead of as shades of gray. By changing to a digital version for manipulation on a digital computer, each disadvantage is overcome.

4.3.2. digital computer-aided version of the overlay method. The initial steps in the use of the modern, numerically-valued digital version of the overlay method are similar to the first steps in the traditional graphic overlay method, but with some added steps:

4.3.2.1. Deliberate goals, objectives and constraints.

4.3.2.2. Decide what parameter to map for each of the above.

4.3.2.3. Deliberate Criterion Functions for each of the above, and choose the numeric scale by which Beta-value judgments will be shown on each map. A useful scale is a scale from one through nine, where 9 = maximum desirability, 1 = maximum undesirability, and 5 = neutral in terms of desirability.

4.3.2.4. Map the parameter data for each of the above.

4.3.2.5. Convert the map to a grid of cells, and fill each cell with a numeric or alphabetic symbol for what data fits that cell.

4.3.2.6. Convert the data grid-map to a grid-map of Beta-value judgments on the 1-5-9 scale, using the appropriate Beta-values, derived from the Criterion Function for each parameter-map.

4.3.2.7. Deliberate Alpha-value judgments for each parameter-map, indicating the judged relative importance of each.

4.3.2.8. Instruct the computer to multiply all cells of each grid-map by that parameter's Alpha-value relative importance judgment.

4.3.2.9. Having done this for all parameter-maps, instruct the computer to construct a new grid-map containing the sums of all the Alpha-Beta-products for each cell from all of the individual parameters' Alpha-weighted-Beta-value grid-maps.

4.3.2.10. Divide every cell by the sum of the Alpha-values for all parameters. The result is a grid-map occupied by values from 1 to 5 to 9, with 1 = maximum undesirability, and 9 = maximum desirability.

4.3.2.11. Seek sites of the appropriate size and shape that are occupied by high values like 9's and 8's. In a large grid-map with many thousands of cells, it is helpful to ask the computer for selective print outs, the first printout containing only 9's, the second printout containing only 9's and 8's, and so on. This pattern recognition task could be done by the computer alone, but would require the computer to exhaustively enumerate all possible sites of the desired size and shape and compute the values occupying the cells in each site. This is technically possible but for most present day computers is not feasible because of the number of operations necessary to do this exhaustive enumeration. For the present, for most computers, the human eye is still a more efficient pattern recognizer, and by asking for selective printouts, the eye is presented with fields of numbers in which to seek the desired size and shape of site.

4.3.2.12. The final map of 1-9 values, resulting from a series of Alpha-value and Beta-value judgments, is a map of judgments about the relative desirabilities of every grid cell from one specific point of view, at one time, for one purpose. The uniqueness of point of view corresponds to the uniqueness of the Alpha-value and Beta-value judgments made from that point of view. See Figure 4.3.3.

4.3.3. Modeling Multiple Points of View. If the entire process described in the preceding section 4.3.2 is carried out separately for each of several points of view, the product is a separate output map for each point of view, each occupied by numbers on the original 1-9 Beta-value desirability scale. See Figure 4.3.4.

If the same cell has a high number for each of two points of view, then those two points of view agree on the desirability of that cell for the purpose at hand. Likewise, two low numbers would indicate agreement on the undesirability of that cell. If one client point of view yielded a high value and another client point of view yielded a low number, that would indicate disagreement as to the desirability of that cell.

4.3.3.1. Sensitivity Analysis If, after deriving the output grid cell-map for a given client's point of view, a particular cell had a low output number in it, for example, a "one," then that would indicate undesirability. It might be that if that client were willing to change the degree of relative importance (Alpha-value) of one parameter, then the output value would be changed. If that were so, then the outcome would be said to be "sensitive" to that change in Alpha-value. The same applies to changes in output values from any change or changes in Alpha- or Beta-values. Sensitivity analysis in this manner requires a complete re-iteration of the entire process with the changed Alpha- and/or Beta-values, to see how sensitive the output is to such changes. For this reason, sensitivity analysis was seldom if ever economically feasible using the traditional graphic overlay method, due to the high cost of each labor-intensive iteration. In the digitized version, however, sensitivity analysis is easy and economical, requiring only the input of changed Alpha-values for relative importances and Beta-values for relative desirability. The changed Alpha-values are input as single numbers applicable to each parameter's grid-map. The changed Beta-values are input in the form of altered Criterion Functions to instruct the computer.

Parameter A: Zoning of Land	Parameter B: Quality of School	Parameter C: Proximity to Low-Income Housing:

Map of Param.A:

R3	R3	R4	R4
R2	R3	R3	R3
R1	R1	R2	R2

Map of Param.B:

F	F	P	P
VG	VG	F	P
Ex	Ex	VG	F

Map of Param.C:

C	D	E	E
B	B	C	E
A	A	C	D

Criterion Function for Each Parameter:

Descri-tion:	Beta-Value:	Descri-tion:	Beta-Value:	Descri-tion:	Beta-Value:
R-1	= 4	Excellent	= 9	A=>=1.0 km.	= 9
R-2	= 6	Very Good	= 7	B=>= .5 km.	= 8
R-3	= 8	Fair	= 5	C=>= .3 km.	= 6
R-4	= 9	Poor	= 1	D=>= .2 km.	= 3
other	= 1			E=<= .2 km.	= 1

For all Criterion Functions:
9=excellent 7=good 5=Neutral 3=poor 1=very poor

Map of Beta-Values for A:

8	8	9	9
6	8	8	8
4	4	6	6

Map of Beta-Values for B:

5	5	1	1
7	7	5	1
9	9	7	5

Map of Beta-Values for C:

6	3	1	1
8	8	6	1
9	9	6	3

Alpha-Values (Relative Importance) for each Parameter:

Alpha:A = 3 Alpha:B = 5 Alpha:C = 2

Alpha x Beta Product Matrix for Each Parameter:

(Alpha)x(Beta)= (A)

24	24	27	27
18	24	24	24
12	12	18	18

(Alpha)x(Beta)= (B)

25	25	5	5
35	35	25	5
45	45	35	25

(Alpha)x(Beta)= (C)

12	6	2	2
16	16	12	2
18	18	12	6

Sum of the Alpha x Beta Products for All Parameters:

61	55	34	34
69	75	61	31
75	75	65	49

Sum of the Alpha x Beta Products Divided by Sum of Alpha's:

OUTCOME:
Sum of the Alpha's = 10

6	6	3	3
7	8	6	3
8	8	7	5

= Judged Quality on a Scale from 9 to 1

Pattern Recognition: Print only 9's; then 9's and 8's; then 9's, 8's and 7's; etc.

Print 9's

	8		
8	8		

Print 9's,8's

7	8		
8	8	7	

Print 9,8,7

Figure 4.3.3

Outcome Matrices for Clients R, S, and T:

Outcome for R:

6	8	9	9
4	6	8	9
3	2	6	8

Outcome for S:

6	3	2	1
8	7	4	2
9	8	6	3

Outcome for T:

7	9	9	6
5	8	7	3
2	5	4	2

Iteration 1: Each client weighted equally

Gamma:R=1

Weighted R:

6	8	9	9
4	6	8	9
3	2	6	8

Gamma:S=1

Weighted R:

6	3	2	1
8	7	4	2
9	8	6	3

Gamma:T=1

Weighted R:

7	9	9	6
5	8	7	3
2	5	4	2

Sum of the weighted matrices:

19	20	20	16
17	21	19	14
14	15	16	13

Sum of the weighted matrices divided by Sum of the Gamma-Values = 3

6	7	7	5
6	7	6	5
5	5	5	4

The best sites are "7's"

Iteration 2: Each client weighted differently:

Gamma:R=1

Weighted R:

6	8	9	9
4	6	8	9
3	2	6	8

Gamma:S=5

Weighted R:

30	15	10	5
40	35	20	10
45	40	30	15

Gamma:T=4

Weighted R:

28	36	36	24
20	32	28	12
8	20	16	8

Sum of the weighted matrices:

64	59	55	38
64	73	56	31
56	62	52	31

Sum of the weighted matrices divided by Sum of the Gamma-Values = 1 + 5 + 4 = 10

6	6	6	4
6	7	6	3
6	6	5	3

The best site is a "7" given the new weighting

Figure 4.3.4

4.3.3.2. Simulating Conflicts of Interest Over Site Selection

The form in which conflicts of interest are simulated in this approach is the occurence of different values in corresponding cells for different clients' points of view.

4.3.3.3. Generating Compromise Proposals

A compromise proposal consists of a two-part proposed trade-off. For example, client A might be asked if they would be willing to place less importance on some parameter in return for Client B placing less importance on some other parameter, in order to arrive at one or more cells on which they would consequently agree. Before the designer would make such a proposal to either client, however, sensitivity analysis should be carried out to insure that if the compromise proposal were accepted by both sides, that some useful agreement would result. It is of no use making proposals to either or both sides in a disagreement if the acceptance of the proposals would not result in a useful outcome.

4.3.3.4. DMUC, DMUR, and DMUU Decision Methods

The method as described above is in the decision-making framework of DMUC, Decision-Making Under Certainty. The assumption is that a chosen cell will have the Beta-value desirability that the Criterion Function specifies, with a probability of 1.0. A future direction for developing this method is to make probability estimates for each possible outcome, instead of assuming a probability of 1.0. This would move the decision process into the DMUR context, that is, Decision-Making Under Risk, with probability estimates. Yet a further development will be to carry out many iterations with random changes and to study the distribution of the output values in a DMUU or Decision-Making Under Uncertainty context, in the manner of modern, large-scale simulations.

4.3.3.5. Glass Box versus Black Box

This method is an attempt to develop a glass box method for site selection. All the elements that go into making a decision are retraceable, including the listing of goals, objectives and constraints; the ranking and weighting of objectives; the construction of Criterion Functions and the resulting Beta-value assignments; the Aggregation Function chosen; and the process of pattern recognition by which possible sites are discovered. The method clearly identifies where human value judgments must be made and input, what kind of data manipulation occurs, and what pattern recognition tasks result and how they are undertaken. It is for these reasons that this method can be seen as an attempt at a glass box method for generating proposals for site selection. Similar transparency of procedure characterizes the way in which conflicting points of view are simulated, and in which compromise proposals are posed and tested for sensitivity.

4.3.4. Scale of Application. Early work with the method, as reported in Cole, Crescione, Morse and Schweitzer (1969), Ward and Grant (1970a; 1970b), Ward, Grant and Chapman (1970); Grant and Thompson (1971); Grant and Chapman (1972), Grant (1972b; 1972c; 1973c; 1973i) and Grant (1974a), was based on mapping an entire community and then seeking potential sites out of the whole array of potential sites in the city. The mapping of even a small city for several parameters was labor-intensive and expensive in those past years, and still is, even with the great advances in machine capacity and speed and in the ease of digitizing graphic data that are available today. Further, it is seldom that in seeking a site for any new building or buildings that the designer can choose any site that is desirable for the purpose. The sites occupied by the post office or the mission church are not likely to be available for housing location no matter how good such sites look on paper. In real world applications, there are usually a finite number of sites that might realistically be available for housing location, and it is these sites that the designer wishes to investigate. Thus, instead of a grid-map of an entire city, it is more frequently useful to determine a list of available sites and to carry out the same method on this finite list instead of for every potential site in the city. This makes it feasible to implement the method using any good spread sheet program, and that is what is feasible at present in applying the method to low-income housing location.

4.3.5. Nature of the Method. The method described here is essentially an application of the Design by Objective Methods, with added steps for generating solution proposals and for simulating different points of views, carrying out sensitivity analysis and generating compromise solution proposals.

4.3.6. Applications. The method described was applied by Grant and Thompson (1971) to low-income housing location in Oakland, California, in which there were five client groups whose points of view had to be considered. It has subsequently been applied by the author in his role as public housing commissioner in a small California city, and is at present applied by the author in the planning of projects to be carried out by the not-for-profit housing corporation of which he is currently president. One observation based on past efforts is that actual clients involved in decision processes about low-income housing location often do not want their motives and interests revealed, let alone modeled in explicit and transparent fashion. This can be the source of considerable resistance to the use of this or any other design method that attempts to achieve objectification, transparency and retraceability in the decision process.

5. Science And Design Methodology

So how does the work in design methodology described here relate to science? Obviously great benefits are derived from quantification, orderliness and the transparent traceability of decisions, all of which seem to relate to the subculture of science. The method is probably efficient and effective compared to less explicit methods, but that cannot be proven. Perhaps it might be possible to estimate relative efficiency and effectiveness between this method and other methods by conducting parallel studies under controlled conditions, but at present there are no plans for doing such studies.

The way in which the method differs substantially from science is in the explicit inclusion of values in the process. The values are identified and objectified and are entered into the process in a transparent manner, but are nevertheless subjective values, which must be excluded from the processes of science, but must be included in the processes of design. That is perhaps a succinct statement of how any design methodology, while borrowing from science, must remain essentially a-scientific, that is, qualitatively different from science in its intentions and its processes.

References

Ackerknecht, Dieter (1972) "Swiss Environmental Information Grid [Informationsraster]" in DMG-DRS JOURNAL: DESIGN RESEARCH AND METHODS, Volume 6 Number 3, July-September 1972, pp.108-111.

Alexander, Christopher (1964) NOTES ON THE SYNTHESIS OF FORM. Cambridge, Massachusetts, USA: Harvard University Press, 1964.

Alger, John R.M., and Carl V. Hays (1964) CREATIVE SYNTHESIS IN DESIGN. Englewood Cliffs, N.J.: Prentice Hall, 1964.

Alinsky, Saul (1972) RULES FOR RADICALS: A PRACTICAL PRIMER FOR REALISTIC RADICALS. N.Y.: Vintage, 1972. Confrontational techiques for bringing disagreements into open conflict, with attention paid to effects on third-party viewers.

Alonso, William (1972) "Beyond the Interdisciplinary Approach." in DMG-DRS JOURNAL: DESIGN RESEARCH AND METHODS, Volume 6 Number 3, July-September 1972, pp.99-104.

Archer, Bruce (1963) "Systematic Methods for Designers." DESIGN. series beginning in issue no. 172, April 1963, and continuing through issue no. 188, August, 1964.

Archer, Bruce (1971) TECHNOLOGIAL INNOVATION - A METHODOLOGY. London: Science Policy Foundation, 1971.

Asimow, Morris (1962) INTRODUCTION TO DESIGN. Englewood Cliffs, N.J.: Prentice Hall, 1962.

Bellman, Richard E.; Kenneth L. Cooke, and Jo Ann Lockett (1970) ALGORITHMS, GRAPHS AND COMPUTERS. N.Y.: Academic Press, 1970. A simple introduction to dynamic programming and graph theory requiring only algebra, rudimentary matrix algebra and an introduction to computers. Games and puzzles are the form used to describe the subjects presented.

Bellman, Richare E.; and Lofteh A. Zadeh (1971) "Decision Making in a Fuzzy Environment." MANAGEMENT SCIENCE, Vol.17 No.4, December 1971. Fuzzy goals and fuzzy constraints can be described as fuzzy sets, that is, sets for which the boundaries are not precisely defined. Abstracted in DMG NEWLSETTER, Vol.5 No.12, December 1971, p.9. See also, Zadeh (1973), this reference list.

Bertalanffy, Ludwig von (1968) GENERAL SYSTEM THEORY: FOUNDATIONS, DEVELOPMENT, APPLICATIONS. New York: George Braziller, 1968.

Broadbent, Geoffrey (1973) DESIGN IN ARCHITECTURE: ARCHITECTURE AND THE HUMAN SCIENCES. N.Y.: Wiley, 1973.

Churchman, C. West; Russell L. Ackoff and E. Leonard Arnoff (1957) INTRODUCTION TO OPERATIONS RESEARCH. N.Y.: Wiley, 1957.

Churchman, C. West (1969) THE SYSTEMS APPROACH. New York: Dell, 1969.

Churchman, C. West (1972) "Architecture and Operations Research: Is Collaboration Possible?" in DMG-DRS JOURNAL: DESIGN RESEARCH AND METHODS, Volume 6 Number 3, July-September 1972, pp.105-107.

Cole, Gregory; Joseph Crescione, Samuel F.B. Morse III, and Stephen Schweitzer (1969) Unpublished student project on site selection and planning based on ecological factors, using a computerized version of the McHarg technique. San Luis Obispo, CA: California Polytechnic State University.

Coser, Lewis (1956) THE FUNCTIONS OF SOCIAL CONFLICT. N.Y.: The Free Press, 1956, The sociological study of conflict as a useful form of social interaction. For related work, see: Simmel, Georg (1955) CONFLICT AND THE WEB OF GROUP AFFILIATIONS. New York: The Free Press, 1955, on the role of conflict in group behavior. See also, Simmel, Georg (1968) THE CONFLICT IN MODERN SOCIETY. New York: Teachers College Press, 1968. See also, Mills, C. Wright (1962) THE MARXISTS. New York: Dell, 1962. The desirable role of conflict in resolving social issues. Critique of American middle class tendency to suppress conflicts, even when they are necessary and desirable for the resolution of significant issues.

Cross, Nigel, ed. (1984) DEVELOPMENTS IN DESIGN METHODOLOGY. New York: John Wiley, 1984. DESIGN METHODS: THEORIES, RESEARCH, EDUCATION AND PRACTICE. A quarterly journal published by the Design Methods Institute, Donald P. Grant, Editor; P.O.Box 3, San Luis Obispo, California 93406 USA. 1992 is Volume 26 of the journal. Annual subscription $30 in the USA, $34 outside the USA ($38 via Air Mail).

DESIGN STUDIES. A quarterly journal published by the Design Research Society, Nigel Cross, Editor; Faculty of Technology, Open University, Walton Hall, Milton Keynes, MK7 6 AA United Kingdom. 1992 is Volume 13 of the journal. Annual subscription in the UK (for DRS members), outside the UK.

Franklin, Benjamin (1961) THE AUTOBRIOGRAPHY AND OTHER WRITINGS. New York: The New American Library, 1961. Introducing and managing conflict into civic life in order to stimulate needed action.

Grant, Donald P.; and Barbara Thompson (1971) "Simulating Conflicts of Interest Over Low-Income Housing Location Using a Computer-Aided Space Allocation Technique." PROCEEDINGS OF THE SIXTH ANNUAL URBAN SYMPOSIUM OF THE ASSOCIATION FOR COMPUTING MACHINERY. New York: Association for Computing Machinery, 1971, pp.80-93. Grant, Donald P.; and Arthur J. Chapman (1972) "A Comprehensive Approach to User Participation in the Space Planning Process." in PROCEEDINGS OF THE SEVENTH ANNUAL URBAN SYMPOSIUM OF THE ASSOCIATION FOR COMPUTING MACHINERY. N.Y.: Assoc.for Computing Machinery, 1972,pp.107-122. Grant, Donald P. (1972a) SYSTEMATIC METHODS IN ENVIRONMENTAL DESIGN: AN INTRODUCTORY BIBLIOGRAPHY. Monticello, Illi- nois, USA: Council of Planning Librarians, Exchange Bibliography Number 302, August 1972. 8-1/2" x 11", 62pp.

Grant, Donald P. (1972b) "A Compromise-generating Planning Information System for Public Housing Location." in THE DMG FIFTH ANNIVERSARY REPORT: DMG OCCASIONAL PAPER NUMBER ONE. Berkeley, California, USA: The Design Methods Group, 1972, pp.23-28.

Grant, Donald P. (1972c) "Combining Proximity Criteria with Nature-of-the-Spot . . . Space Planning Problems." in NINTH ANNUAL DESIGN AUTOMATION WORKSHOP PROCEEDINGS. N.Y.: Assoc. for Computing Machinery and IEEE, 1972, pp.197-202. Grant, Donald P. (1973c) "User Participation in Housing Location Patterns." in Kouskoulas, V., ed., URBAN HOUSING. Detroit, Michigan, USA: Wayne State University, Civil Engineering Dept., 1973, pp. 161-166. Grant, Donald P. (1973i) "The Computer and the Siting of Prefabricated Buildings - Involving the User." INDUSTRIALIZATION FORUM, Volume 4 Number 5, 1973, pp.43-48.

Grant, Donald P. (1974a) AN APPROACH TO USER PARTICIPATION IN THE SPACE PLANNING PROCESS. Ph.D.Dissertation, University of California at Berkeley. Available from Xerox University Microfilms, 2000 Zeeb Road, Ann Arbor, Michigan, xxx, 285 pp., 8-1/2" x 11".

Grant, Donald P. (1974b) "The Problem of Weighting." in DMG-DRS JOURNAL: DESIGN RESEARCH AND METHODS, Vol.8 No.3, July-Sept.1974, pp. 136-141.

Grant, Donald P. (1975; 1983) THE DUAL GRAPH METHOD: HOW TO USE PLANAR GRAPHS AND THEIR DUALS AS DESIGN AIDS. San

Luis Obispo, California, USA: The Small-Scale Master Builder, P.O.Box 5, 93406.
Grant, Donald P. (1976e) "How to Use the Method of Paired Comparisons: A Systematic Approach to the Task of Ranking Things." in DESIGN METHODS AND THEORIES, Volume 10 Number 4, October-December 1976, pp.212-218.
Grant, Donald P. (1976f) "How to Weight Objectives Using the Churchman-Ackoff Method." in DESIGN METHODS AND THEORIES, Volume 10 Number 4, Oct-Dec 1976, pp.219-237.
Grant, Donald P. (1976j) "How to Use an Alpha-Beta Model for Decision-Making with Multiple Objectives." DESIGN METHODS AND THEORIES, Vol.10 No.4, Oct-Dec 1976, pp.200-211.
Grant, Donald P. (1977a) "How to Use the IBIS as a Procedure for Deliberation and Argument in Environmental Design and Planning." in DESIGN METHODS AND THEORIES, Volume 11 Number 4, October- December 1977, pp. 185-220.
Grant, Donald P. (1979) THE DUAL GRAPH APPROACH TO PLAN GENERATION AND EVALUATION. San Luis Obispo, California, USA: The Design Methods Group, DMG Reference Sheet 6, 1979.
Grant, Donald P. (1982) "Issue-Based Information System (IBIS)," in Olsen, Shirley A. (1982) GROUP PLANNING AND PROBLEM-SOLVING METHODS IN ENGINEERING MANAGEMENT. New York: Wiley-Interscience, 1982, pp. 203-246.
Grant, Donald P. (1992a) "Argumentative Information and Decision Systems, Part One: Their Emergence." in DESIGN METHODS: THEORIES, RESEARCH, EDUCATION AND PRACTICE, Volume 26 Number 1, January-March 1992, pp.1524-1540.
Grant, Donald P. (1992b) "Argumentative Information and Decision Systems, Part Two: Guidelines Drawn from the Social Psychology and Sociology of Conflict and Bargaining." in DESIGN METHODS: THEORIES, EDUCATION, RESEARCH AND PRACTICE, Volume 26 Number 2, April-June 1992, pp.1576-1588.
Grant, Donald P. (1992c) "Argumentative Information and Decision Systems, Part Three: Utility Field Analysis and Environmental Problems as Conflict Systems." pp.1630-1640; Part Four: Exploring Symptomatic Levels of Problem Definition in Search of Potentially Benign Problem Landscapes." pp.1641-1646; Part Five: Indifference Curves in Utility Fields." pp.1647-1668; in DESIGN METHODS: THEORIES, EDUCATION, RESEARCH AND PRACTICE, Volume 26 Number 3, July-September 1992, pp.1630-1668, forthcoming.
Grant, Eugene L.; and W. Grant Ireson (1970) PRINCIPLES OF ENGINEERING ECONOMY, Fifth Edition. New York: Ronald Press, 1970.
Grason, John (1970) "A Dual Linear Graph Representation for Space- Filling Location Problems of the Floor Plan Type." in Moore, Gary T., ed.

EMERGING METHODS IN ENVIRONMENTAL DESIGN AND PLANNING. Cambridge, Mass., USA: MIT Press, 1970.

Harrisberger, Lee (1966) ENGINEERSMANSHIP. Belmont, California,USA: Brooks/Cole, 1966.

Heath, Thomas (1984) METHOD IN ARCHITECTURE. New York: John Wiley, 1984.

Jones, J. Christopher (1970) DESIGN METHODS: SEEDS OF HUMAN FUTURES. N.Y.: Wiley-Interscience, 1970.

Kaufmann, Arnold (1968) THE SCIENCE OF DECISION MAKING: AN INTRODUCTION TO PRAXEOLOGY. N.Y.: McGraw Hill, 1968.

Kuhn, Thomas (1970) THE STRUCTURE OF SCIENTIFIC REVOLUTIONS, 2nd Edition. Chicago: University of Chicago Press, 1970.

Levin, P.H. (1964) "The Use of Graphs to Decide the Optimum Layout of Buildings." THE ARCHITECTS JOURNAL, London, October 7, 1964, pp. 809-816.

March, Lionel; and Philip Steadman (1971) THE GEOMETRY OF ENVIRONMENT: AN INTRODUCTION TO SPATIAL ORGANIZATION IN DESIGN. Cambridge, Mass., USA: MIT Press, 1971.

Mark, Robert (1982) EXPERIMENTS IN GOTHIC STRUCTURE. Cambridge, Mass.,USA: MIT Press, 1982.

Osgood, Charles E.; Percy H. Tannenbaum and George J. Suci (1957) THE MEASUREMENT OF MEANING. Champaign, Illinois, USA: University of Illinois Press, 1957.

Osgood, C. (1971) "Exploration in Semantic Space: A Personal Diary." JOURNAL OF SOCIAL ISSUES, Vol.27 No.4, pp.5-64.

Rittel, Horst W. J. (1970; 1971) The Design of an Education System for Design. DMG NEWSLETTER. Vol.4 No.12, Dec.1970, pp.3-10; and Vol.5 No.1, Jan.1971, pp.2-11.

Rubin, Jeffrey Z.; and Bert R. Brown (1975) THE SOCIAL PSYCHOLOGY OF BARGAINING AND NEGOTIATION. New York: SOCIAL CONFLICT: ESCALATION, STALEMATE, AND SETTLEMENT. New York: Random House, 1986.

Steadman, Philip (1983) ARCHITECTURAL MORPHOLOGY: AN INTRODUCTION TO THE GEOMETRY OF BUILDING PLANS. London: Pion Ltd., 1983.

Ward, Wesley S.; and Donald P. Grant (1970a) "A Weighted Parameter, Numerically-Valued Adaptation of the McHarg Technique for Spatial Planning and Analysis." in DMG NEWSLETTER, Volume 4 Number 3, March 1970, pp. 2-3.

Ward, Wesley S.; and Donald P. Grant (1970b) "A Computer-Aided Space Allocation Technique." in PROCEEDINGS OF THE KENTUCKY WORKSHOP ON COMPUTER APPLICATIONS TO ENVIRONMENTAL DESIGN. Lexington, Kentucky, USA: The University of Kentucky, 1970, pp. 180-185.

Ward, Wesley S.; Donald P. Grant and Arthur J. Chapman (1970) "A PL/1 Program for Computer-Aided Architectural and Planning Space Allocation." in PROCEEDINGS OF THE FIFTH ANNUAL URBAN SYMPOSIUM OF THE ASSOCIATION FOR COMPUTING MACHINERY. N.Y.: Assoc.for Computing Machinery, 1970, pp.122-141.

Yu Da (1992) "The Relation Between Form and Function: Principles and Application Using Topological Methods." in DESIGN METHODS: THEORIES, RESEARCH, EDUCATION AND PRACTICE, Vol.26 No.1, January-March 1992, pp.1541-1557.

Zadeh, Lofteh A. (1973) "A Partially Annotated Bibliography of Material on the Concept of Fuzzy Sets." DMG-DRS JOURNAL: DESIGN RESEARCH AND METHODS. Vol. 7, No. 1, Jan-Dec 1973, pp. 11-12.

ENGINEERING DESIGN, CONCEPTUAL DESIGN, AND DESIGN THEORY: A REPORT

W.R. SPILLERS
NJ Institute of Technology
USA

S.L. NEWSOME
IBM Corporation
USA

ABSTRACT. This paper discusses the results of a research project funded by the US National Science Foundation. It is divided into three parts:

Part 1. A Review of the Formal Development of Engineering Design. The paper begins by tracing the formal (mathematical) development of engineering design. While there are roots of engineering design which go back at least to the 19th century, it is argued that the real roots of the activity we see today began in the work following World War II driven by the advent of the electronic computer. Parallels are drawn between the developments of formal theories of engineering design and the development of both mathematical programming and artificial intelligence.

Part 2. Elements of Design Theory. If design theory is to involve more than just a clever play on works, it is necessary to be quite specific about the kinds of activities to be included in it. It is argued that ambiguity is central to any discussion of conceptual design as is the ongoing work in knowledge representation. At this point, there is a division of activities into a very theoretical branch concerned with how the mind works, formal theories of knowledge,... and a more practical branch (an engineering or even human factors branch) which is not specifically concerned with how people are creative but concerned with providing the best possible environment (computer workstation) for the designer.

Part 3. Engineering Design versus Design in the Arts: Case Studies. It is finally argued that conceptual (creative) design in engineering is closely related to other creative design activities such as sculpture. The is done using interviews with a designer of tall buildings, a designer of long span bridges, and a sculptor whose work is rather geometric.

1. Formal Development Of Engineering Design

1.1. INTRODUCTION

This paper is concerned with the formal aspects of engineering design. It, for the most part, deals with examples from structural design which has a long and rich tradition. Structural design examples are robust since they range over topics

M. J. de Vries et al. (eds.), Design Methodology and Relationships with Science, 103–120.

from the environment to human behavior but are clearly not all inclusive and seem to have little in common with VLSI, machine design,...

The history of structural design is the history of mechanics which is several hundred years old in its modern version. A milestone commonly cited is the 1904 paper of Michell who dealt in a very elegant manner with structures which could and could not be built. Michell's problem was one of finding a minimum weight structure to carry a given load.

From 1904 to 1958 little work of the Michell type is to be found in the literature. The year 1958 is cited here since it is the date Arrow, Hurwicz, and Uzawa published their seminal work on mathematical programming. Michell's work is in modern terms mathematical programming. What is most interesting is that Michell constructed his solutions without the assistance of a computer.

In the '60s structural engineers discovered mathematical programming, thought it to be a panacea and began the flurry of activity which now has structural optimization as a discipline. When structural engineers discovered mathematical programming, claims of complete automation were put forward. If it were possible to describe structural design as an optimization problem for some computer to solve, the engineer could simply sit back or turn to other activities. This , of course, has not happened. Mathematical programming algorithms did not develop as predicted; what is easy to formulate as a mathematical programming problem turns out to be difficult to deal with numerically. Difficult to solve or not, it will be argued below that activity of this type, which can be handled by rote, plugging into some algorithm, is not design.

There is a parallel here in software design. As early as 1965 Dreyfus noted the gullibility of scientists who had predicted significant steps over a short period of time in the automation of intelligent behavior. It may simply be the case that intelligent behavior is difficult to define. In any case it seems common to extrapolate too quickly in dealing with software related issues.

The other side of the story is that while software has not lived up to expectations, opposite is the case for computer hardware which has commonly exceeded performance predictions. There should be a theorem in here somewhere.

1.2. STATE-OF-THE-ART

On the practical side, automation continues to progress. Analysis is now highly automated and much less a topic of interest on both the practical and theoretical sides. The automation of design using mathematical programming algorithms is moving forward slowly. What is most common is the automation of ad hoc design procedures. That is, what ever engineers do by hand to produce designs (usually not optimal designs) is becoming highly automated. The best examples of this are some cases of simple industrial buildings whose design is fully automated. In these cases it is only necessary to specify a few parameters such as size, location (for environmental loads),... and the computer will automatically produce a set of drawings and specifications. This really is design at the push of a button. Again, it will be argued below that this type of activity is not what is meant by design in the formal sense. Engineers themselves make the distinction between "detail design" and "conceptual design".

With regard to formal design activities, architecture has been a lively area. George Stiny, for example, has been on the scene working for some 20 years and by now has his own literature in place. In a recent piece (Stiny, 1990) he tasks designs as "....used to describe things for making and to show how they work." This allows him to use an algebraic approach in discussing designs. From the point of view of this paper Stiny's approach does not deal with the central issue of conceptual design which is creativity. It is analytical and allows existing designs to be compared but does not support creative activity. How, for example, is a radically new design to be generated as some kind of combination of existing designs? This point will be returned to below.

The US National Science Foundation has supported formal studies of engineering design over the years. {(Spillers, 1974; Newsome, Spillers, Finger, 1989) can be used to gain insight into the type of work supported.} This work is wide-ranging and will not be summarized here. But it is perhaps appropriate to comment on the developing maturity of the field. The 1974 NSF Symposium on Basic Questions of Design Theory came out of an era of early automation which was concerned largely with analysis and permeated with the idea of common ground among the engineering disciplines. The 1988 NSF Workshop came out of a much later period of automation and addressed design more in its own terms. The extension of this has engineering design appearing as the state-of-the-art issue in the forefront of technology (e.g. DARPA, 1989).

There is one final segment of the literature which is not well represented by the work described above. This is human factors driven work which attempts to provide an optimal working environment (Rouse, 1991; Rouse and Cody, 1989). While work of this type does not really deal with the basic issues of knowledge representation and creativity , it has the practical advantage of providing immediate support for engineers doing real design.

1.3. TRENDS

When the US National Science Foundation started its program in Design Theory and Methodology in 1984 within its Directorate of Engineering, it was argued that the program represented one of the truly new programs within the Foundation. It now seems clear that design theory is simply part of a greater and on-going study of cognitive processes.

2. Elements Of Design Theory

2.1. CONCEPTUAL DESIGN VERSUS DETAIL DESIGN

Engineers often make a distinction between the early stages of the design process in which a task is being discussed without having a specific resolution in mind and later stages of the design process in which it is a matter of specifying the details of an already identified design. The problem with this distinction is that a design will ocassionally fail in its detail. (A mathematical programming problem may not have a solution). Nevertheless, for the purposed here, detail design will be regarded to be a matter of rote and not the subject of design theory. More broadly, the idea is to distinguish between bookkeeping type activities which can easily by automated and cognitive activities which need to be treated more carefully. The problem again is that this distinction may not always be clear: Sometimes a difficult task upon completion appears routine. And obviously, once a design breakthrough has been achieved, its repetition becomes routine.

In these terms ambiguity becomes a central element of design theory: That is, once the task to be done (design or build something) is well-understood and it is simply a matter of going through well-defined steps, doing so is not called design here.

2.2. ATTRIBUTES OF DESIGN THEORY

The arguments above allow some general attributes of design theory to be identified:

o Design must have a heuristic component. If any fully automated process is not it be called design, certainly one way to exclude it is to require heuristics.

o Optimization is not design. As a corollary of the above, once a design has been reduced to parametric studies of a given model, design activity has ceased.

o Enumeration is not design. Once the design process has been reduced to simply examining all possibilities, design activity has ceased. This argues against the use of grammars of some kind to generate designs since grammars do not respond to demand. Put another way,

o Design implies a demand.

o Design theory is a metatheory. It is assumed that existing theories of structures, machine design, aesthetics,... stand alone and that design theory is a theory about these theories or a metatheory.

2.3. COGNITIVE SCIENCES

As matters turn out, there are groups now in place within the cognitive sciences which are attempting to deal with specific issues of engineering design or problems equivalent to engineering design. Some of the more popular topics to appear include:

o Knowledge representation. To some extent engineering design can be thought of as conflict resolution in the presence of ambiguity. The manner in which knowledge is a represented and manipulated is then basic. Sowa (1991) gives a good overview of this area.

o Neural Networks. Since the human brain is the best cognitive machine now available, there has been great interest in modelling its working (Touretzky, et.al, 1991) Perhaps neural networks can provide a good structure for design theory.

o Case-Based Reasoning (DARPA, 1989). Do creative designers work from known examples to solve new design problems with which they are faced? If so, how is this done formally?

o Machine learning (Shrager and Langley, 1990). Is it possible to turn a computer into a creative designer? What formalism should be used to do so?There are several points to be made here. First of all, there exists a strong literature relating to artificial intelligence or more broadly cognitive sciences. Problems of engineering design fall into the realm of this activity. But that does not settle matters:

o Domain. If a group were assembled to study engineering design, what would its composition be? Clearly the choices would include computer scientists, information theorists, mathematicians, educators, psychologists, engineers.....

o Domain specificity. Just how universial is case-based reasoning? Are the human factors of architectural design the same as the human factors of VLSI design?

o Generality. Does a breakthrough in the cognitive sciences imply a concomitant breakthrough in engineering design?

o Role of engineering. What should be the structure of the modern university? Why the present interest in engineering?

There will probably be no easy resolution of the larger cognitive issues of engineering design but rather a long and on-going resolution of them as would be expected to occur in a developing area of science. In the short run, the idea of providing an optimal working environment or even design aids for the engineer can be quite productive.

2.4. AN INFORMATION SYSTEM TO AID DESIGNERS

The idea of attempting to help the designer work through the maze of design requirements and even be creative is hardly new (e.g. Rouse, 1990). Information systems come in many shapes and forms and are quite application specific. This section will discuss how such a system should be constructed.

First of all, when engineers discuss conceptual design they do so in a top-down, hierarchical manner (Figures 1,2). That is they start with the concept itself and provide additional detail as the discussion moves forward. (As an aside, this is the opposite of the manner in which engineering design is taught and suggest that conceptual design might be addressed early in the engineering curriculum.)

Figures 1 and 2 refer to the work of the late structural engineer David Geiger. Figure 1 has Geiger designing the US Pavilion at the 1970 Worlds Fair in Osaka where he created the first large pavilion to be covered by an air-supported, fabric-covered cable net. Geiger went on to build some 12 major structures of this type in the following 10-15 years and is now regarded to be the father of this type of structure. The logic of Figure 1 goes something like this:

1. The US federal government initiates a design competition for the US Pavilion at the 1970 Worlds Fair in Osaka.

2. It is stipulated that there are floor space requirements, a fixed and restrictive budget, and considerable interest in the "image" projected by the structure.

3. Size invokes considerations of environmental loading, floor space requirements, and the advantage of an unobstructed interior.

Image invokes considerations of new materials and new configurations.

Budget invokes considerations of the low costs of single story and temporary structures.

4. ......

Figure 2 derives from a 1986 article (ENR, 1986) which has Geiger describing how he came up with a new type of structure for the then forthcoming Olympic games in Korea:

o Image. There were pressures from the project architect to develop "innovative concepts."

o Air supported, fabric covered cable nets. While air supported structures had served Geiger well in the past, they had nagging mechanical problems involving both maintaining internal pressure and melting snow. How then could the mechanical problems be eliminated while maintaining the other features of air supported, fabric-covered cable nets?

o Tensegrity. Geiger was familiar with the work of the inventor Buckminster Fuller.

o ....

A comparison of Figures 1 and 2 shows how subjective the hierarchical representation can be. Clearly, early in his career Geiger was looking for something quite different form what he was looking for later. The point is that an information system or design aid and should not anticipate a designer.

It is possible to formalize this discussion some (Spillers and Newsome, 1989). First of all, the hierarchical representation of information described above is a kind of "dictionary" approach. That is, an object is described by listing its properties; if more information is required, properties of properties are listed. The properties then form a partial order which leads to the idea of a chain decomposition. Without going into details, Figure 3 attempts to show the usefulness of such a decomposition. The point is that a designer is faced with a complex set of requirements as indicated by the original heirachy. The chain decomposition is a) more simple to deal with and b) supplies a symantics for the hierarchy when names are assigned to the elements of the decomposition.

More generally, the idea of a decomposition of information into a simple representation is a traditional one from analysis. When it works it can be very effective. Unfortunately, it seems that a simple chair decomposition is not sophisticated enough but there are of course other variations (weighted decompositions, contractions,...). How this matter is resolved can be of some importance.

This section is closed with two gratitious examples of how information should be represented (Figures 4,5). The problem with both of these figures is, of course, that they don't lend themselves to any sort of formal analysis. Their meaning then remains quite subjective.

3. Engineering Design Versus Design In The Arts: Case Studies

To determine the characteristics and representations used in conceptual design, four expert, creative designers were interviewed about their own conceptual design process. Their responses revealed a mental process that closely resembled the mental processes of experts in other problem-solving domains. The implications of the interviews for computer-aided design tools are discussed.

3.1. INTRODUCTION

One intriguing aspect of engineering design is the complex, cognitive processes involved in the early stages in the design process. This portion of the design process is important since a single concept is often selected during this early stage of processing and eventually implemented. Past evidence (Ullman, Staufer, & Dietterich, 1987) has shown that designers are very unlikely to radically alter their ideas once this stage of the design process is complete. Hence, one could argue that many of the successes and/or failures in the design process have their origin in this early conceptual stage of design.

Despite the importance of conceptual early portions of the design process, there are few tools that aid the designer in this stage of design. Most CAD packages are designed for and used in later, more routine stages of the design process. The prevalence of computer support tools intended to support routine stages of design is not surprising since these routine activities are characterized by their predictability and are thus, the easiest activities to automate. However, the cognitive processes most difficult to automate such as conceptual design could benefit the most from computer tools designed to facilitate the process.

With the goal of providing tools to be used in early complex conceptual stages of engineering design, Newsome and Spillers (1988) and Meister (1987) described characteristics of a CAD package that would potentially support the conceptual stage of design. Both Newsome and Spillers (1988) and Meister (1987) began their analysis of conceptual design with the premise that the process is a problem-solving activity. Both papers presented ideas concerning the nature of this problem-solving process.

Meister (1987) drew upon the general literature on problem solving to conclude that the design support system should incorporate the mental model of the designer and careful emphasis should be given to formulating the information the system provides. Newsome and Spillers (1988) drew on the literature on the influence of expertise in problem-solving to conclude that computer-aided design systems should support abstract, broad representations of the design alternatives. However, the investigators in both cases were forced to admit that their conclusions were speculative because of the sparse empirical data available to describe conceptual processes in design.

The work described below is a beginning attempt to provide information on the actual practice of conceptual design. Towards this goal, we interviewed for expert designers about the processes and behaviors that characterized conceptual design. The interviews were structured to capture the mental and concrete representations that are used by expert designers in early conceptual stages of design. It can be argued that by discovering the representations used in conceptual design we indirectly provide evidence on the similarity of expertise in design to other problem-solving domains (Newsome&Spillers' concern).

3.2. THE INTERVIEWEES

Three of the interviewees are internationally known engineers with expertise in three different areas of civil engineering: Horst Berger is a recognized expert in a fabric roof design; Les Robertson is internationally recognized for his work on the structure of tall buildings; Gerard Fox is an expert in the design and structure of bridges. Our fourth interviewee is a well-known sculpture, Brower Hatcher, who is recognized for his innovative large figures. All of the interviewees could be characterized as expert designers with 20 years or more experience in their field. Although we were primarily interested in engineering designers, we interviewed a sculptor as a comparison case to determine if the design process of a less constrained creative artist is the same as the more constrained (albeit creative) design process of the engineer.

3.3. THE INTERVIEW QUESTIONS

We asked each of the interviewees six questions that were intended to elicit when ideas in the conceptual design stage were generated, the types of mental and concrete representations used to express conceptual ideas, how alternatives were selected, and the potential usefulness of computer tools in the conceptual design phase. Each interview lasted approximately an hour. The interviewees responded to the questions with both concrete examples and with general information about the process of "design". The six questions were:

1. Do you conceive of an idea in the abstract and then modify it to meet specific constraints or do you get the constraints first and then generate ideas?
2. In the early stages of conceiving of the design do you see ideas in your mind and then transfer them to a sketch or if you sketch, do you start to sketch immediately as you are thinking of ideas?
3. Are your sketches in the early design phase rough or detailed? How do the sketches change over time?
4. Do you think of many ideas in your head before finding the one you want to go with or do you work through one idea at a time to see how it fits the constraints?
5. When do you work out the details of the design?
6. If a computer tool were available that allowed you to draw a design without details would you use it in the early stages of design? Is the use of your hands important in working through design ideas?

3.4. RESULTS

Although each of the interviewees were asked all six of the questions, they were allowed to expand on each of the questions in a relatively unconstrained manner. Their responses and comments are reported here as paraphrases and in some cases summaries of the answer.

Question #1: Abstract or constraint-driven ideas.

All three of the engineers reported that their design ideas are constrained by external requirements. Hatcher, the sculptor, reported that in some cases he is able to use an idea developed in the abstract but he too (for large commissioned sculptures) must deal with the client's social concerns and the physical environment in which the sculpture will be erected. All of the interviewees saw the constraints and their role in dealing with the constraints as slightly different.

Burger reported that he starts with the goal (i.e., with what he needs to achieve) and often finds that the answer to how to achieve the goal is simply there. Hatcher also talked about ideas bubbling up from the unconscious and an idea simply coming to him while studying a model of the space for the sculpture. Both Robertson and Fox saw their role more as providing alternative solutions to the constraints detailed by the client. Robertson was sensitive to the engineer taking a too dominant role in the design and emphasized the responsiblity of both providing alternatives and determining consequences of design decisions. In a similar vein, Fox talked about initially generating (in a group meting) up to twelve alternatives in response to the initial site and functional requirements. For all the interviewees this initial idea generating phase appeared to be a mental interaction between understanding the goals and requirements while sorting through previously stored information on how various global structures met the constraints.

Question #2: The role of sketches.

All of the interviewees sketched ideas and alternatives in this early phase of the design process. However, the degree to which the sketch was an integral part of generating the idea varied. Hatcher, the artist talked about clearly seeing a sculpture in his mind and then starting to sketch. Nevertheless, both he and Burger reported that the sketches were part of the thinking process. Fox and Robertson see the role of sketching as partially for communication purposed and partially as part of thinking about the design idea. Robertson reported that the sketches allowed him to overcome his preconceptions about the design. Fox talked about the usefulness of the sketches in selecting among alternative ideas.

Question #3: The form of the sketches.

Once again the interviewees were unanimous in their agreement that the initial sketches were indeed rough sketches. Robertson claimed that he often made poor sketches himself and thought universities should teach young engineers to sketch. Fox reported that within a group of engineers they may come up with many different rough sketches very rapidly while discussing design ideas. Hatcher, the artist, compares the early sketches to notes. Burger agreed that his early sketches were rough and added that the sketches were difficult to do because of the need for the third dimension.

Question #4: Many versus a single idea.

The interviewees disagreed on whether they had one or many ideas in this early stage of the design process. Hatcher and Burger talked about having a single idea that they could visualize in their mind. Burger talked about starting with a good concept and improving it. He admitted to not being aware of any supermarket shopping (for ideas) that he may go through. Hatcher reported that he may make up to twenty sketches but they are based on transformations of one or two insights. Roberston, like Fox, may work with a group of engineers and architects in generating and selecting alternatives but he reported knowing immediately when an idea is a good one. Fox was the only one of the interviewees that reported having several alternatives occur to him very rapidly.

Question #5: Design details.

All of the interviewees reported that design details emerged only after the initial rough sketching and often after narrowing the alternatives. Fox reported that although twelve sketched might be generated in an initial group meeting, almost half of the ideas will be quickly dismissed. The remaining half are drawn again and rated in a matrix on a number of dimensions that require a more detailed analysis. Hatcher selects the best from up to twenty sketches and draws these forms to scale and in the actual site. Burger reported that the details emerge when the design is modeled. Robertson reported some variation in when details might emerge. In some cases he described going from a blank sheet of paper to a completed project very fast. In other cases, he talked about preparing small color sketches of several ideas that may incorporate varying details.

Question #6: A computer tool for conceptual design.

The interviewees seemed to unanimously agree that a computer tool could potentially be used in the early stages of conceptual design. Fox and Burger saw an advantage in such a tool because it could allow them to represent the design in three dimensions. Robertson mentioned the usefulness of simply tidying up a rough sketch on a computer to present later to a client. Hatcher believed that he could use such a tool in visualization of the sculpture and in testing his hypotheses. However, both Burger and Hatcher mentioned some reservations about a computer tool. Burger pointed about that beautiful computer pictures do not always translate into beautiful or feasible buildings. Hatcher believed that the crisp, mechanical properties of computer drawings may be too sterile to capture aesthetic qualities.

3.5. DISCUSSION

There are several conclusions that can be drawn from the responses of these expert designers. First, it should be noted that despite the fact that each interviewee has a seperate area of expertise and experience (and one is not an engineer) their descriptions of the conceptual design phase are quite similar. These similarities suggest that a fundamental model of the conceptual design process can be developed and computer tools can be used to facilitate the process in more that one domain. Although clearly each of the interviewees utilizes a highly individualized knowledge base, the general process or framework of conceptual design appears to be the same across domains.

A second conclusion drawn from the interviewees' responses concerns their similarities to experts in problem-solving domains other than design. Newsome and Spillers (1988) had hypothesized that if expert designers resembled experts in other domains, their initial approach to a design problem would be characterized by abstract, global representations of the design. The verbal reports of the interviewees describe the initial phase of conceptual design as one in which rough non-detailed sketches were most useful. These sketches appeared to be representations that allowed the designer to study the form in a global manner instead of focusing on the details.

The approach of these designers is very similar to expert computer programmers, for example, who represent a computer problem in terms of the semantics instead of the syntax (Adelson, 1981). Similarly, both expert computer programmers and the expert designers appeared to take a breadth approach in the early stages of the process. For programmers a breadth approach may be characterized by their attempt to identify all sub-problems (e.g., procedures, sub-routines, etc.) before completely specifying the steps for any one sub-problem (Anderson, 1985). In the expert designers we interviewed this breadth approach is characterized by rough sketches of the whole form prior to determining the details on any part of the form.

A final conclusion based on these interviews concerns the usefulness and properties of a computer-aided design tool to support conceptual design. Despite a couple of reservations, interviewees agreed that a computer tool that allowed the designer to globally rough sketch their ideas would be useful. However, most CAD tools currently available fail to support this early design activity. The primary problem with the majority of CAD tools is that they force the designer to specify the design in too great a detail in early design stages. The CAD tool needed and desired by the interviewees would incorporate the rough sketch function while still allowing the designer to depict the form in three-dimensional space. In addition, the designer should be able to manipulates the design in space while it is in a rough global representation.

The advantages of such a tool not only include the ability to picture an object in three-dimensions but with windowing or multiple video monitors, simultaneous comparisons of alternatives could be made. As details of a selected design(s) emerged, the representation could be transformed to illustrate the effects of the details.

These interviews coupled with evidence from other expert problem-solving domains (Newsome & Spillers, 1988) provide preliminary data about the mental processes of an expert designer. Further research is required to determine if CAD tools designed with the characteristics of these expert designers would facilitate or inhibit the conceptual design process.

References

Adelson, B. (1981) 'Problem solving and the development of abstract in programming languages', Memory & Cognition 9, 42-433.

Anderson, J.R. (1985) Cognitive Psychology and Its Implications, W.H. Freeman and Company, New York.

Arrow, K.J., Hurwicz, L., and , H. (1958), Studies in Linear and Nonlinear Programming, Stanford University Press, Conneticut.

DARPA (1989), 'Defense Advanced Research Projects Agency, Peoceedings: Case Based Reasoning Workshop', Morgan Kaufmann Publishers, San Mateo, California.

Dreyfus, H.L. (1965) 'Alchemy and Artificial Intelligence' report P-3244, RAND Corp., Santa Monica, California.

ENR (1986) 'Olympic Domes First of Their Kind', Engineering News Record cover story, March 6, 1986, pp. 24-27.

Meister, D. (1987) 'A cognitive theory of design and requirements for a behavioral design aid' in Rouse, W.B. & Boff, K.R. (Eds) System Design, New York:North Holland, pp. 229-244.

Michell, A.G.M. (1904) 'The Limits of Economy of Material in Frame-structures',Philosophical Magazine, 8, No. 47, November 1904.

Newsome, S.L., Spillers, W.R., and Finger, S. (1989) Design Theory '88, Springer Verlag, New York (These proceedings are summarized in W.R. Spillers and S. L. Newsome 'Another Look at Design Theory' (1990)IEEE Transactions on Systems, Man, and Cybernetics, Vol 20, No. 2pp. 528-530.

Newsome, S.L., and Spillers, W.R. (1988) 'Tools for expert designers: Supporting conceptual design' Newsome, S.L., Spillers, W.R., and Finger, S., (1989)op cit.

Rouse, W.B. (1990) 'Information Systems for Supporting Design' Collection of work available from Search Technology, 4725 Peachtree Corners Circle,Suite 200, Norcross, Georgia 30092, USA.

Rouse, W.B., (1991) Design for Success, Wiley-Interscience, New York.

Rouse, W.B., (1989) 'Information systems for design support', Information and Decision Technologies, 15, pp. 281-289.

Shrager, J., and Langley, P., editors (1991), Computational Models of Scientific Discovery and Theory Formation, Morgan Kaufmann Publishers,California.

Sowa, J. F., (editor) (1991) Principles of Semantic Networks: Explorations in the Representation of Knowledge, Morgan Kaufmann Publishers, San Mateo, California.

Spillers, W.R., (editor) (1974) 'Basic Questions of Design Theory, North Holland Asterdam. (These proceedings are summarized in W.R. Spillers (1977) 'Design Theory', IEEE Transactions on Systems, Man, and Cybernetics, 7,pp. 201-204).

Spillers, W.R., and Newsome, S., (1989) 'Design Theory: A Model for Conceptual Design, op cit Newsome, et. al. (1989).

Stiny, G., (1990) 'What is a design', Environment and Planning B: Planning and Design, 17, pp 97-103.
Touretazy, D.S., Elman, J.L., Sejnowski, T.J., and Hinton, G.E. editors (1991) 'Connectionist Models', Morgan Kaufmann Publishers, San Mateo, California.
Ullman, D.G., Staufffer, L.A., and Dietterich, T.G., (1987) 'Preliminary results of an experimental study of the mechanical design process', in Waldron, M.B. Results from the NSF Workshop on the Design Process, February 1987.

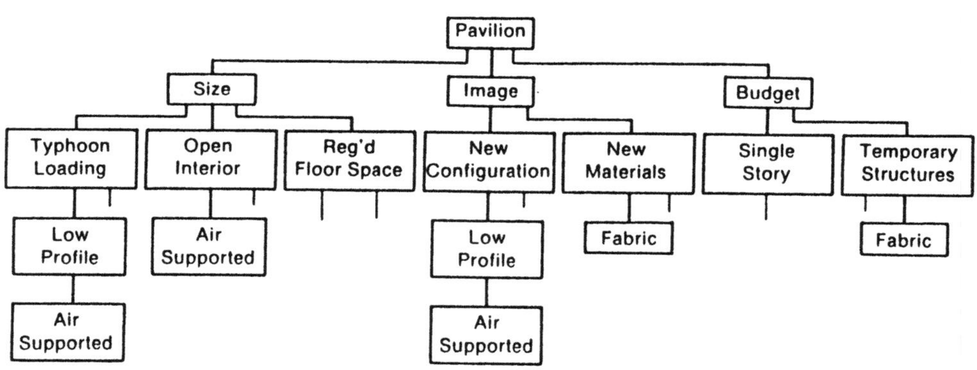

Figure 1. A hierarchical tree structure

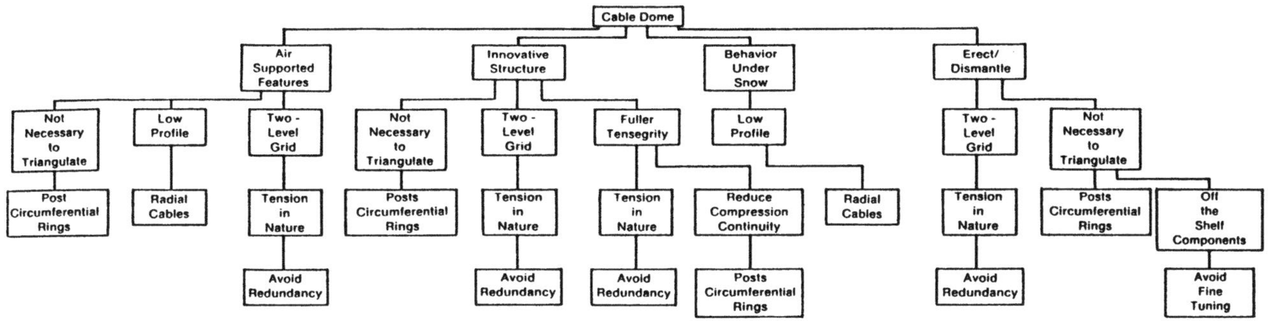

Figure 2. Design of cable dome

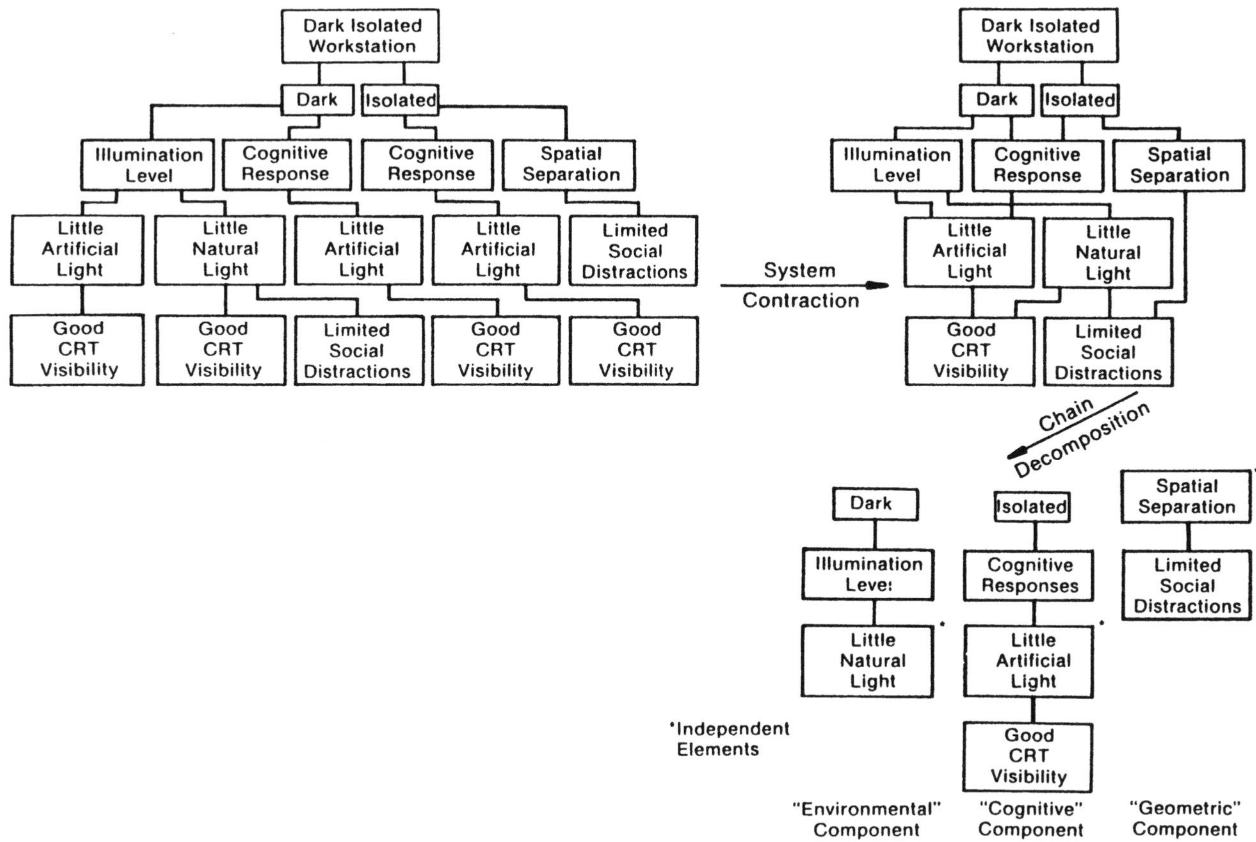

Figure 3. Analysis of a simple hierarchy

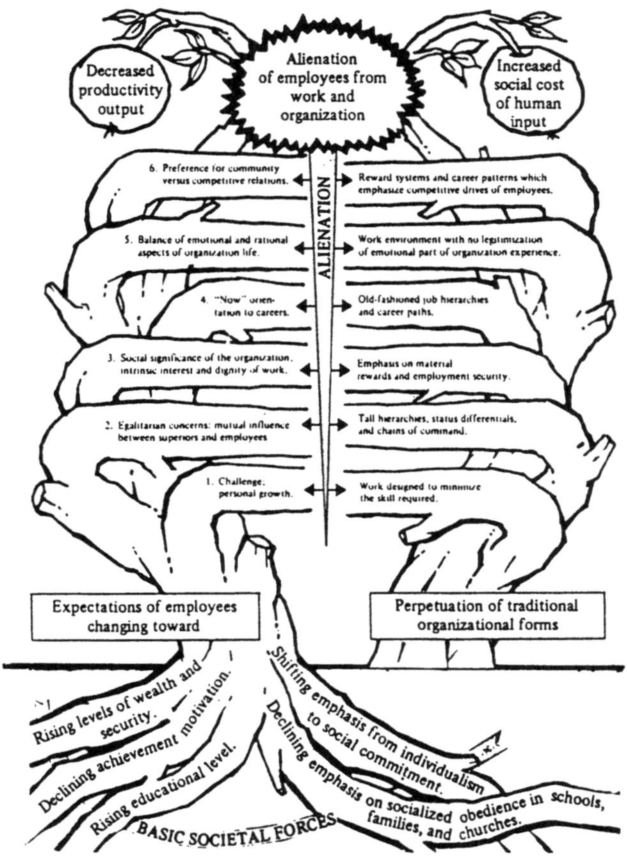

Figure 4. Diagnosis of alienation

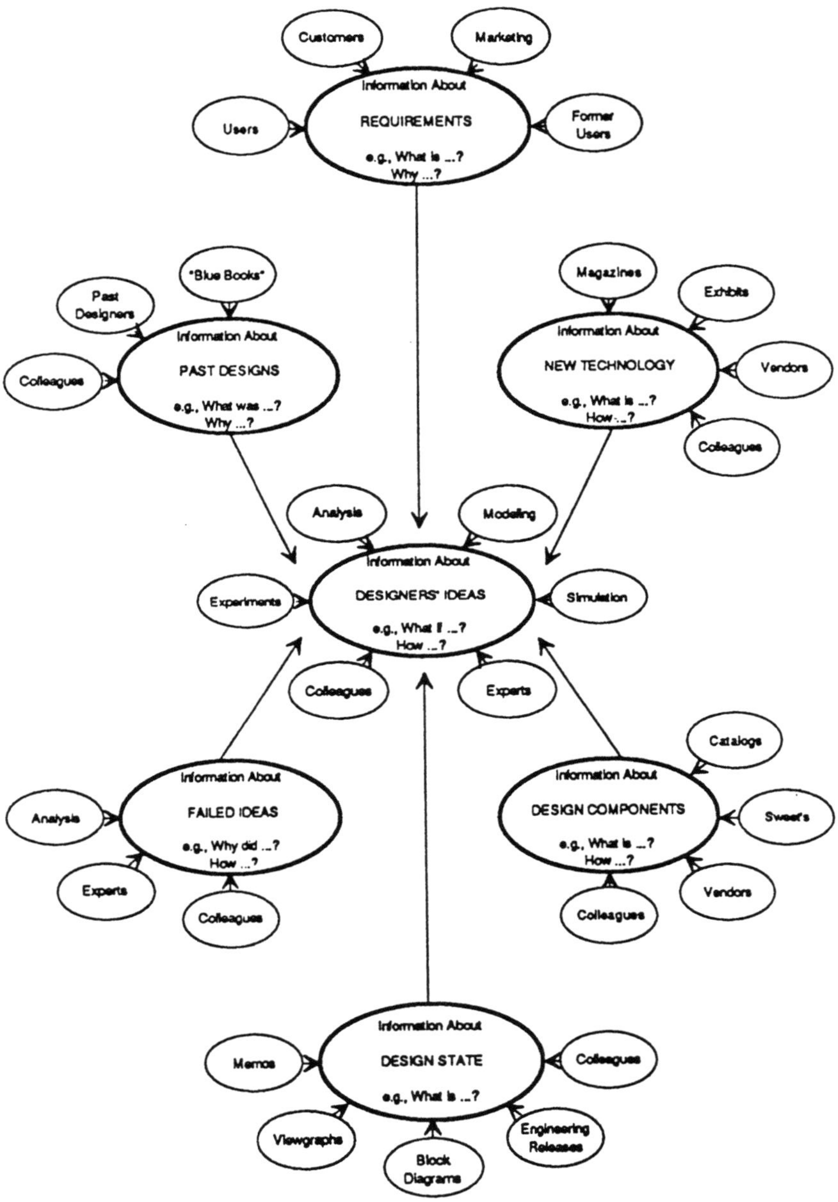

Figure 5. Information environment of designers

DESIGNING: DESIGN KNOWLEDGE: DESIGN RESEARCH: RELATED SCIENCES

N. BAYAZIT
Istanbul Technical University
Turkey

ABSTRACT. At present we are questioning design methods from a different point of view. We are working with computers and trying to decode the methods of the professional practice into computers to be used by the professional practitioners in their turn. These approaches lead us to the investigation of the design knowledge of designers. The understanding of the design knowledge, a part of design research, requires investigation into designing activity. Design research obligates us to apply different approaches and the methods of different disciplines. But in the elicitation of the design knowledge we have to consider the information known by the designers, how that information is gained, how that information is processed, how and when it is used while solving such a design problem. Several disciplines are used in the design process, according to their relevance to the problem area. When we look at a multi-disciplinary area such as designing, we must clarify the disciplines or sciences involved, and their relation to design must be studied.

1. Design

Designing in all the disciplines is demanding collaborate and interdisciplinary team works to investigate nature of design and to understand the way it reaches effective solutions. A decade ago we were trying to investigate the theories and the approaches of the other sciences to designing. The complexity of designing activity acquainted us with the variety of sciences. We were trying to make speculations about the design methods and applications of methods developed in the paradigms of the other sciences to the designing activity. The first generation design methodologists developed design methods depending on their own experiences. We needed the apprehension of the methods applied by the experts like ourselves or others working in design. We gained many experiences in this field while trying to interpret the knowledge of designers into a formalized representation: At present we are looking at design scientifically.

M. J. de Vries et al. (eds.), Design Methodology and Relationships with Science, 121–136.

The concept of "design science" was used for the first time by Gregory in 1965 "The Design Method" Conference in Birmingham. He was defining the concept as follows. 'Design Science is concerned with the study, investigation and accumulation of knowledge about the design process and its constituent operations. It aims to collect, organize and improve those aspects of thought and information which are available concerning design and to specify and carry out research in those areas of design which are likely to be of value to practical designers and design organizations' (Gregory, 1966).

This thinking is only valid for the investigation of designing activity and design methods put to use by the experts who develop their own methods through the experiences gained in the practice in their life-span throughout the technological domain. This does not try to describe the accumulation of the methods developed, some how by some people. It is the in- situ investigation of the methods practiced in a specific field by a particular expert designer on a certain problem. Only in this case we can speak about the investigation of the activity of designing scientifically. From this point of view we can say that we are at present at the very beginning of the design science.

2. Design Knowledge

Nowadays we are working with computers and trying to interpret the methods of the professional practice into computers to be used by the professionals in the practice. These approaches lead us to the investigation of the design knowledge of designers in view of the procedures they apply while they are working in the domain of professional practice. Knowledge was defined by Wiig (1990) as, 'Truths, approaches, judgements and methodologies that are available to handle specific situations. Knowledge is used to interpret "information" about a particular circumstance or case.' Design knowledge is the knowledge hold by designers who can behave in different ways and can be classified as professional practitioners, practical knowledge workers, performers or communicating negotiators.

Design knowledge is the knowledge utilized by the designer during the design process, and it is related to the perception of the information, the organization of the knowledge in the memory, the evolution process of the new knowledge for the specific design situation. It is necessary to examine at some length the relation between the known knowledge by designers and its consequence as the design. 'Some parts of designers knowledge such as validity of the syllogism depend on intuition, and practical skill. Every profession is to some extend a craft, and their development and progress depends on the advancements in the craft.... Practitioners knows "what to do" but not "what he/she does" '(Heath, 1984).

Even though design as a profession is well defined, design knowledge is not understood definitely. The understanding of the design knowledge requires investigation efforts into design which is a part of design research.

3. Patterns of Knowledge In Design

We can talk about the patterns of knowledge and we can talk about the individual as well as cooperate knowledge. The types of knowledge which are owned by an individual can be split into various groups. Design domain knowledge has many facets such as procedural knowledge, normative knowledge, concrete knowledge, positive knowledge, craft knowledge, knowledge of design discourse. Some academicians define knowledge only on account of procedural and declarative. This common classification does not comprise the whole space of design knowledge which needs much deeper explorations to be done into the mind of the people.

The knowledge of the expertise area of a specific design system is called domain knowledge. Domain is the professional environment which comprises structural, mechanical, electrical engineers and other specialist experts. Architects feel that they have been doing something which is against the background of the others. The other experts have been doing different parts of design in collaboration with architects as well as the other experts. Therefore we can not talk about the single domain of an architect as an expert and his knowledge of the domain. We have to consider the participants of the design team as expert designers having different roles. Normative values of the designers are more effective in individual designs but less consequential in team work designs. We can classify designers' knowledge into two main groups, such as procedural and declarative.

3.1. PROCEDURAL KNOWLEDGE

It is concerned with design methodology, the study of the process of designing. 'Procedural knowledge and information about courses of action that may be sequential in nature' (Wiig, 1990). Procedural knowledge is concerned with descriptions and explanations of the process. Procedural knowledge can be composed of reasoning to derive information about a design problem under analysis, reasoning to derive knowledge about the existing or available knowledge, reasoning to generate hypotheses based on design domain knowledge and information which can be assumptions, statements and facts. Procedural knowledge can be about the direction of the reasoning process which can be chained forward or backward or can be bidirectional in nature. Procedural knowledge can have a control structure for the reasoning operations by selecting the rules to be used, accessing and executing those rules, and determining when an acceptable solution has been found. There can be search patterns in procedural knowledge to lead from an initial state to a goal state.

Every profession is still to a great extent a craft, and its practice and progress in most cases depend on the development of these craft elements. 'In the more technically advanced professions, research and practice are in a symbiotic relationship, each conferring benefits on the other but nevertheless quite separate, and having their lives and character' (Heath, 1984). Architecture possesses craft skills of many kinds. As Heath explained in his book (1984) craft knowledge is operative knowledge (or as we defined here is procedural knowledge); knowing how, or "know-how" rather than knowing what. 'To the extent that it is figurative, i.e. called up before the mind's eye, it is sparsely verbalized; it need not be verbalized at all. Hence the reliance we have seen, on demonstration rather than explanation. Insofar as verbal rules are used they are proverbial; formulae used to release the stored knowledge rather than the knowledge itself......Craft knowledge depends on faith; the mysteries of the craft are what the masters knows, and they lead to mysticism' (Heat, 1984). This inexplicit knowledge is hard to differentiate from the other kind of explicit knowledges. Craft knowledge is the accumulation of various skills which are gained throughout the professional practice. The difficulty of obtaining craft knowledge is obvious and needs behavior studies.

Study of the process of designing is at present made depending on the empirical cognitive studies which are forced by the representation requirements of the knowledge of the expert designers into computers.

3.2. DECLARATIVE KNOWLEDGE

This group of knowledge is composed of design discourse, positive knowledge, symbolic knowledge, concrete knowledge. The following paragraphs contain the definitions of different kinds of declarative knowledge.

3.2.1. Positive Knowledge. The basic definition of the positive knowledge can be made as the knowledge to enable people to derive a large number of descriptive statements from a single explanatory statement. Positive knowledge theory in design is an attempt to explain the accumulation of facts about the world. Positive knowledge theory is intrinsically tentative and subject to revision in the face of the first deviant case that does not conform to its explanation and prediction (Roberts, 1969). Positive knowledge of designers is related to the human beings assembled under this heading, such as ecological psychology, environmental psychology, ergonomics, etc.

The purpose of environmental design knowledge is to enable interior designers, architects, landscape architects, and urban designers to better understand the nature of design process and the present nature of the built environment-how it is experienced and used. This can be gained by doing design research on the users' behavior in the relevant domain. Most of the men-environment-relation (MER) studies, architectural psychology studies, environmental design research association (EDRA) studies fall into this category of knowledge.

3.2.2. Concrete (Substantive) Scientific Knowledge. Substantive or scientific knowledge in architecture is concerned with the description and explanation of the physical nature of the products and the built environment-its material, space and structure-. It comprises the segments of the environmental design knowledge. Materials, space, the facts of geometry in service of products for human purposes are the subjects of the concrete knowledge. This knowledge relates to the physical, social as well as psychological aspects of the built environment. The emergence of building science in this century influenced the use of scientific information in architecture. For most of the architects, science is a useful source for facts and generalizations. The enormous amount of increasing information and the necessary scientific knowledge is hard to keep in mind and requires different information devices and structures to be able to utilize them in the building design. There are organizations, standards institutes, laboratories and schools of architecture which work on the scientific and technological aspects of building design to produce concrete scientific knowledge. There are some skepticisms about the role of this scientific knowledge available to the designers to increase the quality of the building during the design process. This information must be put into a form easy enough to be learned and used by the people who do not know scientific contents or implications.

3.2.3. Knowledge Of Design Discourse. 'For the moment, a discourse can be defined as a formation constituted by all that is said, written or thought in a (more-or-less) determinate field.... A discourse is a formation that consists of all that expressed, represented or meant, (that is 'statements' which may or may not have been said or written) around some objects. It exists under the positive conditions of a complex group of relations (Faucault, 1972). Yet these relations are neither simply internal (i.e. between concepts, words, etc.) nor simply external (i.e. limitations, impositions, etc.), but are relations that establish discourse as a practice. This begs the initial question of exactly what the analysis of a discourse actually refers to: to a referent , to an object outside it, or to things? .....a) discourses are practices that "systematically form the object of which they speak" (Foucault, 1972). b) a discourse analysis should not be seen as a substitute for the "concrete analysis of concrete situations". c) it must not be assumed that "the knowledge of an object might ultimately replace the object or dissipate its existence", for the knowledge of this ideology.... is simultaneously the knowledge of the conditions of its necessity (Althusser, 1969).' (Teymur, 1982). The concept of knowledge of design discourse is an important term which comprises design practices, design studies, design theories -what exactly they are dealing with, discursive rules and formations. The subject of the design discourse is the objects, structure of the objects and the mechanisms of the discourse itself, the formations of the discourse as a network of relations as indicated by Teymur (1982).

3.3. DESIGN NORMATIVE KNOWLEDGE

This is an ambiguous term which relates to preferences, values, tastes, attitudes of designers. To some it means, "what has been consensually agreed upon, the norms for a given time"; to others it consists of statements on "what ought to be-what a good world is". Normative knowledge theory of design consists of value-laden statements of philosophers, politicians, and architects, among others , on what ought to be. Some people have described their normative statements as scientific. This is a contradiction in terms. The scientific method provides rules for description, and explanation, not for creation. The acquisition and the structuring of the normative knowledge is challenging as much as the knowledge itself. There are two main design research areas in the study of normative design knowledge. One of them is to make introspective analysis of individual designers; the other is the interpretive analysis of the design itself. The normative knowledge of designers varies from society to society. 'Ordinary observations show the infinite variety of individual value systems that develop among members of a complex society' (Harding, 1966).

Most of the studies on normative knowledge of the past designers has been carried out by architecture or art historians not by the design methodologists or behavior scientists. This is why it is impossible to administer an empirical study on the designers of the past. We can not rely on the weak studies of the art historians who generally develop speculative theories about the knowledge of the designers, their political attitudes, their culture and the environment that shape their thinking. Normative knowledge of the present designers can be the subject of study by the design methodologists and the behavior scientists.

3.4. COLLABORATIVE DESIGN KNOWLEDGE

Collaborate and individual design work are two different methodological approaches to design. The difference originates from the group structure and the distributed responsibilities of the work and work flow. Individual design works can be considered more powerful than group design work from creativity point of view. For some psychologists individuals can be more creative when they are apart from the others. On the other hand 'Groups play a crucial role in the organization theory because they influence and are influenced by organization structure, and because they affect their members' behavior and compliance. ....Groups can influence their members' perception, judgement, and behavior, in part because cohesive groups can enable their individual members to better withstand complaints, abuse, or pressure emanating from the groups' environment' (Dessler, 1986).

At present collaborate design work is recognized as more powerful than the individual design work with its various qualities. There are several significant characteristics which make the collaborate work powerful, such as co-ordination between the people, a cooperated goal shared by the participants, goal-directed behavior of the participants, a reward system for the participants, a shared responsibility between the participants, an organic learning process between the participants in the group, belonging to a social group, mutual interaction between the group and the individual.

Communication with others whether in the same group or not is getting easier in this era by means of new computer technologies and international standards. Electronic mail, facsimile correspondence, video supported telephone conversations, tele-conferencing are contributing to our daily lives. Communication for design can not be behind the technological advancements. The successes of Japanese industry and business became a reference for other nations' industrial as well as business development. Image processing and high definition television studies will assist to improve the much cheaper technological developments for tele-conferencing and decision making. Some western groups already started to study the remote "Computer Supported Co-operative Design Working" such as MIT and ROCOCO project at the Loghborough University.

Knowledge developments of the individuals in the groups working in collaboration with a computer screen are different from the individuals working alone with or without computers. In this multi-media design environment some participants' roles are undertaken
by the AI programs which cover the knowledge bases of an unknown expert in the system. This area of knowledge study is new and needs experimental investigations.

4. Design Research

"Design research is systematic enquiry whose goal is knowledge of, or in, the embodiment of configuration, composition, structure, purpose, value and meaning in man-made things and systems" (Archer, 1981).

The study of design and the designing is one of the subjects of the design research which is the scientific investigation of design. We are speaking about -as Cross, et al (1980) explained-"knowing that" nature of designing and we are talking about the "knowing how" characteristics of designing as the human activity. These two aspects of design research obligate us to apply different approaches and the methods of different disciplines. Design knowledge consists of both areas. But in the investigation of the design knowledge we have to consider the information known by the designers, how that information is gained, how that information is processed, how it is used, and how it is decided to be necessary while solving a problem. Design research can be conducted in various directions utilizing approaches of various scientific disciplines developed for basic physical, human, etc. sciences.

5. Knowledge Acquisition

Knowledge acquisition is an approach developed for the purpose of building expert systems in various fields, can be accepted as an empirical investigation tool for design research, in the exploration of design methods of expert living designers. It is not a new approach but its acceptance scientifically is new. Knowledge acquisition is a process which involves eliciting, analyzing, and interpreting the knowledge that a human expert uses when solving a particular problem and transforming this knowledge into a suitable machine representation.

What knowledge acquisition involves:

- o. It employs a technique to elicit data (usually verbal) from the expert
- o. It interprets these verbal data in order to infer what might be the expert's underlying knowledge and reasoning process
- o. It uses this interpretation to guide construction of some model or language that describes (more or less accurately) the expert's knowledge and performance. Interpretation of further data is guided in turn by this evolving model (Johnson, 1985).

What knowledge engineer does:

'The professional activities associated with eliciting (or acquiring), codifying and encoding knowledge, conceptualizing and implementing knowledge-based systems, and engaging in activities to formalize knowledge and its use, particularly through application of AI' (Wiig, 1990). Design researcher in this case acts like a knowledge engineer who organizes whole research activity during the knowledge acquisition.

We can talk about the method of knowledge acquisition which applies the methods of cognitive science. The problems of the research on the experts can be written as below (after Breuker and Wielinga, 1987):

1. Transforming acquired knowledge into an implementation formalism is the main problem.
   a) Eliciting data from the written documents is one of the problems which can have a structure.
   b) Acquired verbal data can be incomplete and unstructured as well as unreliable and conflicting.
   c) It is necessary to interpret verbal data.

2. Expert knowledge bases need retrieval in the time being, knowledge structure of the experts are not well structured and are not same, this makes the updates difficult.
3. The performance of the expertise may be hard to define, and this knowledge may not well fit to the requirements of its user. Experts do not only solve problems but they are competent in the communication with the environment (user, client, etc.) in order to elicit well specified problem statements and to carry out solutions. The recordings of this communication is difficult to obtain.
4. The complexity of the selected domain to capture knowledge of the experts is one of the main problems. The feasibility and the cost of the study must be socially wise and economical.
5. Especially the normative knowledge is the most difficult part of the knowledge to be elicited from the experts. It is hard to define the concepts of the normative knowledge.

Methodology of the knowledge acquisition is the crucial problem in the investigation of designers' knowledge. There are various approaches developed in this field in relation to the characteristics of the domain and the purpose of the expert system. In knowledge acquisition we can formulate three main functions:

a) knowledge elicitation
b) interpretation of knowledge
c) structuring design knowledge

5.1. KNOWLEDGE ELICITATION TECHNIQUES

Knowledge elicitation techniques are various and depend on the cognition and the knowledge of designers about the knowledge acquisition procedure. The above stages are empirical in general. In each new design knowledge acquisition procedure design domain is observed, data are collected, refined, rejected until reached to a satisfactory position. Then the knowledge is interpreted and concepts are structured. In this procedure review and analysis of design discourse is made, concepts are identified, depending on the initial approaches the expert knowledge is studied applying different techniques such as work observations, questionnaires for experts, interviews, free talks, conversations, protocol analysis, questionnaires for users, team discussions.

5.2. KNOWLEDGE INTERPRETATION

After the collection of large data, knowledge structuring and modelling is completed for the analysis and interpretation of the data. Interpretation and structuring are the most complicated stages in the knowledge acquisition procedure. We need a model in order to interpret verbal, written, and design data. Data interpretation model should be simple enough and robust (i.e., compatible with a wide range of alternative assumptions about human information processing) as Ericsson and Simon (1984) indicated in their book. 'We must consider verbalization in the context of our general model of memory. Information may reach, and be stored in, memory in a variety of encodings-visual, auditory, tactile' (Ericsson, Simon, 1984). As they defined the 'information is stored in several memories having different capacities and accessing characteristics: several sensory stores of very short duration, a short-term memory (STM) with limited capacity and/or intermediate duration, and a long-term memory (LTM) with very large capacity and relatively permanent storage, but slow fixation and access times compared with the other memories' (Ericsson, Simon, 1984). They define the two forms of verbal report to reflect the cognitive process. 'Foremost are concurrent reports - "talk aloud" and "think aloud "reports- where the cognitive processes, described as successive states of heeded information, are verbalized directly. We claim that verbal processes are not modified by these verbal reports, and that task -directed cognitive processes determine what information is heeded and verbalized' (Ericsson, Simon, 1984).

These encodings can be obtained orally and interpreted on account of the developed model. During the verbalization procedure loss of information is recognized by Chafe (1984) and the written texts are preferred because of the slowness of writing which leads to quality which is called "integration". 'Integration as he indicated, shows up packing of more information into written idea units, through typical written devices such as normalization, the increased use of attributive adjectives (properties of entities) and particles, and so on' (Bayazit, 1990). Written reports are easy to analyze compared to oral reports. A method based on "Kinaestheic Image-Schemas" was developed by Bayazit, (1990), principles taken from "Cognitive Semantics" (Lakoff, 1986) which depends on Mark Johnson's "The Body in the Mind".

The knowledge behind the logical structure of a design as an artifact is considered to include the knowledge of the culture and the creative components inherited from the designer. This approach assumes a model of artifact as a theorem which constitutes the program, the rules and the laws about the principles of design. 'In order to understand an artifact, such as a building, it is desirable to talk about attributes not immediately evident from its description. Implicit in a linguistic model is the notion of semantics (the study of meaning). The interpretation of design concerns the discovery of meaning' (Coyne, 1988).

5.3. KNOWLEDGE STRUCTURING

Knowledge interpretations can not be completed without giving enough consideration to the knowledge structure of the referent expert. 'The basic assumption in expert system studies is to form a mental model that accurately represents the design as well as the process that can be performed by it and on it' (Sowa, 1984). The structure of the design knowledge is analyzed depending on the concepts their understanding and meaning. Structure of concepts generates from the structure of the experience which take place at two levels. These levels are named by Lakoff (1986) as "the basic level" and "the image-schematic level". 'Image-schematic concepts and basic level concepts for physical objects, actions, and states are understood directly in terms of structuring of experience. Very general innate imaginative capacities (for schematization, categorization, metaphor, metanymy, etc.) characterize abstract concepts by linking them to image-schematic and basic-level physical concepts. Cognitive models are built up by these imaginative process. Mental spaces provide a medium for reasoning using cognitive models' (Lakoff, 1986). Four major schemas are proposed by Bayazit (1990) after Lakoff (1986) such as container, part-whole, source-path-goal, link. Complex events like design activity is understood in terms of a source-path-goal schema. Concepts related to design can be easily structured with the rest of the schemas.

6. Related Sciences

Designers (design teams) in practice utilize the methods of several disciplines in the design process, according to their relevance to the domain. As design is a multi- disciplinary process, the methods and techniques of various sciences are applied, and various theories are developed.

The sciences can be considered in two groups such as classical sciences and transclassical sciences. 'Classical sciences such as formal (abstract) sciences (logic, mathematics), concrete (descriptive) sciences (physics, chemistry, biology), normative sciences (ethics, aesthetics) each use the methods of, resources and rules of the sciences below them in order to become scientific. This is also true for P and D which uses the methods, techniques and rules of other sciences' (Bayazit, et al, 1981). The group of transclassical sciences are based on either formal as well as concrete sciences or normative as well as human sciences, for practical purposes. Building physics, environmental psychology, housing sociology, environmental geometry, building economics, morphology design, structural anthropology, cognitive design, ergonomics, etc. are some examples.

Many theories have been developed and new sciences are brought into existence in association with the design research since the beginning of the studies o explicit design methodologies. Some of the existing sciences are put to use in the area of design research. 'Design research bases like many other discipline on the systematization of collected data, gained by the results of the case studies in which objective research methods were applied with critical evaluation procedures. Design research can be considered to be wholly scientific in this respect. Popper defines the phases of scientific research as establishing a scientific theory in three groups: scientific observations and experimentation, recording observed regularities establishing systematic explanatory schemas'

(Bayazit, et al 1981). Design process deals with the objective as well as subjective or normative aspects of design related to the human beings in the form of human preferences, attitudes, evaluations, values, etc. Information gathering and development technologies are using already accepted quite common and basic techniques such as searching documentary evidence, getting direct information from user, observation, experimentation to be used for the design research. A long list of these techniques are given in Bayazit's article (Bayazit, et al 1981) in regard to related sciences. These techniques are widely used for the knowledge acquisition by the design researchers for miscellaneous purposes. In the same study Bayazit tried to establish the relevance of the other tools to design research in relation to the different phases of design.

7. Theories Of Design

Looking at the design theories developed in the last three decades of this century from methodology point of view, we can recognize some specific characteristics. We can not refute any of them for the reason that they are not convincing. They can be beneficial for particular design issues and present themselves in specific circumstances. Although many of them originated from other disciplines they are accepted as design theories because of their applicability to the stages of design.

Two appendices for the list of the relevant sciences (Appendix 1.) and theories (Appendix 2.) are prepared to design on account of design knowledge. The related patterns of design knowledge to the theories and sciences are indicated in the lists such as procedural knowledge (PR), positive knowledge (P), normative knowledge (N), concrete knowledge (C), collaborate knowledge (CL), knowledge of design discourse (D).

Positive theories are related to the explanation of the user behavior and its influences on the design. Procedural theories define the phases or sometimes the methodological structure of the design process and most of them try to investigate the designers' ways of thinking. Intuitive approach, system approach, expert systems, cognitive science, decision theory, game theory, etc. are some of these. Theories and sciences are utilized in design research for different purposes. Some of the theories are developed to investigate designers' mental processes or to model these processes which are profession oriented theories. The other group of theories deal with users' attitude patterns and decision making roles in design and design processes. Participatory theory, environmental ecology, performance theory are typical examples of these. There are some theories which concern both users and professionals at the same time. Structuralist theory, performance theory, expert systems, artificial intelligence are of this kind. Most of the theories of architecture and art historians have been concerned with the appearance of the buildings related to the economic, cultural, social characteristics of the social context. The examples to these theories are post modernism, new brutalism, deconstructivism, symbolism, etc. These theories try to investigate the normative values of the designers.

8. Conclusions

The new technological advancements are forcing us to work on the new design methodologies of our age from a different perspective. The use of computers as a part of our daily life force us to develop new approaches to the investigation of design process. Some points of discussion follow. To what extent can we utilize them in design? To what extent can we comply and/or confront with the computers? What can design research achieve?

We can not refuse the past theories and methods as they have their own truths and the knowledge of the developers. If there are so many theories and sciences related to this area of design methodologies, it means there are problems of our concern and there is a lot to be done in the area of design theories and methods.

There are big gaps between the research and practice of design, but most of the theories and methods fit to the problems of the professional practice. There is a selection procedure by professional practitioners about these approaches. If they find something useful then they use it. This is the fact. Therefore I will conclude with the sentence, " We have to make more empirical studies on designs and designers."

References

Archer, L.B., (1981), 'A view of the Nature of Design Research', in: R. Jacques and J. A. Powell (Eds.), Design: Science: Method, IPC Business Press Ltd.

Bayazit, N., Esin, N., Özsoy, A., (1981), 'An Integrative Approach to Design Techniques', Design Studies, Vol 2, No: 4, pp. 215-223.

Bayazit, N. (1982), 'Gelecege yonelik tasarlama (Design for the future)', in N. Bayazit, M. Tapan, N. Ayiran, N. Esin (Eds.), Tasarlama (Dizayn) I. Ulusal Kongresi (I. National Design Conference), 24-26 May 1982, I.T.U. Mimarlik Fakultesi, Istanbul. (in Turkish)

Bayazit, N., (1990), 'Development of a knowledge acquisition model for computer aided design', in V. Hubka, and A. Kostelic (Eds.), Proceedings of the 1990International Conference on Engineering Design, ICED 90, Heurista and Yudeko, Dubrovnik-Zagrep.

Breuker, J., Wielinga, B., (1987), 'Use of models' in A. L. Kidd (Eds.), The Interpretation of Verbal Data, in Knowledge Acquisitions for Expert Systems: A Practical Handbook, Plenum Press, New York.

Chafe, W., (1984), Cognitive Constraints on Information Flow, Institute of Cognitive Studies, University of California at Berkeley, July.

Coyne, R., (1988), Logic Models of Design, Pitman Publishing London.

Dessler, G., (1986), Organization Theory: Integrating Structure and Behavior, Prentice- Hall, Inc., New Jersey-Englewoods Cliffs.

Ericsson, K. A., Simon, H. A., (1984), Protocol Analysis: Verbal Reports as Data, The MIT Press, Cambridge, Massachusetts.

Gregory, S.A., (1966), The Design Method, Butterworth and Co. Ltd.,London.

Harding, D. W., Social Psychology and Individual Values, Hutchinson University Library, London.

Hayes-Roth, F., Waterman, D.A., Lenat, D.B., (1983), Building Expert Systems, Addison- Wesley Publishing Company Inc., Massachusetts.

Heath, T., (1984), Method in Architecture, John Wiley and Sons Ltd.

Johnson, L., (1985), 'The need for competence models in the design of expert systems', International Journal in Systems Research, and Informational Science, No:1.

Kidd, A.L., (1987), 'Knowledge acquisition an introductory framework', in A. L. Kidd, (Eds.), Knowledge Acquisition for Expert Systems, Plenum Press, New York.

Lakoff, G., (1986), Cognitive Semantics, Institute of Cognitive Studies, University of California at Berkeley, March.

Lang, J., (1987), Creating Architectural Theory, Van Nostrand Reinhold, New York.

Roberts, E. (1969), Theory Building, New York: Free Press.

Sowa, J.F., (1984), Conceptual Structures, Information Processing in Mind and Machine, Addison-Wesley Publishing Company.

Uluoglu, B., (1990), Mimari Tasarim Egitimi: Tasarim Bilgisi Baglaminda Stüdyo Elestirileri (Architectural Design Education: Design Knowledge Communicated in Studio Critiques), I.T.U. Fen Bilimleri Enstitusu, Istanbul. (Ph.D. Thesis in Turkish)

Wiig, K., (1990), Expert Systems: A Manager's Guide, Geneva: International Labor Office, Management Development Series No:28.

APPENDIX 1.

s. Acoustics (C-P)	
s. Aesthetic psychology	(N)
s. Aesthetics	(N)
Empirical Aesthetics	(N)
Formal Aesthetics	(N)
Sensory Aesthetics	(N)
Symbolic Aesthetics	(N)
s. Anthropometry	(P)
s. Architectural History	(P-N-D)
s. Behavioral Science	(P)
Behavior circuits	(P)
s. Behaviorism	(P)
Behavior Setting	(P)
s. Cognitive Science	(P)
s. Computer Science	(C-PR)
s. Cost-benefit Analysis	(C-PR)
s. Cybernetics	(PR)
s. Database (PR)	
s. Design Science	(PR)
s. Design Methodology	(PR)
s. Ecological Psychology	(P)
s. Environmental Psychology	(P)
s. Environmental Sociology	(P)
s. Ergonomics	(P)
s. Ethics (N)	
s. Human Ecology	(P)
s. Linguistics	(P-N)
s. Management	(P)
s. Materials	(C)
s. Morphology	(C)
s. Operations Research	(C)
s. Problem Solving Psychology	(P)
s. Psychology of Creativity	(P)
s. Psychobiology	(P)
s. Regional Planning Science	(PR-P)
s. Sociobiology	(P)
s. Structural Anthropology	(P)
s. Synectics	(P)
s. Urban Planning Science	(PR-P)

APPENDIX 2.

t. Analogical Design Theory	(PR)
t. Architectural Theory	(N-D)
International Style	(N-D)
New Brutalism	(N-D)
Post Modernism	(N-D)
Archigram	(N-D)
Deconstructivism	(N-D)
Symbolism	(N-D)
Functionalism	(N-D)
Constructivism	(N-D)
Rasyonalism	(N-D)
Romanticism	(N-D)
t. Artificial Intelligence	(PR-P)
t. Expert Systems	(PR-P)
t. Balance Theory	(N)
t. Behavioral mapping	(P)
t. Canonic Design Theory	(PR)
t. Creativity	(PR-P)
t. Cognitive Maps	(P)
t. Communication Theory	(PR-C)
t. Information Theory	(C)
t. Decision Theory	(PR-C)
t. Empricisism	(P)
t. Functional Theory in Sociology	(N-P)
t. Game Theory	(PR)
t. Graph Theory	(PR)
t. Gestalt Theory	(N-P)
t. Information Theory	(C)
t. Mental Mapping	(P)
t. Morphological Theory	(PR)
t. Nativisim	(N-P)
t. Organizational Theory	(P)
t. Performance Theory	(C)
t. Participation Theory	(P)
t. Pragmatic Design Theory	(PR)
t. Proxemics Theory	(P)
t. Psychoanalitical Theory of Memory	(P)
t. Relativity	(C)
t. Semiotics	(N)
t. Systems Approach	(PR)
t. Typology	(N)
t. Undermanning Theory	(P)
t. Utility Theory	(P)
t. Value Theory	(P)

SCIENCE IN ENGINEERING, ONE COMPONENT OF THE SCIENCE OF ENGINEERING DESIGN

W.E. EDER
Royal Military College of Canada,
Canada

ABSTRACT. Human actions are driven by knowledge, methods and goals. Science is defined as systematized knowledge, and its scope and contents are discussed. Design science to underpin engineering design is introduced. Design science is based on concepts of descriptive and prescriptive knowledge related to objects and processes, which must form systems of knowledge. Some of the typical content of these knowledge types is discussed, especially from the viewpoint of applicability to engineering design for problems requiring new solutions. This knowledge must be derived from scientific and experience knowledge and adapted to suit the forms of questioning used by designers. The content of object knowledge relating to a technical system covers various ways of modelling, its properties and life cycle, and the concept of quality. Design process knowledge covers the management and design process to develop a new technical system, and the embedded strategic and tactical tools and problem-solving processes. The role of the operators of design processes is mentioned.

1. Introduction

1.1. ACTIONS

Any human action is based on a combination of knowledge and method of procedure, regardless of whether that action is unaided (the human acting alone) or assisted by tools.

Admittedly the *knowledge* may be faulty and it is never complete. Parts of this paper will therefore explore various aspects of the nature of knowledge used for a particular kind of human action, namely engineering design in its most usual contexts. Equally, *method* of procedure can range from a purely random set of unconsciously applied action elements to a fully explainable sequence of actions. 'Lack of method' is nevertheless itself a method of proceeding. Parts of this paper will explore several aspects of methods and procedures. Sequencing of methods and the logic supporting those procedures to *methodology*.

M. J. de Vries et al. (eds.), Design Methodology and Relationships with Science, 137–164.

Any action must also have a *goal*, and is performed with the intention of reaching that goal. It is not necessary, and at times even not possible to reach the envisaged goal. Approaching the goal frequently needs a reappraisal or new acquisition of knowledge and method.

1.2. GOALS AND GOAL-DIRECTED ACTIONS

The goal may be to achieve some sort of change in a *transformation process*, starting from an available object (e.g. material, energy, and/or information), and the product of that change is to be the end-result. Alternatively the goal may be to achieve a change, and the search is for a *set of tools* (the end-result) to assist in achieving the change. The tools may exist, or they may need to be *designed* and made. We can now recognize that there are two categories of knowledge: *knowledge about processes* including methods of procedure and *knowledge about objects*, products, artifacts or technical systems.

The end-result of designing, the usual activity of engineering designers in an industrial context, should consist of the instructions to someone about what tools (products, artifacts, technical systems) to make and how to use them so that yet another person may achieve a change or make a product.

This paper should therefore outline the background knowledge for bringing all of the above ideas into a describable context, a science of engineering design. For this purpose, we need to explore the context and contents of science and the types and forms of knowledge needed to aid and enable engineering design.

1.3. SCIENCE DEFINED

According to a well-known dictionary [1], **science** is: *from Latin 'scientia' -- having knowledge; a branch of study concerned with observation and classification of facts and esp. with the establishment of verifiable general laws; accumulated systematized knowledge, esp. when it relates to the physical world*, and **theory** denotes *the general principles drawn from any body of facts (as in science); a plausible or scientifically acceptable general principle offered to explain observed facts; hypothesis -- an assumption made esp. in order to test its logical or empirical consequences (without assumption of its truth [2]).*

Gregory [3] states that the maturity of science about a body of knowledge is characterized by a gradation of behaviour which may be expected, and which causes that knowledge to be reckoned as scientific. This behaviour and the state of organization of the body of knowledge may range through:

(1) Description of phenomena (natural history phase).
(2) Categorization in terms of apparently significant concepts.
(3) Ordered categorization whose pattern may be deemed a model (the evolutionary taxonomy or periodic table phase).

(4) Isolation and test of phenomena, with implied reproducibility by independent observers (foundation of 'research').
(5) Quantification (classical physics phase).

In most areas of knowledge, even the first of these is considered to belong to science and is a basis for being able to use knowledge. To quote Virginia M. Satir [4], *'I am aware that as long as anything remains in the area of magic, it cannot fruitfully be used.'*

The descriptions of a phenomenon resulting from each grade will almost invariably conform to the currently available *paradigms* and *disciplinary matrices* [5,6]. In other words, no science is isolated from all others, all sciences fit into a world picture. Knowledge, even if it is 'acquired for the sake only of knowledge', can only exist in the context of the society in which it resides -- research is not 'value-free' [7]. Research tries to validate the existing world picture and its disciplinary matrices -- until the acquired knowledge and insights can no longer be made to fit into that pattern, when (over the time of about one or two generations) a revolution of thought and paradigms takes place.

Experience, practice and activity leads into formulating the appropriate science in most areas of knowledge. Applying the developed scientific knowledge to the practical tasks can help to verify the scientific formulations and can help to improve the practice. Whether the scientific knowledge is applied in practice depends on many circumstances, including the mutual awareness between the different people who formulate science and who apply knowledge for designing, about the scope and contents of the other side.

Only a few areas of knowledge can ever reach Gregory's fifth stage of quantification, in which mathematically 'exact' models of the phenomena are available. Many others may not even reach the fourth stage, where 'reproducible experimental results' are achieved.

As human knowledge has been developed, the tendency to concentrate in depth on smaller units of study has been strong. Even so, various attempts have been made to co-ordinate its broad range by formulating a hierarchy of sciences and meta-sciences. For instance, taxonomies permit classification of knowledge, and various models (verbal, graphical and symbolic) permit description of phenomena, processes, objects and factual insights. The natural and pure sciences have generally benefitted from interactions recognized by the meta-sciences. In contrast, engineering has been slow to develop a theory to describe what it does, how it achieves its successes, and how the underlying knowledge is interrelated.

1.4. SYSTEMS

Again referring to a dictionary [2], a system is a *complex whole, set of connected things or parts, organized body of material or immaterial things; method, organization, considered principles of procedure, (principle of) classification; body of theory or practice pertaining to a particular form of government, religion, etc.* We can (and should) therefore be able to speak about *systems of knowledge*, as well as about *object systems* and *process systems*.

Each system has a *boundary* which separates it from everything else -- but this boundary may be arbitrary, assigned by a person for a purpose. Within the boundary, a system has *elements* (parts, components) which show certain *relationships* among each other and across the boundary to outside elements. These elements and relationships reveal a *structure*. When the term 'element' is defined in different ways, different relationships and structures can be recognized within the same system.

Systems exist within a *hierarchy*, each system is an element of a larger system, and each element of a system can be regarded as a system in its own rights.

1.5. ART AND SCIENCE

There are fundamental differences in outlook, philosophy, procedures and constraints in various human activities. Dichotomies seem to exist between 'science', 'art' and 'engineering' and must be explored.

Science investigates existing phenomena, with the purpose of providing explanations and models of why and how they happen, and predicting what would happen (under the usually very limiting assumptions of the model) under different circumstances. The usual procedures are to set up a speculative hypothesis, test it in an attempt to show its validity, impartially observe a set of phenomena, and derive an improved hypothesis.

Art aims to allow free expression in order to appeal to the aesthetic senses. It is a very intuitive activity, which involves human feeling.

Engineering (particularly through design) aims to produce the necessary information to realize a manufacturable product that can potentially satisfy a human need. In fact, for human progress and improvement of the quality of life, engineering design plays a fundamental role.

Engineering designers must, of course, use the available scientific knowledge, but must also be concerned with aesthetic, ergonomic, economic and other factors. Designers can (and do) select ways of realizing the duties of a technical system which science has not yet explored -- typical inventions, which may work even though we do not fully know why or how.

Some unpredicted happenings occur in the products of engineering (e.g. failures) that can initiate further scientific research. Failures, when they occur, help to redirect the practical applications, and can influence scientific activities [8]. 'Technology transfer' from science to application is only one of the paths linking science and practice, and in most areas of knowledge it is probably the least important one -- the noteworthy exceptions are the glamour areas of 'high technology'.

Research delivers mainly verified and co-ordinated knowledge, some of which can lead an observant and creative individual to new insights that can result in inventions and new products. The boundaries between science, engineering and the arts are at best fuzzy (see also [9]).

The highest human achievement is 'creating something new' that is of some benefit to some part of mankind. Its form can be that of original artistic works (termed 'aesthetic expression'), new products and processes that are useful to humanity ('designing' or 'engineering'), or new scientific knowledge ('research'). Every human being has some capability in creative work, but whether it is recognized and enhanced depends on circumstances. It is also useful to discover to what levels of activity this creativity can be applied, and what mental operations are needed [10] -- this will help us to establish (but not completely define) what should be the objectives of our investigations of design science, design practice and teaching.

1.6. DESIGN AND SCIENCE

Designing has been recognized as a human activity, one that is essential for planning and describing the processes and products that are to be made. Historically, the processes of designing and making have been performed by the same persons -- the division of tasks only took place during and as a result of the first industrial revolution.

In this paper, I studiously avoid using the word 'design' alone -- it is ambiguous, it may mean the process of designing, or the (appearance of the) resulting artifact (product, technical system). My references are to 'designing' or 'engineering design' as the process, and 'product' or 'technical system' as the artifact.

Designing is thus one phenomenon that can be studied, a body of knowledge can be defined and categorized -- *design science* [11]. This study has been taking place in many countries since about 1950, with almost continuous developments in some parts of Europe and sporadic advances in other regions. A major impetus has been given in the U.S.A. by the National Science Foundation in 1985, but with limited scope. We are convinced that designing, especially for engineering, can be taught in a rational way, because design science can be (and has been) formulated. Design practitioners will still need adequate knowledge, suitable attitudes, and favourable working conditions.

2. Knowledge

2.1. TYPES OF KNOWLEDGE

Engineering design demands a wide range of knowledge of different kinds. The ranges of knowledge have recently been formulated within Design Science [11].

Some of the existing knowledge has been codified (classified, categorized into its elements and the relationships among them) to generate the framework of Design Science. The four major sections of this knowledge are (see figure 1):

descriptive knowledge (theories) about the *objects* -- artifacts (processes and products) being designed;

prescriptive knowledge about the *objects*, how to achieve their properties -- recommendations about what to design for and how to achieve each property of a product;

descriptive knowledge about *design processes* -- theories about methods, procedures, appropriate modelling, etc.;

prescriptive knowledge about *design processes* -- recommendations for action and activity in designing.

The descriptive knowledge about objects (artifacts, technical systems) [12] contains theoretical considerations and generalizations about human needs, and about the nature of general system that perform transformations useful to humans, processes (including manufacturing), technical systems (products), ways of classifying systems, properties, evaluation, representation (e.g. in drawings), life cycle, development in time, state of the art, etc.

The prescriptive knowledge about objects contains practical and theoretical advice about the generalized properties of products and processes and how they may be achieved, life cycle (including typical progress during designing), complexity, difficulty of designing, development in time, etc. (as above), and the implications for designing of all the descriptive (theory) knowledge. Some of this knowledge is based on experience and empirical observations (parts of it not even in retrievable form, e.g. embedded in someone's experience), other parts relate to sciences. An impression of the range of object knowledge needed for engineering design is shown in figure 2. This knowledge should be re-worked to give direct advice to designers' questions -- knowledge as presented in sciences is not generally suitable for searching for solutions, and has only limited applicability for analysis tasks.

Descriptive knowledge about design processes relates designing to the theory about objects, to explain why a recommended procedure should be the most rational. An outline of this theory may be found in [13].

Prescriptive knowledge about design processes gives advice on actual stages and procedures of designing [14], when and how to use design methods, and what other alternative procedures there are. This knowledge also contains paradigmatic solutions (exemplars) of solving problems of conceptual design using known and recommended methods [14,15].

A further sub-division yields knowledge for *analyzing* (especially mathematical modeling and predicting), and knowledge for *synthesizing* which are related to the tactical aspects of problem-solving, see section 4.3 -- the latter requires familiarity with (and feel for) individual phenomena and their relationships, and alternative ways in which an outcome can be achieved. Formal (i.e. mathematics-based) analysis is necessary for designing, but is neither sufficient in itself, nor a prerequisite to designing.

2.2. OBTAINING KNOWLEDGE

Quoting Aristotle: *It is through knowledge that I gain understanding -- and understanding lets me do by choice what others do by constraint of fear.* Knowledge about design processes can be obtained in various ways:

theorizing on the basis of prior experience (usually design experience gained in industry) -- including self-observation and abstracting from that;
abstracting from independent observations (of engineering designers at work) -- e.g. scientific research and action research; and
attempting to generate useful tools for engineering design (usually with computers).

The first approach may suffer from personal bias, and is difficult (but not impossible) to perform by an active engineering designer. The second approach may lead to wrong conclusions about the importance of some aspects of designing (e.g. the common myth that designing consists almost exclusively of producing engineering drawings), and cannot adequately observe mental (brain-internal) processes. The third approach is usually too narrowly focussed to produce a generalized theory. These three approaches are not mutually exclusive. In order to develop *systems* of knowledge, the knowledge obtained from the various sources must be categorized and coordinated, i.e. an underlying theory must be proposed that explains the observations and the recommendations.

Much of the European work, including the efforts of WDK--Workshop Design-Konstruktion, falls into the first category, and has a history of over 35 years. Most of the recent work in the U.S.A. and elsewhere concentrates on the latter two aspects. The computer tools are mainly applicable to the most concrete design stages (layout and detail, see section 4.1), or to analysis, simulation and/or decision-making in the problem-solving cycle. Some headway is being made towards bringing computer tools into the more conceptual higher levels of abstraction. The most comprehensive framework for these efforts relevant to engineering design exists in the references [11,12,13,14,15].

3. Object Knowledge

Social and technical systems (viewed as objects) are complex, and this complexity should be investigated from all possible aspects -- this is the duty of the Theory of Technical Systems [12]. The search is for those aspects (and the related knowledge) that are common to all technical objects, whereby the most important questions refer to:

- the purpose and usage of technical systems (TS) within hierarchically higher systems such as the socio-technical transformation system;
- the nature of technical systems;
- structures recognizable in TS, and modelling of these structures;
- classifications of TS;
- properties of technical systems, and their relationships within and external to the TS;
- evaluation of TS;
- representation of TS, in verbal, symbolic and graphical models;
- phases of ontogenesis (development of a TS from idea to manufactures product) and operation (usage);
- development in TS over periods of time, current state of the art.

A dynamic model of a general 'transformation system' has been proposed [16,17] in order to show the purposes of technical systems. This model should correspond to the essentials of all artificial processes in society, and all natural processes. The purpose of any transformation is to add value to the properties of the materials, energy and information that undergo the transformation from an existing initial state to a desired state in a process. Technical, economic, aesthetic, human, utilization or other measures of value (individually or in combination) may be applicable. Figure 3 shows such a model, and indicates the essential elements of every technical process systems, and their most important relationships, especially its operators. The transformation system receives the operands as inputs from the environment, and delivers the transformed operands as outputs. Both the inputs and the outputs contain operands that are desired and those that are unintentional or even harmful. By performing its duty the system changes a part of its environment, and will also change a part of itself [16]. The total system is divided into a process that transforms a set of operands, and the operators of that process, a division that may not always be obvious to the observer. An analogy to the concept of a mathematical function and its operator, operand and result is intentional. Such a division into process and operators is not unique but depends on the viewpoint of the observer.

The *operators* of this process consist of the *human beings* who run the system, the *technical systems* proper that the human being uses as tools to perform the tasks, the *information system*, the *management system* that sets goals and directs towards achieving them, and the active *environment*. They act separately and together, and exert effects on the operands to drive and guide the process. Goals are set in the management system for the transformation system and its process, almost invariably by human beings, and information is generated, stored, retrieved and used. The operators deliver the necessary effects to enable, guide and control the process; this is their *function*, as shown by the vertical connections from operators to processes in figure 3.

The horizontal stream, the process, carries the *operands*, the input that the transformation system should actively *transform* from a less desirable to a more desirable state. This *process*, a transformation of operands, takes place during a period of time. Its characteristics include by implication any feedback functions that help to control the system. The process has an internal *structure* that consists of separable elements (operations) which are bound together by relationships. This process is under the control of the operators, who receive feedback by sensing/measuring from the operands.

In the Theory of Technical Systems (TTS) [12], the transformations are investigated, classified, and a terminology formulated, in order to generalize and abstract from particular systems to the comprehensive models. In particular, the effects that must act on the operands to achieve the desired transformations, and the originators of those effects, the operators, are thoroughly analyzed.

3.1. STRUCTURES AND MODELING

The technical system is recognized as the most important operator of processes in modern times. The main content of the TTS is the nature, classification (taxonomy), properties, evaluation, representation (in various models), stages of development, historic evolution, etc. of technical systems. These contents must therefore include the principles and modes of action of a TS (why and how they work, their internal actions), and their anatomy (how they are built up and constructed). Explanations for these features are contained in the various types of structural diagrams that are available for modelling TS, see figure 4, particularly a function structure (a structure-model composed of the individual functions and their relationships), an organ structure, and a component structure. These structure-models are particularly useful for engineering designers, they support design work by providing starting points for separable stages of design work and rules for transforming one structure into another. Establishing the candidate structures, and selecting a relative optimum among them, is predicated on establishing those characteristics necessary for the TS to be able to perform its duties, namely the design characteristics. It must be recognized that the overwhelming majority of engineering design tasks deal only with the component structure.

Technical systems to be designed can occur at various levels of *novelty*, from almost completely unknown, through major revisions, to minor alterations. These are 'responsible' for break-throughs, evolutionary developments, and adaptations of technical products. Technical systems also appear in different levels of *complexity*, from large-scale plant to single components. Technical systems may be more or less *difficult to design*.

3.2. PROPERTIES

We must 'design for' a vast range of features to get a good and economic product. Every technical system (engineering product and/or process) has a range of properties that make this system suitable for parts of its life, many of which are measurable. These categories are a complete set, but the examples shown within each category need to be adapted to the particular system. The ones that are generally observable by a user are the *'external properties'*, see figure 5. They must be deliberately 'designed for':

Design for functions, it must be able to perform some specific and general duties (property 1) with acceptable performance characteristics (2);
- design for all needed processes on operands;
- design to include all necessary constituents of processes and systems;
- design for all life stages of a product;
- design more or less close to the state of the art in the field;

Design for operation (3), it must be safely operable;
- design for friendliness to the environment;
- design for specific cases, special conditions;

Design for realization (4), it must be manufacturable and assemblable [18];
Design for distribution, packaging, transport (5);
Design and organizational planning for timely delivery (6);
Design for liquidation (7), recycling, and disposal;
Design for human acceptability (8) and appearance (9);
Design for law and standards conformance (10);
Design for economics (11), life-cycle and first costs;
Design for quality and quality assurance (see next section).

These properties are responsible for the behaviour of technical systems throughout their life cycles. Technical systems must be designed to withstand all conditions they will meet during this life cycle, typically as shown in figure 6. The design specifications (figure 4, level I) should therefore cover all these aspects of properties and life cycle, and should contain criteria for evaluating all generated alternative solution proposals.

An engineering designer produces (as output of the design process) a set of detail and assembly drawings, parts lists, operating and maintenance instructions, etc., or their equivalents in computer files. These show the elementary design properties in the central box of figure 5, *internal properties* that are just about invisible to the user. By designing, and deciding on the shape, material, and features of each component, a designer in fact creates the external properties.

The output of the design process can only be achieved if the *design characteristics* (bottom inner box) have been considered, and if the work conforms to the *general design properties* (upper inner box). This is the sort of place where engineering analysis helps a designer to investigate the features of a proposed technical system. Any detected deficiency causes the designer to enter a new iteration towards the eventually accepted solution. Consequently, these internal properties must also be 'designed for':

Design for strength, including fatigue and creep;
Design for stiffness;
Design for (or against) wear;
Design for (or against) corrosion;
Design for .... -- and so on.

Some of the 'design for ...' knowledge is available in a form that is readily accessible for design use, e.g. [18]. One form of this type of information stems from the typical design question: 'What alternative ways can I use to solve this problem?', and is typically available in *Design Catalogues* [19]. Verified data of conventional scientific and experimental form can be found [20].

Much more knowledge is hidden, or not sufficiently classified and cross-referenced, in designers' experienceand heuristics, and in conventional engineering (and other) science, and must be reworked to make it directly accessible for engineering designers. A structure for classifying knowledge [21], and expert systems technology should help in these efforts. Patents information needs to be searched and collected in a suitable way, which can be the source of a useful research effort.

An example of how knowledge from experience must be reworked is shown in figure 7 for a part of manufacturing. Part A gives a description of metal casting and faults, part B derives the requirements for designing and producing castable components, and part C shows typical recommendations for form and dimensions coordinated with parts A and B. Similarly, knowledge from 'pure' science must be brought into a relationship with that of 'applied' science and 'engineering' science to generate applicable knowledge for designing. Science knowledge is generally classified according to (physical) principles, the systems of knowledge try to collect all knowledge within a discipline. In contrast, a major question for designing is 'with what principles and their embodiments can I solve (fulfill) a given function?' For instance, the speed of a machine can be controlled by means of devices using hydraulic, centrifugal, pneumatic, fluidic, electrical, analogue-electronic, digital-electronic or other principles. Comparative data on applicability, performance, cost, ease of integration, repairability, reliability, etc. (all properties of technical systems) should be available for many functions.

3.3. QUALITY

This is a major current concern, defined as [22]: *The quality of an engineering product is the sum of the perceived measures of its properties.*

The most appropriate quality does not necessarily imply the best achievable performance in one or a few properties, but requires the best achievable balance (and compromise) in combination between all properties to satisfy the customer. The customers' (and users') demands for all the external properties in figure 5 are agreed and defined in a design specification. The main aim of recently introduced management techniques such as 'quality function deployment', 'total quality management', Taguchi methods and 'simultaneous engineering' is to achieve this most suitable quality from the perception of the customer at an appropriate cost to the manufacturer.

The quality of a product, as perceived by a customer or user, results from three coordinated aspects:

(a) quality of design, the potential balance of properties from designers' decisions,
(b) quality of conformance, how well the manufactured components conform to the requirements stated in the design documents, and
(c) quality of application, the user's responsibility.

4. Design Process Knowledge

Quoting from Leonardo da Vinci: *Those who give flight to ready and rapid practice before they have learned the theory resemble sailors who go to sea in a vessel without a rudder.* Before a theory is known and established, a 'blind entry' is needed to get the solving process started. A better knowledge of the theory should then be able to improve the practical results. Yet we cannot expect to know ALL the theory before we have to apply it. According to O. Heaviside: *Shall I refuse my dinner because I do not fully understand the process of digestion ?* Experience is, and will continue to be, essential. *Designing* is probably the most difficult task faced by engineers. Creativity in engineering design demands that the engineer shows certain mental attitudes, but creativity cannot be exercised without the appropriate object and process knowledge.

When we consider engineering designers, the norm is that they design intuitively, and are unaware of any theory for executing a design process -- they 'give flight to ready and rapid practice'. A theory and science about engineering design, design science, is increasingly necessary.

Engineering Design has three intrinsic aims:

a) to attain the most appropriate quality of designed product;
b) to attain it in the shortest time;
c) to attain it at minimum costs.

The quality of technical system has absolute priority. The two other aims are often relegated to the background.

Engineering design should deliver full information (a description) of the designed system, ready for possible manufacture if the management decision is favourable. Aims imposed within the manufacturing organization may ask for designing with differing scope of the problem:

a new system without precedents;
a system to be modified or radically re-designed to improve some of its properties (e.g. to modernize it);
a system to be modified to create a range of sizes or capability values;
a system to be adapted to different (manufacturing, operating, climatic, or other) conditions.

These different scopes will also influence the nature and procedures of design process, and the knowledge required to design.

4.1. PROCEDURAL ASPECTS OF DESIGNING - DESIGN STRATEGY

Designing of engineering products (especially technical systems) can take place in various forms:

(a) *trial and error*, or cut and try, an imaginative, creative, but largely random way of getting a solution of questionable quality (it can well be excellent, but there is no way of comparing to other possible solutions);
(b) *intuitive*, spontaneous, where the process is not defined or explainable, the quality of the result depends on experience, and the results need extensive checking and verification;
(c) *methodical*, a systematic procedure according to a plan and methods to support the designer, the resulting system can be nearly optimal;
(d) *partially computer-supported*, some operations or steps are done with the help of computers;
(e) with *integrated computer support*, the human designer is still the key operator;
(f) *fully automated*, only data is needed, a computer does the rest.

Form (a) is obviously inefficient. Actual design experiences use a combination of forms (b) through (e). Form (c) uses a strategic sequence (with many checks and iterations) as outlined in *A--E3* below, supported by various tactical methods [14,23,24]. Form (f) is currently only applicable for a restricted range of technical systems.

Forms (a) and (b) are generally regarded as unstructured, although even when engineers in industrial practice apply them, a management structure is needed. The usual structure (e.g. [25,26]) covers the following stages:

(1) *design specification*, understanding the scope of the problem as given (i.e. recognizing the needs), setting the criteria (including parameters and constraints)

for a solution to be regarded as acceptable, and initial checking of technical and economic feasibility;
(2) *conceptualizing*, thinking and sketching;
(3) *embodying*, layout, arrangement, configuration and parametrization;
(4) *detailing*, generating the instructions for manufacturing.

Especially between stages (1) and (2), and between stages (3) and (4), various *design audits* are performed, and only if the proposals look sufficiently promising is further work (and expenditure) allowed.

Methodical (systematic) procedure -- based on design theory as outlined in [11,12,13,14] -- subdivides the management structure into a set of steps that transform one (more abstract) model of a novel system to be designed into a different (more concrete) model. These steps are obviously coordinated with the TS-structures shown in Figure 4. Each such level gives a progressively more complete description of the future technical system. The steps, after considering alternatives and selecting among them, typically result in:

A -- a *design specification*, a designer's interpretation of the requirements and constraints, stage (1) above and level I in figure 4;
B -- a *process*, the tasks and transformations that are needed by a user, and the effects (outputs from humans and technical systems) that are needed to drive the process, first part of stage (2) and level I;
C -- a *function structure*, the internal tasks of a technical system needed to drive the process, second part of stage (2) and level III;
D -- an *organ structure*, the means (in principle, including their mode of action) to achieve the goals of making the function structure work, third part of stage (2) and level IV;
E -- a *component structure*, level V, generated by:
E1 - a *preliminary layout*, as a sketch of proposed components, arrangements and connections, first part of stage (3);
E2 - a *dimensional layout*, as drawings from which the next stage is accomplished, second part of stage (3);
E3 - full *design documentation*, detail and assembly drawings, parts lists, instructions, etc., stage (4).

Because at each step the designer has a more complete understanding of the initial problem, all prior stages are subject to (periodic or continuous) review -- feasibility checking, refinement and iteration. Progress is largely linear, but with more or less scatter (iteration and back-tracking) depending on the difficulty of the problem, need to refine previously decided features, designers' knowledge, experience and adhesion to the systematic procedural guidelines. Computer aids to designing generally cover levels *E*, the component structures.

In fact, the procedures, knowledge and tools should support the skills and abilities of the engineering designers, and should still allow them to work in whatever way and sequence they choose. The freedom of choice cannot be absolute, there must always be checks and balances. In other words, we can and do not deny the roles of creativity, intuition, iteration, incubation, flair, talent, etc. For engineering design, these can not be ends in themselves, nor can they be sufficient. Creativity can only be effective on the basis of good and verifiable object and process knowledge.

4.2. GENERAL DESIGN TACTICS

In the spirit of the explanations of technical systems we are searching during the design process for a component structure that will be the carrier of the desired properties, the external properties in figure 5. This structure should then be accurately described by the appropriate elementary design properties. The relationship between the given requirements (as input to the design process) and the design properties to be found (as output from design) is complicated, firstly because the number of these relationships is usually large, secondly because quantitative knowledge about some of the relationships is lacking, and thirdly because all phases of the design process are mutually interrelated.

These reasons do not allow directly to solve the problems of achieving particular properties, some additional manipulations must be used. Among the important strategic manipulations in the design process are:

-- *Iteration* (iterating): this is used where a direct solution to the problem is not possible (which is almost always the situation in design problems), usually because the relationships and conditions are too complex. The procedure is similar to that used in mathematics for the approximate solution of a system of equations: initial assumptions are made in order to be able to proceed towards (e.g. calculate) a solution. These results are used in the next stage as improved assumptions and help to determine a more accurate solution. If convergence is sufficiently rapid, a solution is obtained at the desired accuracy after a few iteration cycles.

-- *Abstraction* (abstracting): in this process, attention is concentrated on the important and decisive aspects of a problem situation, and less important ones are initially ignored. This method permits easier entry into a design problem, e.g. by allowing the engineering designers to first concentrate on realizing the main working effects.

-- *Concretization* (concretizing): this procedure leads from rough preliminary and abstract solutions or concepts towards definitive, better defined and more finely tuned ones. It is, in our terminology, the opposite to abstraction.

-- *Improvement* (improving): starting from an existing solution that shows scope for improvement, a satisfactory solution can be achieved by criticism and modification.

-- *Strategy of the Problem Axis*: progressing forward along the problem axis from the problem (or its symptoms) towards physical means may only allow achieving a solution that is palliative, or that paralyze further progress towards a viable solution. Reversal of the search direction towards the causes of the problem can also be attempted ('go back one step'). In many cases, this procedure can permit designers to 'get rid of the problem' by avoiding it from a more abstract starting level.

In practice, each of these strategic manipulations merges with one or more of the others. The strategy of the problem axis can be viewed as a combination of these manipulations.

4.3. PROBLEM SOLVING - DESIGN TACTICS

Problem-solving (as part of engineering design) can also be structured [14,27,28] and these operations occur in two distinct groups:
cyclic steps of problem-solving --

a -- the immediate task should be defined, stating and formulating the problem, elaborating the assigned specification into a full design specification (recognizing a problem, and formulating a problem statement), including developing evaluation criteria for the proposed solutions;
b -- a wide range of solutions to be considered should be generated and synthesized, searching for solutions from existing information and by imagination;
c -- each such solution should be evaluated against suitable criteria, where possible by some form of mathematical 'exact' or approximate analysis based on existing scientific knowledge, and only those one to five solutions considered best should be taken further,
d -- these solutions should be communicated to the next more concrete level of detail and refinement.
auxiliary operations --
x -- preparing and providing information,
y -- verifying all data, checking all work,
z -- representing and documenting the work, data and results including sketching, drawing, modelling (see also [29]).

Again, these procedures may be subconsciously or systematically applied. They may in practice happen so rapidly that they are no longer separable, especially during the most concrete design steps, detailing, etc.

4.4. THE DESIGN SYSTEM AND ITS OPERATORS

The resulting model of a generalized design process is shown in figure 8 [30]. With reference to this model and definition, a holistic study of engineering design should cover a number of factors and aspects:

-- *design process aspects*:
the activity, the design *process* -- its structure (components such as the usable activities of designing, and their relationships), any systematic or procedural aspects (prescriptions to guide individual activities and sequences of activities), creativity and intuitive factors, any other factors affecting design progress, etc. These activities are supported by the knowledge in the top-right quadrant of figure 1, which should be based on the knowledge formulated in both bottom quadrants.
the object to be designed, the *operand* of the design process -- concerning its nature and properties, transformation processes to be performed, necessary influences on the transformation process (from human beings, technical systems and the environment), parts, life stages, evaluations and decisions, etc. -- but also concerning abstractions and models, possible and beneficial transformations between them, representations, etc., both for designing, and for other life stages. This is the subject of the two left quadrants of figure 1.
-- *operators of the design process*:
designers as *operators* of the design process -- their characteristics, working methods, strategies and tactics, etc.
tools and aids that designers may use, including CAD and other computer-based techniques, as operators of the design process, the other main component in North American design research.
their available information and knowledge;
the structure and attitudes of management, and goals set for designers' work and for the company;
their active environment as operator -- the context in which engineering design takes place, industry and management; and the social, moral and political context of designing and using the technical system.

The quality of each of these operators affects the quality of the technical systems they design or cause to be designed. It is not only education, but also the working environment, and the general atmosphere created by management that accounts for the attitudes (particularly motivation and dedication) of design staff. Nevertheless, designers must bring certain positive attitudes to bear on their problems, open-mindedness, a constructively self-critical outlook, willingness to seek and accept advice, etc. The advice (e.g. about manufacturing, laws and economics) must be made readily available -- another management task.

5. Closure

This knowledge is being developed, and has been actively pursued since about 1955. Theories and experiences are being collected by activities of observational research, self-observational investigations, and reflective research (theorizing on the basis of the other two parts).

Design methods and theory can constrain a problem enough to make it comfortable to mess with. These are valuable ways to HELP solve design problems, they are not 'musts', only guidelines; but beware, they can also be used as crutches to 'explain' procrastination. Useful advice is to try to solve it in QUICK AND DIRTY ways to start with, especially for graphical work (sketches) and calculations (using very simple models), and refine later if needed.

Students usually needs more constraint (but not as much as WE give them). Less constraint can be tolerated with increasing experience. -- John J. Rheinfrank (DEED BULLETIN, **14**, No. 1, p. 12)

If engineering designers do not know about these methods (as most people in industry), then they cannot be expected to find time to learn about and use them. Best industry practice can only be a retrospective guide, progress is more likely if these newer insights are adopted.

References

[1] --, (1974) The Merriam-Webster Dictionary, Pocket Books, New York

[2] --, (1984) The Concise Oxford Dictionary (7 ed.), Univ. Press, Oxford

[3] Gregory, S.A., (1966), 'Design Science', Chapter 35 in Gregory, S.A. (ed), The Design Method, Butterworths, London, 323-330

[4] Satir, V. & Baldwin, M., (1983), Satir Step by Step, Science and Behavior Books, Palo Alto CA, 269

[5] Kuhn, T.S., (1970) The Structure of Scientific Revolutions (2 ed.), U. of Chicago Press, Chicago

[6] Kuhn, T.S., (1977) The Essential Tension: Selected Studies in Scientific Tradition and Change, Univ. of Chicago Press, Chicago

[7] Ropohl, G., (1990) 'Die Wertproblematik in der Technik', in Roozenburg, N., & Eekels, J., WDK 17: EVAD -- Evaluation and Decision in Design -- Reading, Heurista, Zürich

[8] Petroski, H., (1985) To Engineer is Human, St. Martin Press, New York

[9] Eder, W.E., (1988) 'Theory vs. Practice in Engineering Design', in L.P. Grayson & J.M. Biedenbach (eds.), Beginning a New Era in Engineering Education, Proc. ASEE Annual Conference, ASEE, Washington, DC 921-925

[10] Guilford, J.P., & Hoepfner, R., (1971) The Analysis of Intelligence, McGraw-Hill, New York

[11] Hubka, V., & Eder, W.E., (1992) Einführung in die Konstruktionswissenschaft (Introduction to Design Science), Springer-Verlag, Berlin/Heidelberg
[12] Hubka, V., & Eder, W.E., (1988) Theory of Technical Systems, Springer-Verlag, New York
[13] Hubka, V., (1976) Theorie der Konstruktionsprozesse, Springer-Verlag, Berlin/Heidelberg
[14] Hubka, V., & Eder, W.E., (1992) Engineering Design, Heurista, Zürich
[15] Hubka, V., Andreasen, M.M., & Eder, W.E., (1988) Practical Studies in Systematic Design, Butterworths, London
[16] Ropohl, G., (1979) Eine Systemtheorie der Technik, Carl Hanser Verlag, München
[17] Hubka, V., & Ropohl, G., (1986) 'Was ist ein technisches System ?', VDI-Z 128 Nr. 22, Nov, 869-874
[18] Andreasen, M.M., Kahler, S., Lund, T., & Swift, K.G., (1988) Design for Assembly (2 ed.), IFS, London
[19] Roth, K., (1982) Konstruieren mit Konstruktionskatalogen, Springer-Verlag, Berlin/Heidelberg
[20] -- , (published annually), ESDU Index & ESDU Catalogue of Engineering Software Engineering Sciences Data Unit, London, England
[21] Eder, W.E., (1989) 'Knowledge Engineering -- An Extended Definition', in Proc. Appl. of Expert Systems in DND, RMC, Kingston, Ont., 11-20
[22] Hubka, V., (1989) 'Design for Quality', in WDK 18: Proc. of the Institution of Mechanical Engineers, International Conference on Engineering Design, ICED 89 Harrogate, I.Mech.E., London, 1321-1333
[23] Adams, J.L., (1980) Conceptual Blockbusting (2 ed), Freeman, San Francisco
[24] Jones, J.Ch., (1980) Design Methods - seeds of human futures (2 ed), Wiley, New York
[25] Pahl, G., & Beitz, W., (1984) Engineering Design, The Design Council, London, & Springer-Verlag, Berlin/Heidelberg (Translated by A.J. Pomerans & K.M. Wallace, edited by K.M. Wallace)
[26] -- , (1987) VDI Guideline 2221: Systematic Approach to the Design of Technical Systems and Products, VDI, Düsseldorf (edited by K.M. Wallace)
[27] Wales, C., Nardi, A., & Stager, R., (1986) Thinking Skills: Making a Choice, Center for Guided Design, Morgantown, WV
[28] Wales, C., Nardi, A., & Stager, R., (1986) Professional Decision Making, Center for Guided Design, Morgantown, WV
[29] Eder, W.E., (1991) 'Role of Graphical Representation in Engineering Design', in Proc. ASEE 1991 Annual Conference, New Orleans, 16-20 June, 1991, ASEE, Washington, DC, 1729-1731
[30] Eder, W.E., (1991) 'Editorial', Design Studies Special Issue for ICED 91 Zürich, Eder, W.E. (guest ed), 12 No 4, October, 194-196

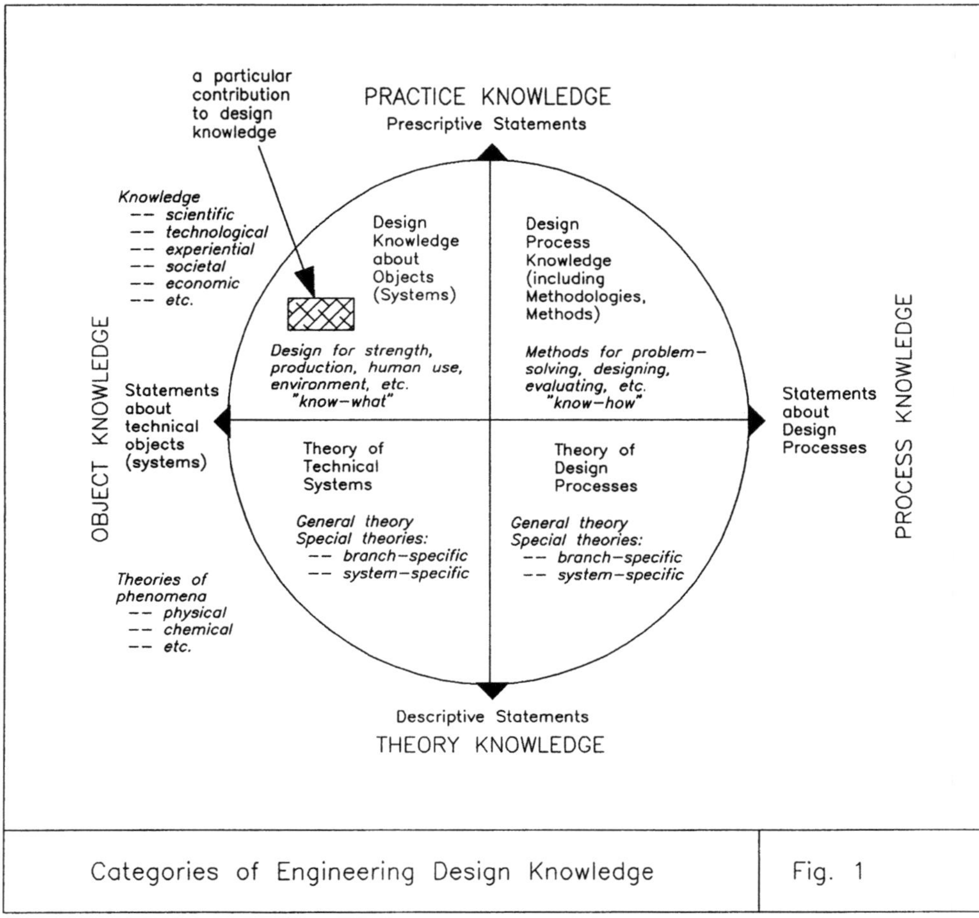

Categories of Engineering Design Knowledge | Fig. 1

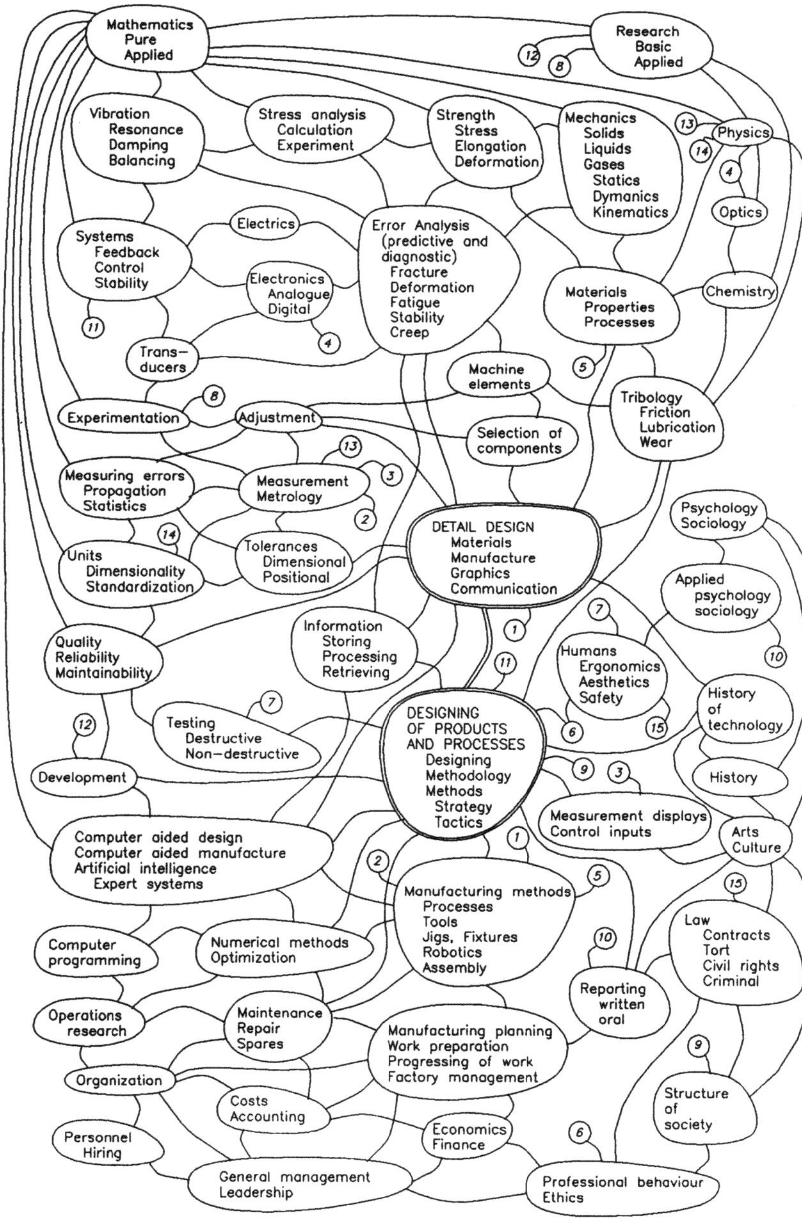

Map of Object-Related Knowledge for Engineering Design

Fig. 2

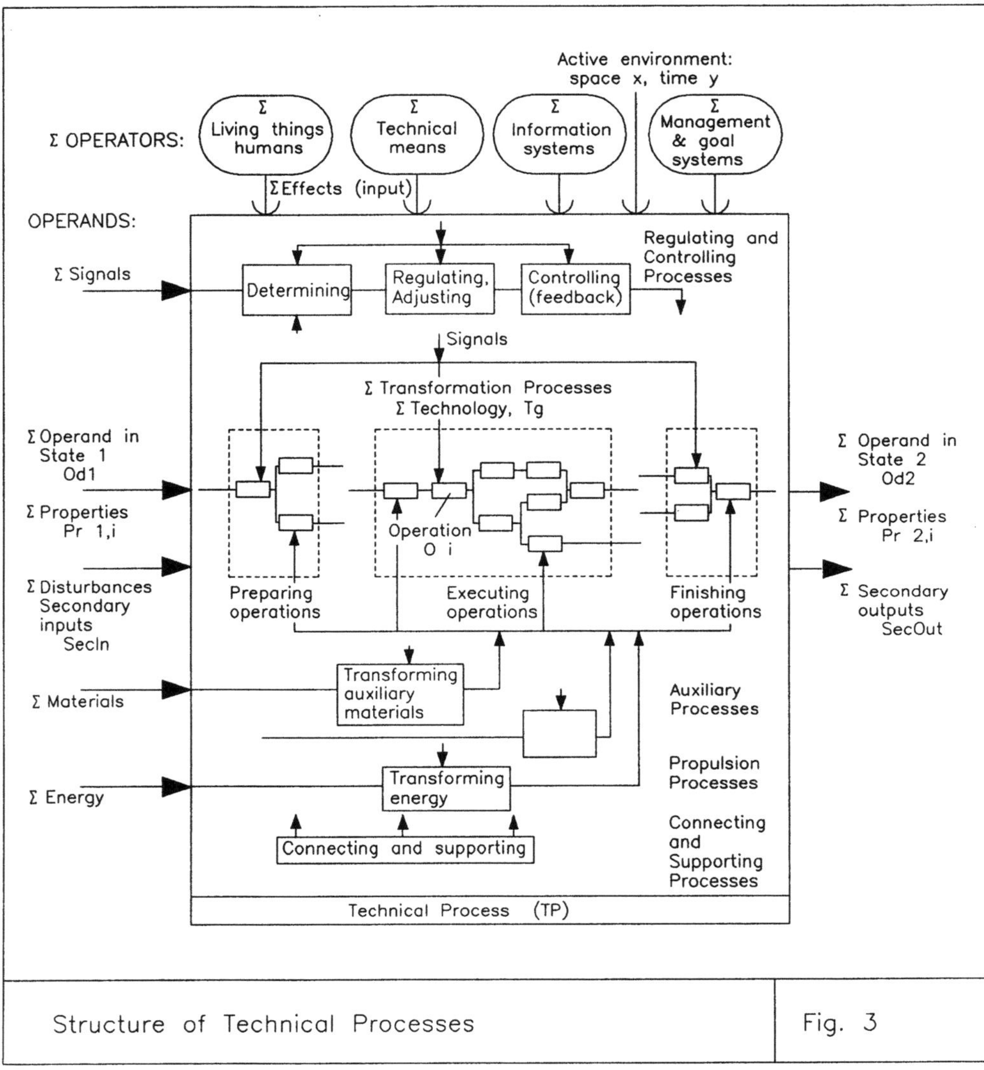

Structure of Technical Processes | Fig. 3

I)

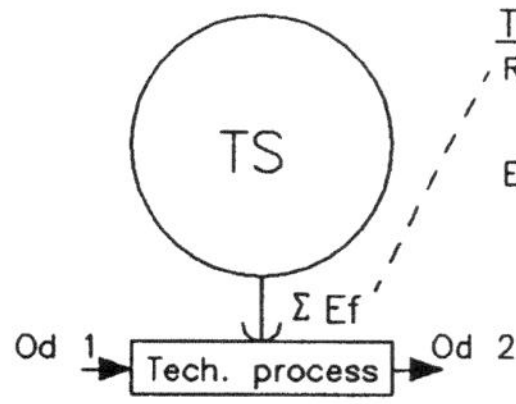

TS Purpose $\Sigma Ef \rightarrow Od$

Realizing the interaction between operand <---> operator, in order to achieve the desired transformation of the operand.
Effects (Ef) are derived from the technology used for the transformation.

Legend:
TS ... technical system
Ef ... effect
Od ... operand
Od 1, Od 2 ... states of the operand

II)

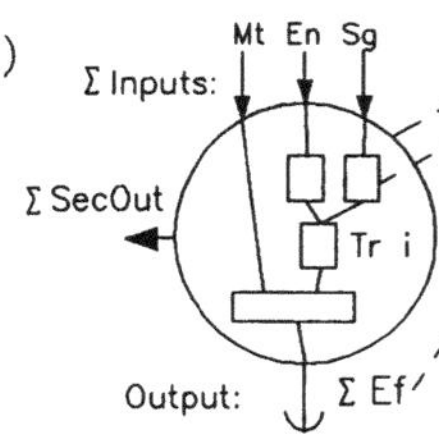

TS Process Model -- Process Structure

TS in working state -- operating.
Internal transformation of inputs to the TS (desired and undesirable) into desired outputs (Ef = effects) plus secondary outputs (SecOut = secondary effects).
Effect --> product (output) of the chain of internal transformations (Tr = mode of action based on laws of nature, e.g. physics, chemistry, biology, etc.).

Legend:
Mt ... material
En ... energy
Sg ... signal (Inf ... information)
Tr ... internal transformation
Tr i ... one of the n transformations of the procss structure
SecOut ... secondary output, secondary effect

III)

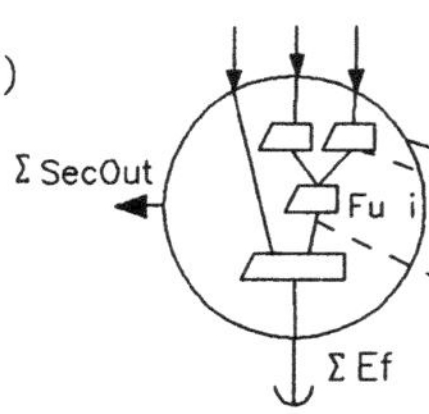

TS Function Model -- Function Structure

TS in state of being capable of working.
Function --> task (capability) of performing the TS-internal transformations, or permitting them.
Function is directly derived from the internal processs.
Relationship between functions.

Legend:
Fu ... function
Fu i ... one of the n functions in the function structure

IV)

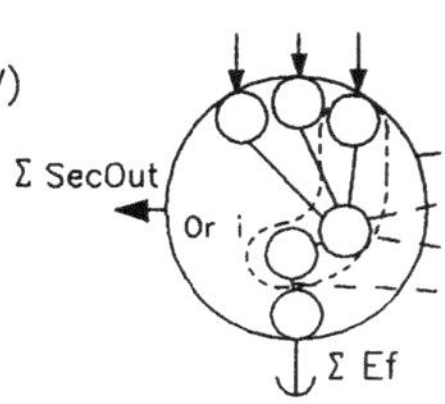

TS Organ Model -- Organ Structure

TS in state of being capable of working.
Organ --> means to realize functions. Main features of organs are the action spaces, surfaces, lines, etc.
Group of organs --> organism.
Relationship between organs.

Legend:
Or ... organ

V)

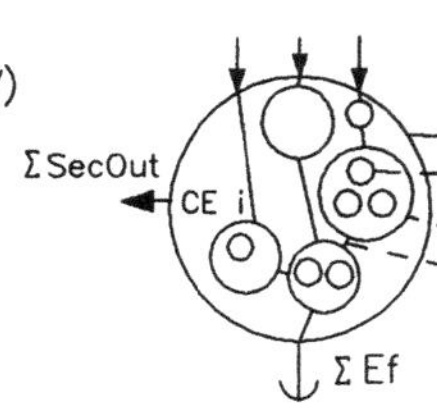

TS Component Model -- Component structure

TS in assembled state.
Constructional element (component) --> means of realizing organs.
Constructional group (sub-assembly, machine element).
Connection between constructional elements.

Legend:
CE ... constructional element

Literature:
Hubka, V., Theorie technischer Systeme, Berlin: Springer-Verlag, 1984, Abb. 5.4
Hubka, V. u. Eder, W.E. Theory of Technical Systems, New York: Springer-Verlag, 1988, Fig. 5.4

Models of Technical Systems (TS Models)	Fig. 4 Part 1 of 2

Example: machine vice in operational state -- state of capability of working.

ABSTRACT MODEL | CONCRETE MODEL

I) PURPOSE

Design specification
Requirements specification
-- from customer, or potential user

Contract specification
-- company level

Designers' working specification
-- designers' approved interpretation

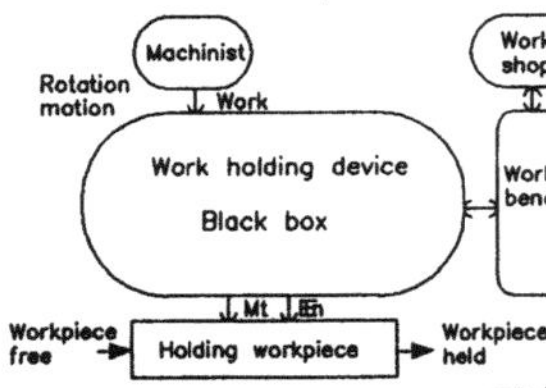

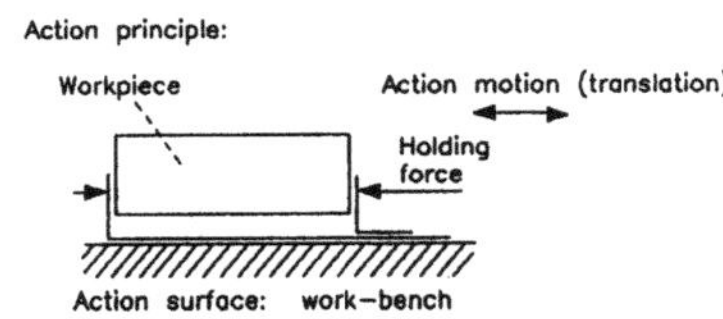

II) PROCESS STRUCTURE

not illustrated -- elements of the function structure and the process structure are in one-to-one correspondence, each function implements the capability to perform the corresponding process.

III) FUNCTION STRUCTURE

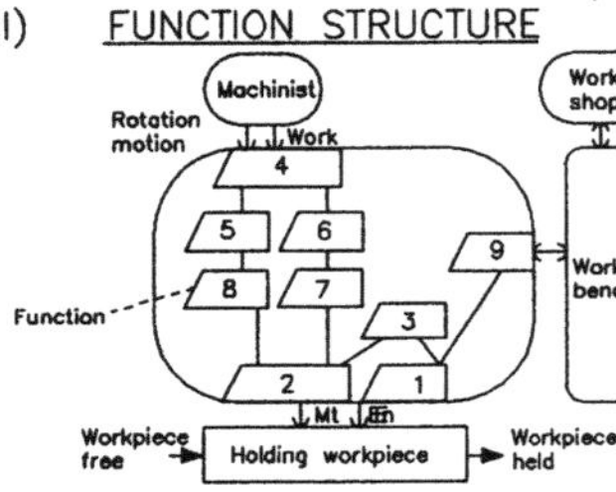

	FUNCTION	ORGAN	
1	FORM fixed holding surface TRANSMIT holding action	Fixed jaw	FJ
2	FORM movable holding surface TRANSMIT holding action	Movable jaw	MJ
3	GUIDE movable holding surface rel. to fixed holding surface	Guidance for movable jaw	MJG
4	FORM the action surface ACCEPT work	Acceptance organ	AcOr
5	TRANSFORM rotation to translation	Motion screw thread organ	MST
6	AMPLIFY force		
7	STORE energy for holding (strain energy)	Storage organ	StOr
8	RETAIN position against holding force	Motion screw guidance organ	MSG
9	FORM contact to fixed system PERMIT making connection	Connection organ	CcOr
10	Evoked addition: REACT radial force from work acceptance	Motion screw guidance organ	MSG

IV) ORGAN STRUCTURE

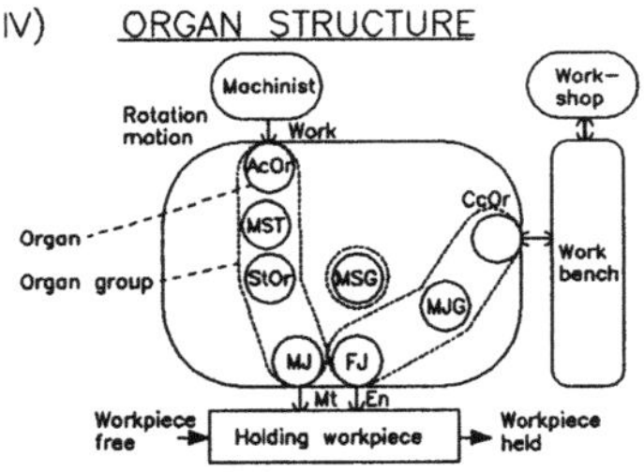

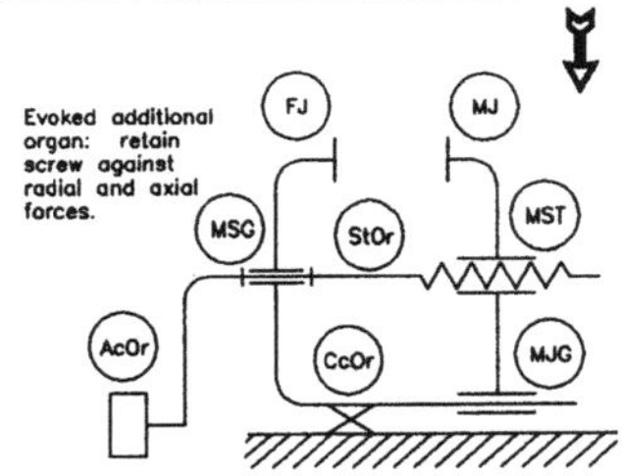

V) COMPONENT STRUCTURE

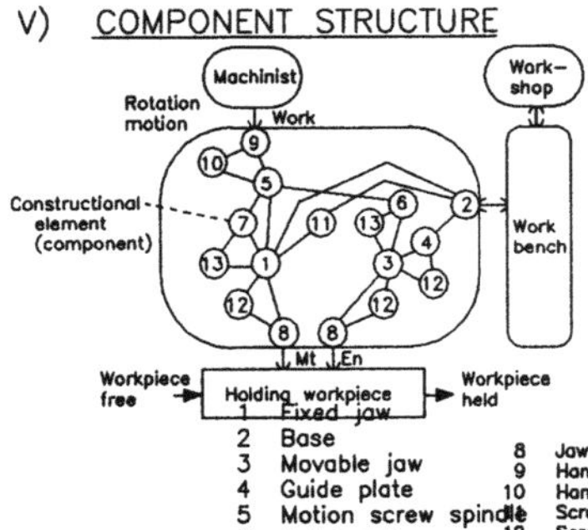

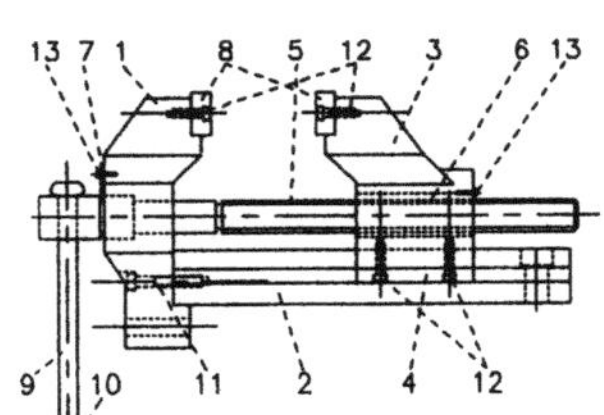

1 Fixed jaw
2 Base
3 Movable jaw
4 Guide plate
5 Motion screw spindle
6 Nut insert
7 Keeper plate
8 Jaw plate (hardened)
9 Handle
10 Handle knob
11 Screw
12 Screw
13 Screw

Detail drawings
Parts lists
Purchased components lists
etc.

Models of Technical Systems (TS Models) | Fig. 4 Part 2 of 2

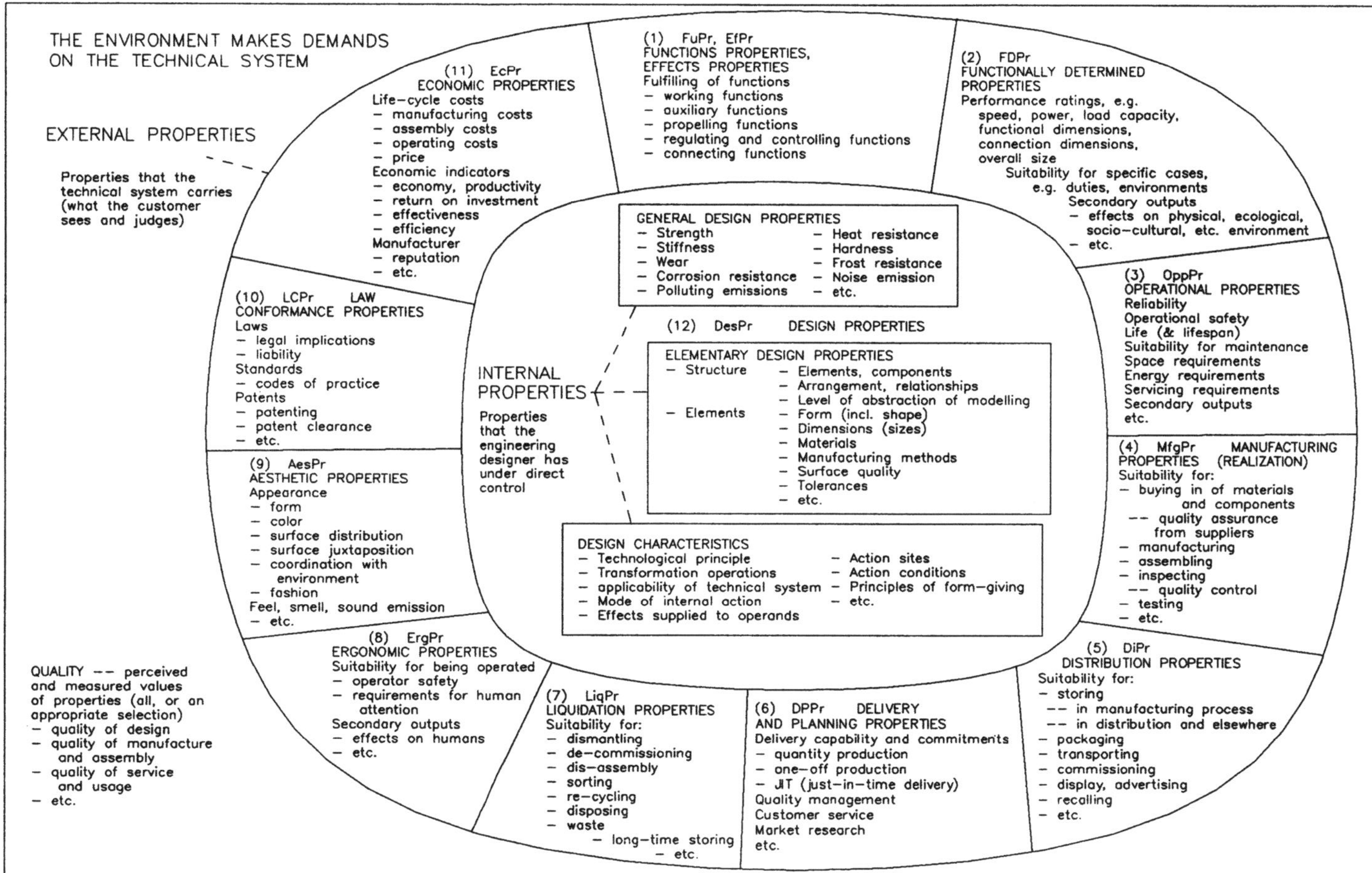

Relationships among Properties of Technical Systems — Fig. 5

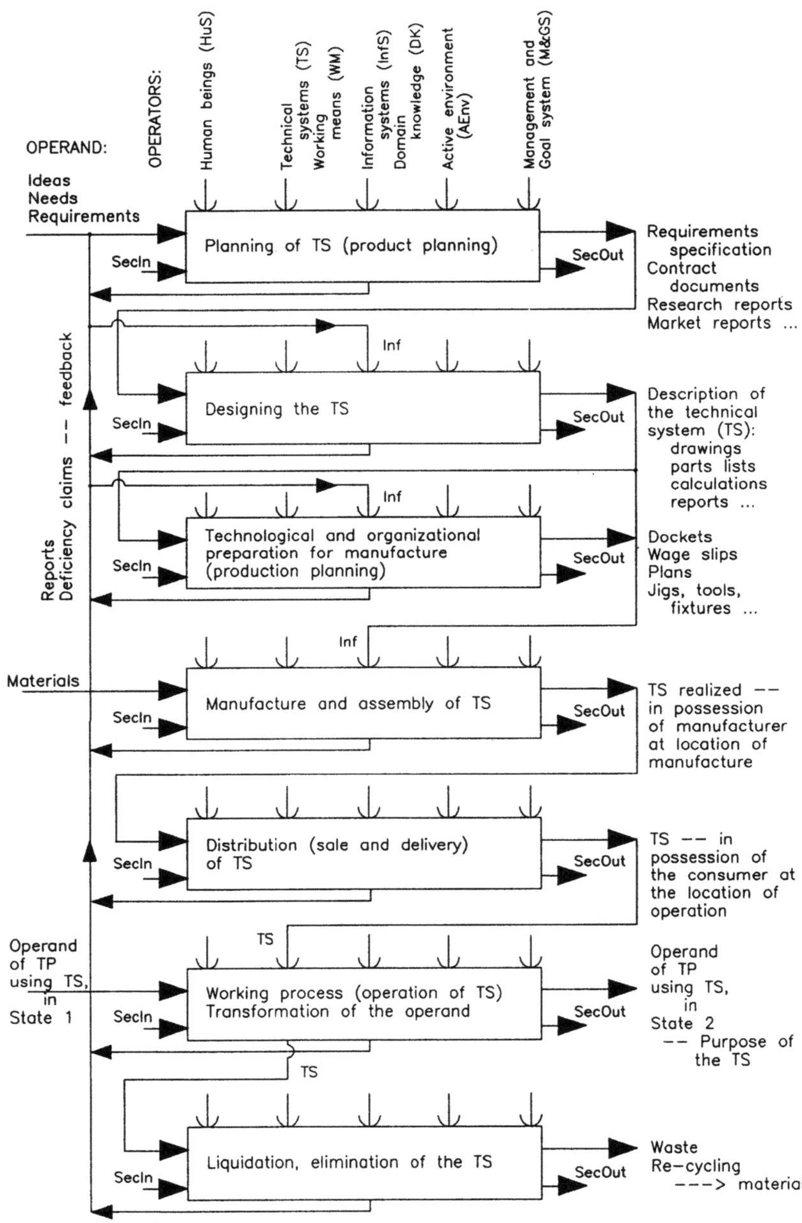

Life Cycle of Technical Systems

Fig. 6

A -- Knowledge of manufacturing technology and faults:

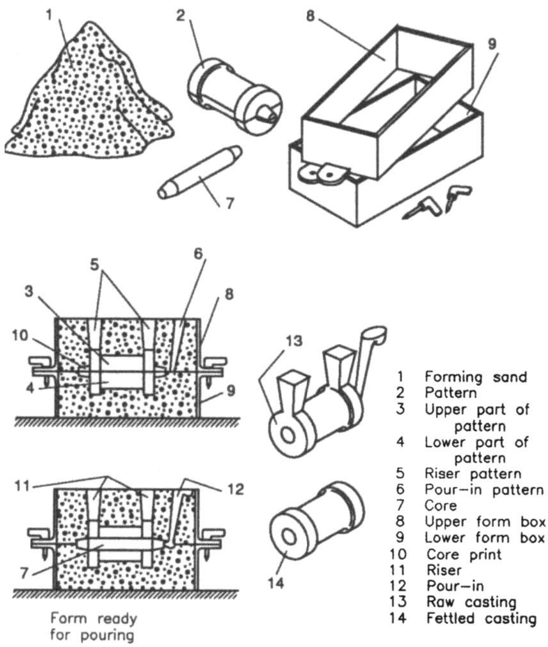

1 Forming sand
2 Pattern
3 Upper part of pattern
4 Lower part of pattern
5 Riser pattern
6 Pour-in pattern
7 Core
8 Upper form box
9 Lower form box
10 Core print
11 Riser
12 Pour-in
13 Raw casting
14 Fettled casting

Defects in castings, fault occurrences on finished cast item. Casting faults are so varied that a complete listing with characteristics and advice for avoidance is not possible in this space. Casting faults can consist of deficiencies in form and dimensions, but can also reduce the strength properties of the cast item, cause leakage, make machining and surface treatments more difficult, influence corrosion resistance and appearance, and lead to scrap. Frequently an interaction between different causes makes recognition and avoidance of casting faults more difficult. Even with careful procedures, the levels of scrap in casting production can vary over wide limits, scrap levels depend on materials to be cast, and forms of castings.

Casting faults include:
- faults in shape due to forming errors;
- shifted cores;
- fettling errors.

B -- lead to requirements for castings:

i.e. regarding structure + form (shape) + dimensions + materials + tolerances + surface

Suitability for casting
- Suitable for pattern-making
- Suitable for forming
- Suitable for pouring
- Suitable for stresses from solidification
- Suitable for fettling
- Suitable for machining

C -- lead to rules for form-design and recommendations for sizing (dimensioning):

- maintain constant wall thicknesses,
- avoid sharp curves,
- avoid collections of material,
- use smooth transitions,
- split up large surfaces for machining,
- provide draft angles for lifting pattern parts.

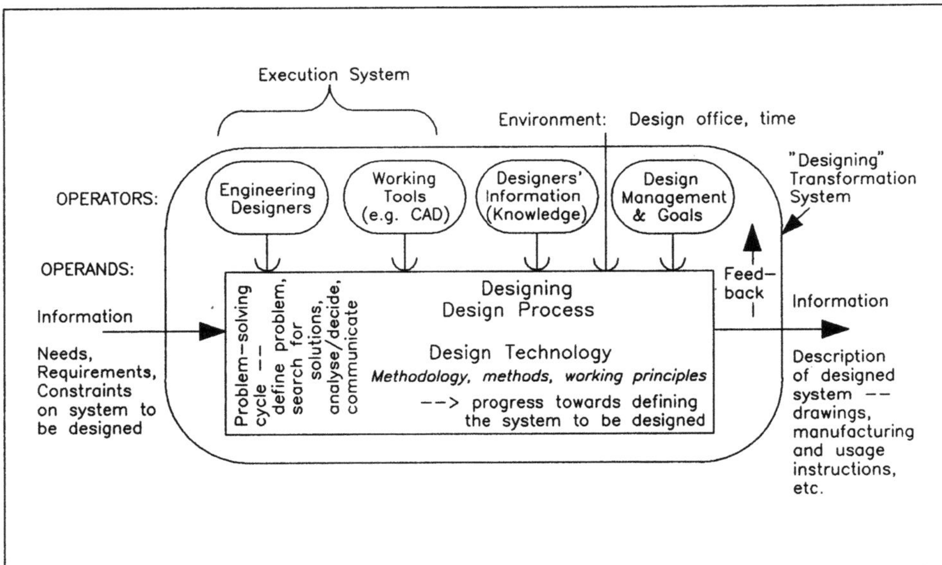

Model of the Design Process — Fig. 8

DESIGN, SCIENCE, AND PHILOSOPHY: THE PRAXIOLOGICAL PERSPECTIVE

W. W. GASPARSKI
Polish Academy of Sciences
Poland

ABSTRACT. This paper examines mutual relationships between members of a triad: design, science (or sciences rather), and philosophy (of science and technology as well as action and value theories, i.e., praxiology and axiology). Part one attempts to recall praxiological issues as related to design. Part two discusses preliminaries related to the concepts of methodology and design. In part three a programme of the methodology of design is presented while part four ventures a collection of general theory (praxiology) of design. In this part some relevant epistemological questions are discussed. Part five taking into account that designing as a meta-action takes the precedence of actions it serves a conceptual preparation should serve a new methodological insight of science activity.

1. Introduction

This paper is concerned with the relationships between *design* and its context as seen from *praxiological* perspective. The perspective refers to praxiology, i.e., *human action theory* (see, e.g., (Bunge, 1989), (Gasparski, 1992)) and/or *general methodology* (see, e.g., (Kotarbiński, 1937), (Gasparski, 1987), (Hall, 1989)).

First, we have to recall the name of a French social philosopher Alfred Victor Espinas who is generally considered an originator of *praxiology* understood as *general technology* (Espinas 1890). Therefore both the praxiologists and the philosophers of technology recognize Espinas as a founding father of their disciplines. Among the contemporary philosophers there are: a German philosopher F. Rapp (1974), an American philosopher S. Toulmin (1977), a Canadian-Argentinean philosopher M. Bunge (1979), and a Chinese philosopher Zou Tsing (1993) who underline the praxiological issues of *technophilosophy* mainly in relation to design.

Second, the praxiological analysis of design comes back to T. Kotarbiński's (a Polish philosopher) study in epistemology published (in Polish) for the first time

M. J. de Vries et al. (eds.), Design Methodology and Relationships with Science, 165–189.

in, 1929 (for English translation see (Kotarbiński, 1965)). According to the study design is the distinguishing methodological mark of *practical sciences*, i.e. disciplines (or skills) dealing with human purposeful activities (see (Kotarbiński, 1971), (Kotarbiński, 1977), (Gasparski, 1983b)). It is worthy to add that the similar (but not identical) point of view was shared by another Polish methodologist K. Krzeczkowski[1] (1986) as well as by an Austro-American economist and philosopher F. A. von Hayek (1952). The H. A. Simon's idea of *the sciences of the artificial* (Simon, 1969) is of similar nature too[2].

Finally, systematic praxiological research on design - considered not only as a method but also as a kind of human activity - were introduced in the sixties by the author of this paper (Gasparski, 1970, 1972, 1973). An Italian philosopher T. Maldonado (1972), an Austro-American methodologist R. Mattessich (1978), and an English design scientist B. Archer (1980) are among other authors who have discussed design issues from praxiological perspective. The perspective has been acknowledged a factor important for the development of contemporary design science and methodology by, e.g., (Gregory, 1973, 1980), (Warfield, 1990), (Reich, 1992).

Both the size and the purpose of the paper prevent its author from summarizing all *designological* (Gasparski, 1981) results achieved under the praxiological umbrella, so more they were recently presented[3] in (Gasparski, 1984, 1988, 1990). What the author is going to offer instead is a collection (a system) of the main theses of general theory (praxiology) of design[4].

---

[1]First published in 1936.

[2]A special Polish annual devoted to methodological problems of practical sciences entitled *Projektowanie i Systemy* (Designing & Systems) has been published since 1978 under co-editorialship of W. Gasparski and D. Miller in co-operation of A. Strzałecki. The annual is sponsored by the Committee of the Science of Science, Polish Academy of Sciences and published by the Ossolineum Publishers, Wrocław, Poland. Among the authors whose articles has been published in the annual are eminent design scientists and philosophers like: A. Bridges, M. Bunge, C. Eastman, W. E. Eder, L. Finkelstein, R. Foque, J. G. Gero, S. A. Gregory, G. J. Klir, H. A. Linstone, T. Mann, G. Nadler, G. Rzevski, H. A. Simon, L. Tondl, J. N. Warfield, L. A. Zadeh, H. Yoshikawa, and many others.

[3]What, after all, forced the author to quote a few passages from those publications in this one.

[4]Earlier versions of the theory were published in: (Gasparski 1983a, 1989, 1990).

2. The Preliminaries

2.1. THE CONCEPT OF METHODOLOGY

No one has put the question what "a fatal ambiguity surrounds the expression 'the methodology of...'" in more clear way than M. Blaug (1982) has. In many cases the term *methodology* is used as a "more impressive-sounding synonym for *methods*", he noticed not without a sense of irony. Let me propose to use the term *methodics* in this context (after, e.g., Bunge (1985)). Methodology, continues Blaug, "denotes an investigation of the concepts, theories, and basic principles of reasoning of a subject" being a synonym of philosophy of science applied to the given discipline. This understanding is congenial with Ajdukiewicz's (1974), Bunge's, Kotarbiński's, and many other's interpretation[5].

Writes P. Checkland: "By 'methodology' I do not mean 'method'. The word does derive from the Greek word for method and this, according to Kotarbiński (1966), originally meant the path of a person pursuing another, then came to mean generally a path, then a way of doing something, and later expert bahaviour in formulating one's thoughts. As a result of this history Kotarbiński distinguishes three current conceptions of methodology, which he calls *praxiological* -'the science of ... ways of expert procedures', *logical* - ' the study of methods of using one's mind', and *epistemological* - 'the study of sciences as historical products and processes'. My sense of the word here is that the outcome of the research is not *a method* but a set of *principles of method* which in any particular situation have to be reduced to a method uniquely suitable to that particular situation. [...]My use of the word is thus nearest to Kotarbiński's 'praxiological' version: methodology not as 'ways of expert procedures' but *the science of* [such] procedures'. [...]I take a methodology to be intermediate in status between a philosophy, using that word in a general rather than a professional sense, and a technique or method (Checkland, 1988, 161-162). Having this in mind let us introduce the following definitions:

Definition 1: A *method* is the way of action applied systematically to achieve successfully the purpose of the action.

---

[5]E.g., Buchler's one expressed in his *The Concept of Method* so characterized by B. J. Singer, an American practical philosopher related to praxiology: "Buchler's study of methods is also guided by his conviction that if methodology (critical study of the presuppositions, implications, conditions, and consequences of methods and their application) and what he calls 'methodography' (description and analysis of methodic process) are to be fruitfully pursued, we require a more adequate understanding than we presently possess of the nature and function of method." (Singer 1983).

Definition 2: A set of methods (i.e., technical procedures of a discipline) related to the same kind of actions (i.e., a discipline) is called *methodics* (of the discipline).

Definition 3: A study of the philosophical foundations (i.e., principles) of methods in any kind of human activity, e.g. science, design, etc., is called *methodology*, e.g. *methodology of science*, *methodology of design*, etc. - shortly: *science methodology*, *design methodology*, etc. - respectively.

Methodology may be *general* (i.e., gender and/or discipline invariant), *generic* (i.e., of a particular gender of action), or *detailed* (i.e., of a particular discipline).

Design methodology being a study, as it was said, is *a theory* of a kind. "It is important to realize -admonishes M. Tribus - that a theory is *constructed*, not *discovered*. That is, a theory can be developed which will contain sufficient concepts to encompass a subject of interest and provide connections among these concepts. A general theory is one which includes many applications without becoming overly complicated by a need for more and more concepts as the domain is extended. When properly understood, a theory defines its own domain of application. Thus, thermostatics applies only to equilibrium - and equilibrium is defined only via thermostatics. This apparent circularity is inherent in all theories. It should not be regarded as a deficiency." (Tribus, 1969). Therefore *design methodology* ought to be constructed first to define (a concept of) *design* as its domain of application.

2.2. THE NAME OF 'DESIGN'

Having in mind the Tribus' warning we have to limit the discussion now to a name of 'design", its equivalents, and they etymology. Such a discussion will help us to sketch the area of inquiry.

In different tongues, i.e., in different cultural environments, words of different etymology are used as labels of a concept of design. In English it is *design* itself based on Latin *designo*='to define', 'to point out', 'to mark', 'to form', and *dissigno*='to unseal', 'to manage'. In many languages (whether Anglosaxonian or not) *design*, taken from English not from Latin, means 'industrial design', i.e., design with an aesthetical flavour, e.g., *disseny* in Catalan, *diseño* in Castilian (Spanish), *esthétique industrielle* in French. By the way, French equivalents for *design* are *dessein* what means 'intention' or *dessin* what means 'pattern' (Polish *deseń*). In Polish (and other Slavic languages) the label is *projektowanie* based on Latin *proicio*='to place something ahead' (like in English 'to project a missile'), which is similar to German *projektirung* of the same Latin origin. In both languages 'engineering design', especially mechanical one, is labeled *konstruowanie*, *konstruirung* resp. from Latin *construo*='to cast', 'to arrange'. In one group of languages different kinds of design are labeled through using one noun (e.g., *tervezes*-in Hungarian) plus different adjectives, while in the other group of tongues different nouns are used. It reflects different beliefs in the unity or disunity of different kinds of the human activity in question. All languages are unified, however, in one common question, namely whether *design* and *planning* are synonims (like *suunnitella in Finnish)* or not (like *sheji*='design' and *jihua*='planning' in Chinese).

Writes G. Nadler in his *The Planning and Design Approach*: "Planning and design are classified together here because their definitions overlap. The words are often used interchangeably as in 'Planning a vacation,' or 'designing a health care delivery system.' No purpose is served by saying that 'planning' is open-ended while 'design' is specific, or that the former has a longer time horizon, or that the latter is project-rather than program-oriented. Whether it be an architect's blueprint, a five-year land-use map, or a family's financial plan, solution specifications are detailed, resource allocations are proposed, innovation is encouraged, and purposes are defined--and this is planning and design." (Nadler, 1981). On the other hand, according to Bunge, design and planning are different though subsequent phases of the technological process. "...technology may be regarded as the field of knowledge concerned with designing artifacts and planning their realization, operation, adjustment, maintenance, and monitoring in the light of scientific knowledge." (Bunge, 1985). Praxiological point of *designing* and *planning* is similar to the Nadlers' one[6], for praxiology recognizes them as different names of the same kind of human action taken from methodological point of view[7].

Praxiology understands *an action* as a human behaviour which is conscious, voluntary and oriented toward a purpose of its agent (actor). Kinds of actions differ between each other because of classes of purposes their agents tend to achieve. Some kinds of actions are *pre-actions*, i.e. actions which general purpose is to prepare other actions; among them are the second order actions or *meta-actions* (Gasparski, 1987), i.e., actions with a purpose to create (using the Simon's word) other actions. Design (planning) belongs to them.

Taking into account what has been said above and anticipating what will be said below, we may follow the advise: <<*"What does this mean?" I asked. I no longer understood anything. "Try to formulate a hypothesis. You must have learned how it is done"*>> (U. Eco, *The Name of the Rose*) and formulate the name of the 'design'.

---

[6]And the Simon's one too: "The second state in decision making is to device or discover possible courses of action. This is the activity that in fields like engineering and architecture is called 'design'; in military affairs 'planning'; in chemistry 'synthesis'; in other contexts 'invention', 'composition; or that most approving of labels--'creation'." (Simon 1977).

[7]Praxiology, however, accepts traditional names of design/planning-like professions: e.g., architectural design vs. urban planning, organizational design vs. economical planning, etc.

Meaning postulate 1: The *gender of 'design'* is the kind of meta-actions (an activity) the supreme purpose of which is a conceptual preparation of an action or any of its elements.

Meaning postulate 2: The *verb of 'design'* is to perform an action (a meta-action) aimed at
formulating a design (a noun).

Meaning postulate 3: The *noun of 'design'* is a description of an artifact thought needed thus worthy to be implemented.

3. The Methodology Of Design

General methodology of sciences (science methodology) is the most experienced branch of general methodological knowledge (general methodology). Its best elaborated programme was formulated by Kazimierz Ajdukiewicz, a Polish logician who - like Kotarbiński - relates to the world famous Lvow-Warsaw school of philosophy. Having that in mind, I suggested, as it was said above, an outline of the methodology of design using the Ajdukiewicz's programme as a background one (Gasparski, 1972). The outline has never been presented in English in its original form (only its shorten version was published in (Gasparski, 1973)), therefore I dare to use this unique opportunity to present it in full. The left column consists on the outline of the methodology of design while the right column consists of the Ajdukiewicz's programme quoted after (Ajdukiewicz, 1974).

The term 'design' is interpreted in two ways. In one interpretation it means the profession of designers, i.e., the totality of the operations performed by designers *qua* designers. In the other hand, it is used to denote the product of those operations, i.e., a system of the designs which designers have come to accept in their striving to change of the reality on the base of learning the facts and on being in an accord with accepted values.

The term 'science' is interpreted in two ways. In one interpretation it means the profession of scientists, i.e., the totality of the operations performed by scientists *qua* scientists. In the other hand, it is used to denote the product of those operations, i.e., a system of the science which scientists have come to accept in their striving to comprehend facts.

Design understood as the profession of designers, i.e., as the totality of the operations undertaken by them when they act *qua* designers, is a process that evolves in the course of time, and as such is a historical phenomenon that may suitably become an object of historical research. Design so conceived is a social phenomenon, too, and as such may be studied by sociologists. Since the operations performed by designers *qua* designers are largely mental, they may also be studied by psychologists.

But, next to history, sociology, and psychology, there is also another discipline which is concerned with the study of design interpreted as the profession of designers. That discipline is the methodology of design, which sets itself a different task from those set themselves by the disciplines mentioned above.

The task of the methodology of design is, first of all, to single out certain types of operations performed in designerly work and to analyse them so as to be able to state in what the operations of a given type consists. Design methodologists thus single out such types of cognitive procedures as problem identification, interpretation, and testing, decomposition of problems and aggregation of solutions, optimization, etc. Such types of cognitive procedures are analysed by methodologists, which results in clear definitions of these procedures.

Science understood as the profession of scientists, i.e., as the totality of the operations undertaken by them when they act *qua* scientists, is a process that evolves in the course of time, and as such is a historical phenomenon that may suitably become an object of historical research. Science so conceived is a social phenomenon, too, and as such may be studied by sociologists. Since the operations performed by scientists *qua* scientists are largely mental, they may also be studied by psychologists.

But, next to history, sociology, and psychology, there is also another discipline which is concerned with the study of science interpreted as the profession of scientists. That discipline is the methodology of sciences, which sets itself a different task from those set themselves by the disciplines mentioned above.

The task of the methodology of sciences is, first of all, to single out certain types of operations performed in research and to analyse them so as to be able to state in what the operations of a given type consists. Methodologists of science thus single out such types of cognitive procedures as substantiation of theorems, inference, solving of problems, description and explanation of facts, formulation and verification (testing) of hypotheses, defining terms, etc. Such types of cognitive procedures are analysed by methodologists, which results in clear definitions of these procedures.

It turns out in this connection that certain types of cognitive procedures are to be found in all kind of design and play the same role in each case, while others are to be found in some kinds of design only to play a certain role in some and a different one in others.The methodology of design is accordingly divided into general and specialized, the latter being in turn subdivided into specialized methodologies of the various types of design which differ from one another by the types of methodological procedures used in them. General methodology works out general methodological concepts, i.e., concepts of those cognitive operations which occur in all design disciplines, even though they may play different roles in different kind of design. These general methodological concepts include those of problem definition, inference, solution, optimization,and many others. Specialized methodologies work out specialized methodological concepts, i.e., concepts of those designerly operations which occur in some kinds of design only. These concepts include such as feasibility, performance measurement, verification (testing) of solution candidates, and many others. These concepts refer to those types of designerly operations which occur in what is termed *engineering design*, i.e., technical disciplines. Endowed with the apparatus of general and specialized methodological concepts, specialized methodologies describe in an outline the whole of the procedures used in those kinds of design.

It turns out in this connection that certain types of cognitive procedures are to be found in all sciences and play the same role in each case, while others are to be found in some sciences only or play different role in different sciences. others. The methodology of sciences is accordingly divided into general and specialized, the latter being in turn subdivided into specialized methodologies of the various types of sciences which differ from one another by the types of methodological procedures used in them. General methodology works out general methodological concepts, i.e., concepts of those cognitive operations which occur in all disciplines, even though they may play different roles in different sciences. These general methodological concepts include those of inference, deduction and induction, definition, classification, and many others. Specialized methodologies work out specialized methodological concepts, i.e., concepts of those cognitive operations which occur in some sciences. These concepts refer to those types of cognitive operations which occur in what is termed *real sciences, i.e., natural and social sciences, but do not occur in what is termed formal sciences*, i.e., mathematical disciplines. Endowed with the apparatus of general and specialized methodological concepts, specialized methodologies describe in an outline the whole of the procedures used in those sciences.

By taking a synthesizing look at the procedures used by designers who represent the various disciplines, a look that covers not only the present, but the past as well, the methodologist forms an image of design procedure which he sees as if it were a striving towards a goal. For instance, the procedures used by engineering designers is presented in a synthetic interpretation, look as if engineers strove to instrumentalize any activity of a man, and to make this instrumantion more and more perfect through their inventions. Similarly methodologist sees the whole image of the activities of agro- and zoo-technicians, with the only difference that they domain of design is limited to the phenomena occurring in living organisms (plants and animals resp.). It is no other with the procedures of physicians, pedagogues, socio-engineers, lawyers, managers, etc. It looks as if they are inspired by the duty of change (modification) of the fragments of reality interesting them. Treating design as a whole, i.e. giving up its divisions according to various objects, the methodology of design takes into account the following differences between the branch and the global treating of an object under design. Each of the branches is interested in the reality on one particular respect, whereas changing of the reality needs examining the object in all respects concerning both the object and its context. Design then, as a whole, is to fulfill as if a superior task: the synthesis of all these respects into an artifact on the grounds of which it would be possible to realize the artifact in practice.

By taking a synthesizing look at the procedures used by scientists who represent the various disciplines, a look that covers not only the present, but the past as well, the methodologist forms an image of scientific procedure which he sees as if it were a striving towards a goal. For instance, the procedures used by physicists, in a synthetic interpretation, look as if physicists strove to formulate the regularities that govern all material nature, whether animate or inanimate, and to explain those regularities by finding higher-level laws from which the lower-level ones follow; to explain those higher-level ones by such which are ranked still higher in the hierarchy of laws; and ultimately to some supreme laws of nature, from which all other laws follow. Likewise, the methodologist take a synthesizing look at what is done by plant or animal physiologists, the only difference being that their field of research is restricted to processes that take place in living organisms, and that is explaining them they avail themselves of physical and chemical laws whose validity covers all natural phenomena. The same applies to those zoologists who work on systematics (...).

It has been said that the procedures used by designers in a given discipline look so, in a synthesizing approach, as if those designers were striving towards a goal. This is not to say that those designers always realize that. Yet they act in the way they would act if they realized what their goal is. If they act so without realizing clearly what they goal is, than it may be said that they are striving towards that goal unconsciously. One of the tasks of the methodologists is to identify those goals towards which designers working in a given field strive, whether consciously or unconsciously.

An explicit listing of those goals makes it possible for a methodologist to formulate the standards of correct procedures in the various disciplines. Such a procedure, in a given discipline, which brings designers closer to their goal, is termed correct, whereas such which does not serve that purpose, and *a fortiori* such which makes it more difficult to reach the goal, is termed incorrect. Incorrect, for instance, would be not only such design of technical object which disregards implementation possibilities but also such which, though possible to realize, disregards the influence of the designed object upon the environment in which it will be exploited and utilized.

The standards of correctness of designerly procedures, as formulated in methodology, are not dictated by it to designers in advance. Such standards are derived from practical activities of competent designers, who approve of some procedures in design while they disapprove of others. Competent designers, when assessing their own and other people's designerly procedures, do so in accordance with such standards, but they usually do not realize them clearly enough to be able to describe them explicitly. In other words, competent designers develop, as a result of their practical activities, what might be termed a design conscience, but they do not always clearly realize the principles by which that 'conscience' of theirs is guided. It is the task of methodologists to codify the principles of that design 'conscience'.

It has been said that the procedures used by scientists in a given discipline look so, in a synthesizing approach, as if those scientists were striving towards a goal. This is not to say that those scientists always realize that. Yet they act in the way they would act if they realized what their goal is. If they act so without realizing clearly what they goal is, than it may be said that they are striving towards that goal unconsciously. One of the tasks of the methodologists is to identify those goals towards which scientists working in a given field strive, whether consciously or unconsciously.

An explicit listing of those goals makes it possible for a methodologist to formulate the standards of correct procedures in the various disciplines. Such a procedure, in a given discipline, which brings scientists closer to their goal, is termed correct, whereas such which does not serve that purpose, and *a fortiori* such which makes it more difficult to reach the goal, is termed incorrect. Incorrect, for instance, would be a classification of animals which would group them in a way that does not refer to any genealogical relationships nor makes it possible to explain the evolutionary process which resulted in the formation of such groups. (...)

The standards of correctness of research procedures, as formulated in methodology, are not dictated by it to researchers in advance. Such standards are derived from practical activities of competent researchers, who approve of some procedures in research while they disapprove of others. Competent researchers, when assessing their own and other people's research procedures, do so in accordance with such standards, but they usually do not realize them clearly enough to be able to describe them explicitly. In other words, competent researchers develop, as a result of their practical activities, what might be termed a research conscience, but they do not always clearly realize the principles by which that 'conscience' of theirs is guided. It is the task of methodologists to codify the principles of that research 'conscience'.

The three main tasks of that branch of methodology which is concerned with design as the profession of designers, i.e., design as an activity, are: (1) singling out the types of activities carried out in design work, and analysis of such types of activities, resulting in definitions which explain in what such activities consist; (2) description of designerly procedures (in general outlines) as used in the various disciplines; (3) finding out the goals for which designers in the various fields strive, whether consciously or not, and the resulting codification of standards of correct design procedures. This branch of methodology is termed *pragmatic design methodology*.

Pragmatic design methodology does not, however, cover the full scope of design methodology. It is not possible to discuss designerly operations without referring to the language in which the results of those procedures are formulated. Thus, for instance, we cannot speak of technical systems design without referring to the way they are formed. It is not possible to discuss the design of a technical system without referring to the system formulated in this connection and used as the paragon of the term in question. It is not possible to discuss the accuracy of a solution without referring to the concept of relevancy of change.

Designs of particular objects are built in different ways, depending on the types of these objects. Wishing to speak of the objects in a more general way than the detailed design methodologies do, the general design methodology discuss these objects in terms of systems. Aiming at the formal analysis of the designed artifacts the detailed methodologies use mathematical apparatus and formal logic.

The three main tasks of that branch of methodology which is concerned with science as the profession of scientists, i.e., science as an activity, are: (1) singling out the types of activities carried out in research work, and analysis of such types of activities, resulting in definitions which explain in what such activities consist; (2) description of research procedures (in general outlines) as used in the various disciplines; (3) finding out the goals for which researchers in the various fields strive, whether consciously or not, and the resulting codification of standards of correct research procedures. This branch of methodology is termed *pragmatic methodology* (from Greek πρᾶγμα, to be read: pragma, which means 'deed').

Pragmatic design methodology does not, however, cover the full scope of methodology. It is not possible to discuss cognitive operations without referring to statements and more complex structures which consist of statements, in which the results of those procedures are formulated. Thus, for instance, it is not possible to discuss the definition of a term without referring to the statement formulated in this connection and used as the definition of the term in question. It is not possible to discuss proofs without referring to constructions consisting of statements which are formulated in this connection and which form the proof as the result of the operation of proving.

Proofs of single theorems are rather simple constructions formed of statements which are elements of more complex constructions that cover proofs of many theorems in a given field. What is termed deductive systems are such more complex constructions consisting of statements. They usually have the form of what is termed axiomatic systems, i.e., sequences of statements all of which, except those few which are called axioms of the system, have in that sequence a proof based on those axioms. The mathematical sciences (interpreted as a product of human activity) in a more advanced stage of evolution become deductive systems, and axiomatic systems in particular.

It is obvious that it would be impossible to discuss the activities of designers as the builders of the systems they design without making reference to the systems which they construct in the process. On the other hand, however, it is possible to discuss systems while disregarding completely man's role in their formation. It is possible to analyse systems as composed of things (elements) forming these systems with regard to given relations and properties if these elements are entitled these properties found in the relations given. We can also (in accordance with the principle of dualism) examine systems as consisting of elements forming these systems with regard to the properties given if these elements appear in relations which are entitled to the given properties. Śo then, we can examine these relations or the properties of these systems which they are entitled to independently of what the people who create these systems think. Thus, e.g., having a system given we can ask whether the system is homogenic, elementary (consisting of elements of which no single one is treated as a system) or minimal (vanishing along with the vanishing of even one element).

It is obvious that it would be impossible to discuss the activities of mathematicians as the builders of their discipline without making reference to the axiomatic systems which they construct in the process. On the other hand, however, it is possible to discuss axiomatic systems while disregarding completely man's role in their formation. It is possible to analyse axiomatic systems as sets (or, rather, as sequences) of statements linked by logical relations, the consequence relation above all, and to examine those properties of the said sets which they have regardless of what the people who formed such sets had in mind. For instance, given an axiomatic system, i.e. a set (sequence) of statements consisting of a number of statements designated as axioms of that system, and of logical consequences of those axioms, we may be interested in whether that system is consistent, i.e., whether the consequences of its axioms include, or not, contradictory statements. We may also be interested in whether that system is maximal, i.e., whether any of two contradictory statements formulated in the language of that system one of these statements is a consequence of its axioms, i.e., in other words, whether every problem formulated in the language of that system can be decided within that system. Now when conducting such research on axiomatic systems we are not interested in any way whatever in people's attitudes toward the statement which are in that system. We are not interested in whether he infers some statements from others, and whether the axioms of the system are self-evident to him, or not. We are interested only and exclusively in the statements which are in that system and in the relations between them; we are also interested in the systems of those objects to which those axioms refer.

When engaging in this kind of research on design, the latter interpreted not as the profession of designers but as a product of their designerly operations (actual or merely possible), we engage in *apragmatic design methodology*. Its most important branch is the theory of systems. The most developed are some sections of apragmatic detailed methodologies of some classes of designs, namely of those which may be described in the language of mathematics. Pragmatic design methodology, which is concerned with operations made in the pursuit of design, is a behavioral science and hence falls under the broad category of social science and the humanities. The methods applied in pragmatic design methodology accordingly are those which are proper to social science and differ signally from those used in mathematics. However, also within pragmatic methodology there are attempts of treating design itself as a system, what could be called meta-design or design of design, enabling the performance of designerly operations by computers. The section of pragmatic design methodology which deals with particular methods and techniques, some of them of an algorithmic form, some of them not, deserves the name of *methodics*.

When engaging in this kind of research on science, the latter interpreted not as the profession of researchers but as a product of their cognitive operations (actual or merely possible), we engage in *apragmatic methodology*. Its most important branch is the theory of deductive systems, which in view of the fact that deductive systems occur mostly in mathematics, is also termed metamathematics. The theory of deductive systems, i.e., metamatematics, may itself be constructed as a deductive system, and then becomes similar, as to its method, to mathematics. Pragmatic methodology, which is concerned with operations made in the pursuit of science, is a behavioral science and hence falls under the broad category of social science and the humanities. The methods applied in pragmatic methodology accordingly are those which are proper to social science and differ signally from those used in mathematics.

The programme for methodology of science of Ajdukiewicz along with the methodological reflections of Kotarbiński provides a strong foundation for the methodology of design. The approach combines praxiology with systems-based concepts of advances in design; it combines the praxiology (a pragmatic methodology in general) with general systems theory (rather apragmatic in its nature) and so-called *science of science* (Gasparski, 1989). The *methodology of design* (MOD) includes two parts:

(1) the *pragmatic* MOD dealing with all questions related to the verbal understanding of *design*;
(2) the *apragmatic* MOD dealing with all questions related to the substantival understanding of *design*.

The way of design-methodological studies consists in *idealization*. It means that the goals, tasks, and procedures involved in design-making are idealized. The idealization consists in singling out, in the course of a methodological reconstruction of the various cases investigated, the invariant features typical of design-making. From such an idealization we can obtain a description of the behaviour of a *perfect designer*[8], as well as a description of various deviations from the ideal type, and also an analysis of the causes of such deviations, together with their consequences (Gasparski, 1983a).

MOD may be (and really is) developed, as any methodology, as: (i) a positive study of past and present design (a part of research & studies *on* design), (ii) as a normative endeavor to re-form present and future design (a part of research & studies *for* design). The umbrella stretched over the MOD is of philosophical nature irrespective of some of design researchers contrary attitude[9]. Writes Robert Rosen (after D. Hawkins): "Philosophy may be ignored but not escaped; and those who most ignore least escape." (Rosen, 1985).

The philosophy in question is predominantly the philosophy of technology[10] that covers all area of man-made products called generally *artifacts*. Contemporary technology *lato sensu*, i.e., from engineering to decision theory, "...is concerned with the ontological problem of the nature of the artificial; with the epistemological problem of the peculiarities of technological knowledge and its relation to basic and applied science; with the pragmatic (or praxiological) problem of defining the concept of rational action, i.e. action guided by designs and plans; with the axiological problem of identifying and analyzing the typical values of technology, such as efficiency and reliability; with the ethical problem of the moral codes that the various branches of technology should abide by - and many more problems. Actually each one of these is an entire problem system with a large number of interrelated components. For example, the ontology of artifacts concerns not only tools and machines but also conceptual tools as designs and plans, and the products of knowledge-guided work, from cheese and computers to domestic animals and artificial social organizations." (Bunge, 1985).

The philosophy is important because it warns that "...there can be no *general design method* enabling one to execute designs in a rule-directed manner and without any substantive knowledge. The idea that such a method exists and can

---

[8]The *perfect designer* or an *ideal designer* is nor a real person, neither a paragon to be followed, but an abstract free of unnecessary (for methodological studies) characteristic. It is a concept similar to, e.g., an ideal gas to be studied by physics.

[9]And some philosophers as well. "In some cases the lack of interest of philosophers in technology has been due to the mistaken belief that technology is far too removed from the lofty realm of ideas. (...)If the philosopher finds technology uninteresting it may be because he fails to see the difference between it and technic, or because he sees only the end product of the research-development-production (or service) process. A more attentive glance cannot fail to locate the areas of maximal conceptual density in this process: those of policy and decision making, research, design, planning, end evaluation of quality, cost, and benefit...However, even the very nature of artifacts is of philosophical interest... " (Bunge 1985).

[10]*Technology* is the body of science-based technical knowledge while *technics* is the body of prescientific knowledge (after Mumford, see: Bunge 1985).

be learned separately from the practice of design is just as absurd as the idea that all one has to do in order to become a scientist is to master the scientific method. Yet, although creativity cannot be taught, it can be informed and disciplined. In particular inventing, like painting or investigating, can be taught... up to a point. This is because there are some general principles and desiderata of design, such as stability and economy." (Bunge, 1985).

4. The Theses Of GTD

4.1. DESIGN APPROACH: FROM HISTORY TO NOW

In my contribution to the earlier NATO conference on systems design[11] I sketched the evolution of *design approach* (Gasparski, 1992).

Definition 4: The *design approach* is an ordered quadruple of: (i) a *body of knowledge* serving as a base for grounded design, (ii) *design problematics*, i.e., a set of non-trivial practical problems to be solved through designerly operations, (iii) a *set of aims* to be achieved through designerly operations, (iv) *design methodics*, i.e., a set of design methods and techniques[12].

Although a combination of thinking and doing is as old as the mankind the very first period of mankind's history is a *Pre-design Era* with an *intuitive quasi-design* as an approach[13]. The next epoch is our epoch which deserves a name of

[11]The NATO Advanced Research Workshop on "Comprehensive Systems Design: A New Educational Technology", Asilomar, Monterey, CA, U. S. A., December 2-7, 1990.

[12]This definition is based on more general on offered by (Bunge 1983).

[13]"All man's actions which are not automatic require reflection. The more complex the action, the more essential is the reflection. Coercive situations require particularly intensive reflection. In everyday life, and in mono-subject actions, reflection is so closely interwoven with action that it would be difficult for the acting individual to differentiate reflective action from implementational action without additional retrospective effort or self-observation planned ahead. Things are however, completely different when more people take part in the action, i.e., when we are dealing with multi-subject actions. In this case, the organization of the action is essential, which in turn requires mutual communication amongst the acting subjects. Communication is essential in defining and agreeing upon goals, and also the way of action.

We should note that in the course of communication amongst the persons engaged in multi-subject action, an externalization of what has been termed above 'reflection' preceding action takes place. This externalization may take various forms: from ordinary conversations accompanying action, to consultation which forms a separate preparatory stage of the multi subject action. Let us further note that the variety of multi-subject action leads to overlapping of one action with another, which gives rise to conflicts; in order to avoid this, prior agreement on intentions is required. Thus, not only must agreement be reached within the group carrying out multi-subject action, but also

the Design Era. One may identify the following stages within the Design Era: (a) Primitive Design Stage (PDS), (b) Design Civilization (DCI), (c) Design Culture (DCU). *Natural approach* is characteristic for the PDS. DCI is subdivided into: (b1) The Craftsman's Phase, with the approach based on experience (*experiential design approach),* (b2) Engineering Phase, with the *science based monodisciplinary approach*, (b3) Systems Engineering Phase, with the *science & arts based multidisciplinary approach*. For the Design Culture being in *statu nascendi* now the so-called *designological approach* is offered. The approach is based on the *proper knowledge on design* formulated, not without difficulties, by all the divisions of design research and studies.

4.2. TRADITIONAL DESIGN VERSUS MODERN DESIGN

An old saying teaches us that the difference between a civilized person and a cultural person consists in their behaviour towards the others: a civilized person who happen to jostle a man expresses an apology while a cultural person does not jostle anybody whatsoever. The same should be said about the difference between Design Civilization and Design Culture. Design Civilization, like a civilized person, expresses its sorrow for side effects it has produced (pollution, greenery effect, ozone hole, etc.). It is doing that through the same way DCI is designing their designs.

Proposition 1: In each particular instance of design, the task takes the following form: for given goal G find such a manner M for its implementation that the goal is attained to the greatest degree, i.e.,G=max, while the costs C which are incurred by use of this manner are at their lowest, i.e., C=min.

The manner is what designers are busy to design. The solving of tasks of this general form is concentrated on objects which represent a materialization of means, i.e. tools *lato sensu*, of attaining goals. As I wrote in (Gasparski, 1990) the designer's task has a strictly defined structure, and the goal of design is an isolated *designed object*, upon which the education of design students is

---

between groups, and possible disagreements must be sorted out.

Thus, design began to take shape from some phase of intersubject communication, externalization, separating off, arbitration etc. It would seem that the decisive moment in the development of design can be found in the point at which for the first time the result of the join reflection referred to above was preserved in the form of a drawing, iconic image or description. Preservation of this kind, which was the first design, made it easier to implement the agreed intention, made it easier to check the conformity of what was realized with what was intended, and hindered possible prevarication on the part of those who were discontented, or who had been forced into agreement. We shall never know whether these first designs were drawn on the walls of caves, or were descriptions written on papyrus: for the term 'design' appeared much later than design itself." (Gasparski 1990).

concentrated reproducing the traditional design paradigm, as is suitable for Kuhn's normal science.

Although the Systems Engineering Phase was introduced a few decades ago it is only a dream yet, as Hall said: "...it exists amorphously in the aggregate of its parts... It is not yet whole, but it is undergoing a process of progressive systematization - like a star coalescing from cosmic dust." (Hall, 1989). The quoted author noticed also that "Action science, developed by the social sciences over many years in relative isolation from the applied physical sciences, and which might otherwise have humanized them and made engineering more productive, were doomed to fail by being on one end of the two-culture problem wherein science and the humanities do not even speak the same language" (Hall *op. cit.*). It is just praxiology[14], the action theory, which is trying to overcome the gap suggesting the modern design as cognitive based activity for solving problems in their general context (Gasparski, 1990). The main differentiating factor of modern design, I wrote in the article, is its opening out in many different respects. Above all, the former designed object has been improved by the addition of a *context*, and has so become the *object of design*. This has modified the task of design from an instrumental search for means to attain the goals set, to a search for means of dealing with *practical situations*. It has also become clear that the matter of *relevance of changes* conceptually prepared by design solutions is of prime importance.

Proposition 2: Designing consists in elaboration of: (i) grounded conceptions of *change* in a given *object of design* and (ii) actions necessary for its implementation.

Proposition 3: An *object of design* is composed of at least a pair of practical situations: a core one and a complement one; the complement situation represents a context ('the rest of the world') of the core situation.

"The development of a new paradigm of design is not intended to wipe out what was valuable in the design governed by the traditional paradigm, i.e., technical expertise, but to limit traditional design to well-defined cases which do not threaten a one-sided solution. The new paradigm does not mean that things will be easier. Design on this model will not be simpler, but it may perhaps be done with more awareness of what it is doing. But what is more aware is more difficult as well, and also more expensive today - but with the hope that

---

[14]"Methodology, or praxiology, refers to the study of human planning, action and behavior. (...)By developing a system methodology, we also obtain knowledge of the art of practical reasoning and common sense problem solving, creating at the same time a modern praxeology." writes Hall (1989) acknowledging my statement that 'generally, systems methodology is becoming understood as a conceptual preparation of any activity or any change' (Gasparski 1982).

tomorrow it will not be necessary to bear the costs of what might not have been taken into account today without this awareness." (Gasparski *op. cit.*).

4.3. OTHER PROPOSITIONS

There is always somebody, real or hypothetical, whose needs are to be met through design. He or she is a user main and direct, e.g., a car driver. There are also another users of the same design. like, e.g., pedestrians who are forced to "use" fumes of the car, they are indirect users. There is a designer responsible for satisfaction or dissatisfaction of the both kinds of users. Three of them: main and direct users, indirect users, and designers are subjects of the object of design. It is because their values are involved in a decision-making process incorporated in the given design enterprise.

Proposition 4: A *practical situation* is the set of *facts* recognized by the *subject* of the situation as satisfactory or unsatisfactory because of the *values* of the subject.

Proposition 5: The *subjects of practical situations* and a designer are *subjects of design*.

Proposition 6: The aim for change is an aim of the subject of a core practical situation, who - when the existing situation does not satisfy him/her - tends to *change* the situation, or - when he/she considers the situation satisfactory - tends to preserve it through *changing* practical situation complementary to the given core situation, or through modifying both situations.

Proposition 7: A *design problem* is an adequate mapping of the object of design in a language of design.

Proposition 8: The general language of design is *systems modeling*; languages of particular kinds of design are languages of relevant practical disciplines (technologies).

Proposition 9: The *principle of modeling adequacy* stands that a design problem maps the essence of the object of design.

Proposition 10: The mapping of an object of design consists of a procedure of *identification* and a procedure of *interpretation*.

Proposition 11: The procedure of identification comprises a sequence of the following operations: (i) measurement of the characteristics (values and facts), (ii) formulation of an hypothesis concerning the essence of the object of design, (iii) testing the hypothesis, (iv) correction of the hypothesis, (v) testing of the corrected hypothesis, (vi) formulation of a description of the object of design in a language of an interpreter.

Proposition 12: The procedure of interpretation comprises a sequence of the following operations: (i) acceptance of the description formulated by an interpreter as a preliminary formulation of the design problem, (ii) confrontation the preliminary design problem formulation with the formulations of known design problems, (iii) assessment of the solvability of the design problem, (iv) correction of the preliminary design problem formulation, (v) assessment of the solvability of the corrected formulation, (vi) formulation of the design problem in the language of a designer.

Proposition 13: The correction-testing operations are performed as many times as (a) one reaches such a formulation which the probability (subjective) of solving it is higher than the *limit probability* of solvability and (b) the probability of finding a formulation with a higher probability at the time limit and means available is very small (almost equal to zero).

Design problems are expressed in a form of verbal representations of objects of design. The adjective 'verbal', however, should not be understood literally, graphical and mathematical representations also fall under this kind of representation. Whatever the form of the verbal copies of the object of design they are interrogative sentences. Some of them, the very primitive ones, are the *'yes-or-no'qustions* or the so-called *tasks*. More complex ones, *sub-problems* of different degree and *design problems*, are *complementations questions* (Ajdukiewicz, 1974). A process of solving a design problem, although creative and inventive from psychological point of view, is - from methodological perspective - a process of finding a proper answer to the question. Therefore methodology is interested in the *inferential structure* of the process. On the other hand, solutions of the design problems are - so to speak - prescriptive verbal copies, i.e. verbal patterns of artifacts, i.e., of what should be done 'in order to' or 'in order that' what is characteristic for *teleological sentences* (Woodfield, 1976). Therefore designs are *teleological descriptions* or structures of such descriptions.

Proposition 14: The procedure of solving a design problem is guided by the principle of the unity of divided design on the bases of which it is possible to synthesize partial solutions into an unified solution of a design problem considered as a system.

Proposition 15: The solution of a design problem, a *design*, is an actualization and a concretization of general or detailed propositions of practical sciences (technologies) to the particular, though sometimes repeatable, conditions outlined by the object of design.

Proposition 16: The inferential structure of a design problem solving process is in the form of a *P-T-S* (Problem-Task-Solution) *Network*, the dimensions of which are determined by: the *rank of the problem*, the *degree of decomposition* of the problem, and the *degree of aggregation* of the solution.

Proposition 17: If the design problem is equivalent to a design task, i.e, a 'yes-or-no' question, than the solution of the problem is the solution of the task; such a problem is called *the elementary problem*.

Proposition 18: If the design problem is not an elementary one and if, moreover, the P-T-S network is a network with complete structure, then the problem can be decomposed directly or indirectly (through sub-problems) into a finite number of design tasks equal to the rank of the problem.

Proposition 19: If the design problem is not an elementary one and if, moreover, the P-T-S network is a network with a complete structure, the design tasks into which the problem has been decomposed have as many true solutions as the rank of problems, while at the same time none of them taken separately, nor all of them taken together, are the solutions of the design problem.

Proposition 20: If the design problem is not an elementary one and if, moreover, the P-T-S network is a network with a complete structure, the solution of the design problem is a solution obtained as a result of aggregation and selection of the solutions of the design tasks into which the problem has been decomposed.

Proposition 21: If the design problem is not an elementary one and if, moreover, the P-T-S network is a network with a continuous structure, the design problem is decomposed into two sub-problems: one with a complete structure and one with a non-continuous structure.

Proposition 22: If the design problem is not an elementary one and if, moreover, the P-T-S network is a network with a non-continuous structure, then the design problem is decomposed into sub-problems with a complete network and the remainder.

Two American design researchers G. Moriarty and A. Stone presented an application of GTD to engineering design (ED) at the First International Congress of Planning and Design Theory held in Boston in 1987. They discussed the linear quadratic optimal control design process as a case of the situation described by the *Proposition 18* and the design of a computer software debugging process as a case of the situation described by the *Proposition 21*. According to the quoted authors "there are, of course, various other ED processes that can subsume the praxiological theory of design. Praxiology, then, provides a very general framework that can cover multitudes of ED processes. These particular processes are deduced from the general theory of praxiology. Even though the general praxiology design theory was originally induced from observation of a number of specific designs, the primary function of the developed theory is deductive. Thus applying the praxiological theory of design to actual ED processes is a deduction, while developing the praxiological theory of design is both a deductive and inductive business (...) we need not only induction and deduction, but also abduction[15]." (Moriarty & Stone, 1990).

Proposition 23: If the design problem is not an elementary one and, moreover, if the P-T-S network is a network with a non-continuous structure, then the solution of the design problem is the solution of that sub-problem with a complete structure which has been selected on the basis of the factor of selection.

Proposition 24: The factor of selection is chosen on the base of the *principle of relevancy of change*.

Changes intended by human beings are results of the processes that belong to a chain of a modificational procedure. The list of criteria to differentiate and/or asses the intended changes are: (i) causative involvement of human being, (ii) authenticity, (iii) rationality, (iv) positive utilitarian values, (v) positive ethical values, (vi) positive aesthetical values[16].

Definition 5: Any change that is at the same time intended, authentic, rational, and positive valued from utilitarian, ethical as well as aesthetical points of view is a *change proper* or a *relevant change*.

The concept of *relevancy* in relation to intentionally man-made changes plays the role similar to the role the concept of *truth* plays in science, both are ideal states designers (and other professionals) resp. scientists tends to achieve in the highest possible degree.

Proposition 25: The *principle of the relevancy of change* stands that in the modificational procedure (design and implementation) the tendency to achieve the highest possible degree of relevancy of change is manifested in longer periods of time.

---

[15]The GDT discusses question of deduction, induction, and abduction in the context of design solution *accuracy* referring to (March 1976) and (Kotarbiński 1962); for detail see (Gasparski 1988).

[16]More on that topic in (Gasparski 1990 and 1991).

Proposition 26: Once the propositions of the practical sciences (which are empirical disciplines), presupposed in design problem solutions, are of hypothetical nature, the nature of a design is also hypothetical.

5. Design Comes First

Although many methodologists agree that plans and designs are conceptual artifacts, it is paradoxical enough that they limit themselves to 'the products of knowledge-guided work' (as it was said in a quotation cited earlier after (Bunge, 1985)). They says nothing about producing pieces of knowledge as *sui generis* conceptual artifacts. On the other hand, they generally accept the concept of *experiment design* saying "that insufficient attention is being paid to the methodology of experiment and, in particular, to its conceptual basis." (Bunge, 1983b). They offer even a methodology (a design methodology in a way) of designing experiments consisting of philosophical and statistical principles (*op. cit.*).

I think that after more than three decades of intensive developing the design methodology, based on the pattern of the methodology of science, it is time to formulate a proposal to reconsider the scientific enterprise from 'designological' perspective. Let this postulate acts as a conclusion of the paper on design, science, and philosophy written from praxiological perspective. For it is praxiology, as general methodology, which was claimed to subsume methodology of science, as detailed methodology (Kotarbiński, 1966), and was referred in the context of the pragmatical theory of research (Radnitzky, 1974).

Proposition 27: Because any theory is constructed and not discovered, the relationship between science (as a set of theories) and design (as a set of construction procedures) is of circular character, first a theory (of any kind) is designed then a design (of any kind) is examined against knowledge (factual, methodological, and axiological) theoretically grounded.

References

Ajdukiewicz, K. (1974) Pragmatic Logic, D. Reidel Publishing Company, Dordrecht.

Archer, B. (1981) 'A View of the Nature of Design Research', in R. Jaques and J. A. Powell (eds.), Design: Science: Method, Westbury House (IPC Science and Technology Press Ltd.), Guildford, pp. 30-47.

Blaug, M. (1982) The Methodology of Economics: Or How Economists Explain, Cambridge University Press, Cambridge.

Buchler, J. (1961) The Concept of Method, Columbia University Press, New York.

Bunge, M. (1979) 'The Five Buds of Technophilosophy', in: Technology in Society 1, 67-74.
Bunge, M. (1983a) Treatise on Basic Philosophy, Vol. 5, "Epistemology & Methodology I: Exploring the World", D. Reidel Publishing Company, Dordrecht.
Bunge, M. (1983b) Treatise on Basic Philosophy, Vol. 6, "Epistemology and Methodology II: Understanding the World", D. Reidel Publishing Company, Dordrecht.
Bunge, M. (1985) Treatise on Basic Philosophy, Vol. 7, "Philosophy of Science and Technology, Part I", D. Reidel Publishing Company, Dordrecht.
Bunge, M. (1989) Treatise on Basic Philosophy, Vol. 8 "Ethics", D. Reidel Publishing Company, Dordrecht.
Checkland, P. (1988) Systems Thinking, Systems Practice, John Wiley & Sons, Chichester.
Espinas, A. V. (1890) 'Les origines de la technologie', Revue Philosophique XXX, 113-135.
Gasparski, W. (1970) Criterion and Method of Selecting a Technical Solution with a Praxiometrical Approach: On Methodology of Design (in Polish), Polish Scientific Publishers (PWN), Warsaw.
Gasparski, W. (1972) 'Introduction to the Design Methodology', in: Prakseologia 41, 5-22.
Gasparski, W. (1973) 'The Design Activity as a Subject of Studying: The Design Methodology', in: Design Research and Methods 7:4, 306-311.
Gasparski, W. (1981) 'Editorial: Designology-Towards a Unified Science of Design', in: Design Methods and Theories 15:2, 31-34.
Gasparski, W. (1982) 'Two Thousand Years Later: From Vitruvius to Systems Engineering', in: R. Trappl (ed.), in: Progress in Cybernetics and Systems Research, Vol. VI, Hemisphere, New York.
Gasparski, W. (1983a) 'A Praxiological Theory of Design', in W. Gasparski and T. Pszczołowski (eds.),in: Praxiological Studies: Polish Contributions to the Science of Efficient Action, D. Reidel Publishing Company, Dordrecht, pp.281-312.
Gasparski, W. (1983b) 'Tadeusz Kotarbiński's Methodology of the Practical Sciences and Its Influence', in P. T. Durbin (ed.), in: Research in Philosophy & Technology, Vol. 6, pp. 93-106.
Gasparski, W. (1984) in: Understanding Design: The Praxiological-Systemic Perspective, Intersystems Publications, Seaside, CA.
Gasparski, W. (1987) 'Praxiology', in M. G. Singh (ed.) Systems & Control Encyclopedia:Theory, Technology, Applications, Pergamon Press, Oxford, pp. 3860-3865.
Gasparski, W. (ed.), (1988) Design Studies: The Elements of Scientific Knowledge on Design (in Polish), Science & Technology Publishers (WNT), Warsaw.

Gasparski, W. (1989) 'Design Methodology: A Personal Statement' in: P. T. Durbin (ed.), Philosophy of Technology: Practical, Historical, and Other Dimensions, Kluwer Academic Publishers, Dordrecht, pp. 153-167.
Gasparski, W. (1990) 'On the General Theory (Praxeology) of Design', in: Design Methods and Theories 24:2, 1195-1215.
Gasparski, W. (1991) 'The Concept of "Relevant Change" and Its Importance for the Assessment of Technology and Innovation', Contribution to the International Seminar on Technology Assessment and Its Role in Processes of Society Transformation in Central and East European Countries, October 7-9, 1991, Prague, Czecho-Slovakia (in press).
Gasparski, W. (1992a) 'Editorial: The Prose of Action', in: J. L. Auspitz, W. W. Gasparski et all. (eds.), Praxiologies and the Philosophy of Economics, Transaction Publishers, New Brunswick, N.J., pp. 3-8.
Gasparski, W. (1992b) 'The Evolution of Design Approach: A Historical Perspective and Its Relevance to the Design of Educational Systems', in: C. M. Reigeluth (ed.), Comprehensive Systems Design: A New Educational Technology, Springer-Verlag, Heidelberg.
Gregory, S. A. (1973) 'The State of the Art in Design Methods', in: DMG-DRS Journal: Design Research and Methods, Vol. 7:2, p. 138.
Gregory, S. A. (1980) 'Deriving a Context', Design Studies 1:3, p. 133-140.
Hall, A. D. III (1989) Metasystems Methodology: A New Synthesis and Unification, Pergamon Press, Oxford.
Hayek, F. A. (1952) The Counter Revolution of Science: Studies on the Abuse of Reason, The Free Press, Glencoe, Ill.
Kotarbiński, T. (1937) 'Idee de la methodologie generale: Praxeologie', in: Actualites Scientifiques et Industrielles 533 (Travaux du Ixe Congres International de Philosophie: Congres Descartes, Vol. IV), 190-194.
Kotarbiński, T. (1962) 'Merkmale eines guten Planes', in: Forta Briefe, n. 4, pp. 82-103.
Kotarbiński, T. (1965) Praxiology:An Introduction to the Sciences of Efficient Action, Pergamon Press, Oxford.
Kotarbiński, T. (1966) Gnosiology: The Scientific Approach to the Theory of Knowledge, Pergamon Press, Oxford.
Kotarbiński, T. (1971) 'The Methodology of Practical Skills: Concepts and Issues', in: Metaphilosophy 2:2,
Kotarbiński, T. (1977) 'Concepts and Problems in General Methodology and Methodology of the Practical Sciences', in M. Przełęcki and R. Wójcicki (eds.), in: Twenty-Five Years of Logical Methodology in Poland, D. Reidel Publishing Company, Dordrecht, pp. 279-289.
Krzeczkowski, K. (1986) 'On Practical Sciences Position' (in Polish), in W. Gasparski and D. Miller (eds.), in: Projektowanie i Systemy (Designing & Systems), Vol. VIII, Ossolineum, Wrocław, pp. 11-35.
Maldonado, T. (1972) Design, Nature, and Revolution: Toward a Critical Ecology, Harper & Row Publishers, New York.

Mattessich, R. (1978) Instrumental Reasoning and Systems Methodology: An Epistemology of the Applied and Social Sciences, D. Reidel Publishing Company, Dordrecht.

Moriarty, G. and Stone, A. (1990) 'Praxiology in Engineering Design', in: Praxiology: Polish Yearbook of Practical Philosophy and Methodology, Vol. 4-5, pp. 71-83.

Nadler, G. (1981) The Planning and Design Approach, J. Wiley & Sons, New York.

Radnitzky, G. (1974) 'From Logic of Science to Theory of Research', in: Communication & Cognition, Vol. 7:1, p. 62.

Rapp, F. (ed.), (1974) Contributions to a Philosophy of Technology: Studies in the Structure of Thinking in the Technological Sciences, D. Reidel Publishing Company, Dordrecht.

Rapp, F. (1980) 'Philosophical Problems of Design', in: Design Methods and Theories 14:2, pp. 42-48.

Reich, Y. (1992) 'Transcending the Theory-Practice Problem of Technology', Engineering Design Research Center, Carnegie Mellon University, Pittsburgh, PA, Report No. 12-51-92.

Rosen, R. (1985) Anticipatory Systems: Philosophical, Mathematical, and Methodological Foundations, Pergamon Press, Oxford.

Simon, H. A. (1969) The Sciences of the Artificial, The MIT Press, Cambridge, Mass. (2nd ed. 1981).

Simon, H. A. (1977) Models of Discovery and Other Topics in the Methods of Science, D. Reidel Publishing Company, Dordrecht.

Singer, B. J. (1983) Ordinal Naturalism: An Introduction to the Philosophy of Justus Buchler, Associated University Press, London.

Toulmin, S. (1977)Human Understanding:The Collective Use and Evolution of Concepts, Princeton University Press, Princeton, N.J.

Tribus, M. (1969) Rational Descriptions, Decisions and Designs, Pergamon Press, New York.

Warfield, J. N. (1990) A Science of Generic Design: Managing Complexity Through Systems Design, Intersystems Publications, Salinas, CA.

Zou Tsing (1993) 'Philosophy of Technology: Epistemological or Praxiological? Some Lessons from the Chinese Philosophy of Technology', in T. Airaksinen and W. Gasparski (eds.), Topics in Practical Philosophy and Action Theory, Transaction Publishers, New Brunswick, N.J. (in press).

DESIGNS ARE CULTURAL ALLOYS, 'STeMPJE' IN DESIGN METHODOLOGY

A. SARLEMIJN,
Eindhoven University of Technology, the Netherlands[1]

" ... the scholar who says he detests any kind of science is not only ridiculous: his attitude is decidedly harmful. Harmful because it encourages those who are responsible for decisions that may determine the fate of mankind to be intentionally ignorant about the material background against which their decisions should be taken. Harmful also because authors and scholars, while gladly using modern commodities, fail to see the philosophical implications of science and tend to deny *scientists and engineers their legitimate place in culture. But we, scientists and engineers, we know that we have not only created material things and above all we know that we contribute to better relations between nations and peoples.* For us it is easy to have understanding of and objective appreciation for the work of others, and from there it is not difficult to arrive also at human understanding and appreciation" [H.B.G. Casimir, 1965; italics added]

Introduction: Technology, melting pot of 'cultures'

In 1978 I was a visiting professor of the University of Bielefeld in West-Germany. During one of my lectures I made a stupid mistake. I used the non-existent expressions *alpha-Wissenschaften*, *beta-Wissenschaften* and *gamma-Wissenschaften* when I explained the influence of scientific research on technological designs. My students' eyes told me that they didn't understand a word. And of course, they did not. I was literally translating Dutch concepts into German. The in Dutch so-called *alpha disciplines* are humanities like literature, law, philosophy. Our *beta disciplines* are sciences, for instance mathematics, physics, chemistry. Examples of our *gamma disciplines* are economics and behaviouristic psychology; they have things in common with sciences as well with humanities. In the context of this Dutch distinction technological disciplines belong to the *beta* sciences.

From this anecdote about my mistake we can learn two things. First, expressions like *science* and *Wissenschaft* have different meanings. Second, several technological arts become sciences (Wissenschaften, Wetenschappen, Videnska-

M. J. de Vries et al. (eds.), Design Methodology and Relationships with Science, 191–248.

ber), if they cross the ocean from the States or if they cross the North Sea from Britain to Germany, to Holland or to the Scandinavian countries. And on the European continent there are differences in meaning too.[2] Due to these two points the discussion about science-based designs is not free of linguistic difficulties. Therefore, I want to make clear from the beginning that I use the expression *science-based designs* in the strict Anglo-Saxon meaning. In our modern industrial society it is hardly possible to find a product that is not a result of mathematical, physical, chemical or metallurgical know-how with regard to its design or its production.

This, however, does not imply that designs are exclusively the results of science-based know-how. The opposite is true. My main thesis states that technology functions as a melting pot; successful designs can be considered as alloys: they are based on sophisticated combinations of many different kinds of know-how. If we remember the title of C.P. Snow's famous book of 1959, then we have to say that technological designs do not belong to only one of his so-called cultures.[3]

My thesis on technological designing as a melting-pot activity also implies that the Dutch distinction, mentioned before, is wrong: the development of technology is not restricted to the field of the *beta* sciences. Look at the activities of mechanical and electrical engineers! Of course, they are active in the fields of mathematics and physics. It is wrong to say that their efforts are restricted only to applications: their R&D often lead to really new mathematical and physical insights. However, mathematics and physics, 'melted' in the context of these technological disciplines, have lost their academic features: fundamental questions about properties of calculi and those about the unity of the basic physical forces are far beyond the horizon of the engineer. Therefore, engineers contribute to the mathematical, physical and chemical growth of knowledge but their contributions have no academic characteristics. This is the reason why their 'scientific' disciplines are considered as arts in the Anglo-Saxon languages.

The same can be said about social insights in the context of technological developments. Business engineering and industrial planning had their origin in academic economics and sociology. But, sophisticated mathematical models of economics and sophisticated sociological theories often lack relevancy in the context of technology. They therefore play no part in the solution of problems of business engineering or of industrial planning. That is why 'social' disciplines like business engineering and industrial planning have to be considered as arts, as well, in the Anglo-Saxon languages (like the fore mentioned 'scientific' disciplines of mechanical and electrotechnical engineering).

There is a gap between academic knowledge and technological know how. This gap leads to frictions. Mathematicians are often convinced that engineers calculate in a very careless way. Physicists are seldom impressed by the technological models of mechanical engineers. Engineers, however, are convinced that they alone can change the world by translating abstract mathematical, physical and chemical knowledge into practical know-how. In the same way as the scientists, the economists and the sociologists usually have contempt for the social insights used in industrial or technological contexts. But these contexts do not form the circumstances for the development and testing of sophisticated social theories. To analyze the different kinds of knowledge and know-how which are combined in designing, I developed the 'STeMPJE' approach. *'STeMPJE'* is formed by the initials of different factors. The success of a design depends on *S- and Te-factors*. Examples of scientific factors (S-factors) will be discussed at the beginning of 1.1. They are phenomena ...

- which are involved in a design,
- which can be explained in the context of a scientific model,
- and which, together with other phenomena or factors, form the Te-factors of the technological system of a product.

This definition seems to be difficult to understand. In fact, the meaning is quite simple: not all the phenomena of a technological system can be predicted scientifically; these aspects remain subjects of what we usually call the *Fingerspitzengefühl* or *intuition* of engineers. And this is the reason why the S-factors have to be distinguished from the other Te-factors of a technological system.

The M-factors (market factors), P-factors (political factors), J-factors (juridical factors) and E-factors (esthetic factors) do not form a part of a technological system; they can, however, influence its design in a strong way. Examples of *M-factors* are the growing diminishing income rate, the changes of sale rates of products, the bank rate, the profit rate. These factors make the successful production of the design possible or impossible. From a methodological point of view T-factors and most of the M-factors have an extensional character: they exist 'objectively'. P-, J- and E-factors and their interpretations have, at the contrary, an intensional character in the language of the logicians: their 'objective' influences on designs are strong; and yet they exist 'subjectively' in the minds of the people. The sale rates of a product in the past has an extensional character; the market demand for a product has, at the contrary, a mixed character: it constructs a relation between people's desires and products. These methodological remarks are important: *it is usually easier to observe and to register in an exact way extensional factors than intensional factors.*

P- and J-factors are political and juridical norms and prescriptions which facilitate the introduction of the product, or which make this introduction difficult or even impossible. A great deal of the designer's reasoning is based on the interpretation of norms. Some of them are exact prescriptions; this is, for instance, true with regard to the American ASA, the German DIN and the Dutch NEN. Others are less exact or global. This aspect, especially, clarifies that designing does not belong to only one of C.P. Snow's cultures. *E-factors*, finally, are esthetic factors which influence the acceptance of the designed product in a positive or negative way. These esthetic factors are extremely important in the context of architecture and industrial designing. The history of architecture illustrates the frequent interactions between S-, Te- and E-factors. Sometimes, E-factors are influenced by ideological changes. In the fifties windows of school buildings, for instance, had to be high enough to make it uninteresting for children to look out of the windows. Later, the windows were lower to demonstrate that there was no gap between the problems in school teaching and those in daily life. This example shows that I do not restrict the meaning of esthetic only to a highly developed sense of beauty; from a modern point of view the schools with the lower windows are nice but not practical for the teachers.

The 'STeMPJE' approach prescribes that designs have to be analyzed as provisional or definitive reactions on the interaction between those factors [see fig. 20]. The multifactorial analysis of the 'STeMPJE' approach can be applied to designs of the past, because the aforementioned factors have always influenced designs in an explicit or implicit way. Historical studies will lead to concrete insights into the interactions of those factors and will help us to define the factors in an adequate way [see §§ 6-8].

The present situation makes such a multifactorial approach very urgent as well. From our contacts with business corporations it became clear to us that an approach such as 'STeMPJE' is needed for a distinctive explanation or prediction of an innovation [see for example § 11]. The reason is that the role of scientists and engineers in business corporations has been changed. During the first half of our century multinationals lured scientists to their laboratories with the promise that pure scientific research could be done in the industrial context as well. The situation is different now. Shell, Akzo, Philips and other multinationals preach the slogan *inventions are not required, innovations are needed*. This means: high-tech sophisticated design concepts alone do not guarantee a solution for the concerns's urgent problems. If scientists or engineers wish to contribute to the solution of the problems of their company, they have to keep in mind the complex situation in which their company has to operate nowadays. This means that a unity of cultures is required by these companies. In turn this means that a kind of 'STeMPJE' approach is necessary.

A. DESIGNS: EXPERIENCE-BASED, MACROSCIENCE-BASED, MICROSCIENCE-BASED

1. Experience-based designing

1.1. STATIC S-MODELS AND TE-SYSTEMS

After this introduction I am going to deal with the question of how we arrive at the combinations of know-how (concerning the different factors) in designing. I will begin with the combination of scientific and technological know-how of S- and Te-factors. My part B analyses dynamic interactions between the S- an Te-factors and the other factors. Speaking about S- and Te-factors we have to point out the following different approaches in science-based engineering.

(a) We encounter a result of the *microtechnological* approach when we are watching a football match on television. What we see physiologically is a football. But from a physical point of view we are looking at little dots shot onto a screen by an electron beam. The design of the technological system of a tv-receiver is based on know-how concerning the behaviour of small particles: in many cases a knowledge of quantum mechanics or electron is needed to achieve innovations.

(b) Machines, airplanes (with regard to their aerodynamic forms), steam engines, and so on, are systems resulting from *macrotechnological* approaches: the know-how to apply classical (macroscientific) theories is required; detailed knowledge about the behaviour of electrons is not necessary.

(c) And finally, we have *experience-based* technological approaches. An example of such an approach is construction engineering in which systems like houses, ships, or bridges are designed.

S-factors in experience-based technology are, for instance, the factors which can be predicted with static models. Static models have a long historical tradition. Aristotle's lectures on mechanics already contained the lever law: the behaviour of the pulleys of figure 1 is analyzed in these lectures. A hundred years later Archimedes discovered the relation between statics and hydrostatics and this relation is illustrated by figure 2 today. An anecdote has us believe that Archimedes discovered this relation while taking a bath. The truth of this story is uncertain. It is, however, certain that Greek engineers used static and hydrostatic laws in designing their houses, ships, and bridges.

The fundamental principles of these experience-based technologies have a long history. However, we still come across certain 'mistakes'. A famous example is the Tacoma bridge which collapsed in 1940. It became famous because an instructive film of the collapse was made. The film clearly illustrates how resonance caused by a storm can lead to the destruction of a bridge. The design of the Tacoma bridge was perfectly alright from the static point of view. However, it is a feature of a static analysis that it does not consider dynamical phenomena like resonance. Since that disaster the behaviour of scale-model bridges is studied in laboratories. And since then, bridge construction has been based on a combination of static, dynamic and aerodynamic know-how. However, the aerodynamic approach in this context is not purely deductive. The opposite is true. Many aspects are left to the so-called *Fingerspintzengefühl*. How then is it possible, that even after the Tacoma bridge disaster static principles still form the starting point for bridge construction? Is it true that engineers have natural blinkers on, which make it difficult for them to analyze all aspects of a technological system scientifically? I will try to find reasons for these blinkers later on. First I would like to deal with another example of an experience-based technology: ballistics.

1.2. S-MODELS AND TE-SYSTEMS IN BALLISTICS

It is an interesting fact that the professor in ballistics of the *Royal Military Academy* in the city of Breda was educated in physics at my university here in Eindhoven. The same can be said for the professor in ballistics of the *Royal Academy of the Navy* in the city of Den Helder.[4] This implies that both professors are graduated in classical mechanics, theories of relativity, quantum theory and quantum mechanics, solid state and surface physics, and so on and so on, and ... that they never apply (in the strict sense of this term) these theories in the field in which they are active now. This means that they use blinkers on purpose. They work in that field of physics in which the methodology and the basic principles were already formulated in the time of Galileo.

I admit that my views on S- and Te-factors, for a part, are based on my readings of Galileo's works, especially of his *Discorsi* of 1638.[5] Therefore, I will pay some attention to the thoughts which I have deduced from this book. In the *Discorsi* the following three tasks are distinguished.

(a) First of all, we have the task of mathematical physicists who reads the book of nature written in a mathematical language. This means as we would say to day: physicists choose the adequate idealized situation which justifies the mathematical formulation of physical laws.

(b) The second task is that of experimentalists testing the physical laws. To confirm these laws they have to search for situations which have the greatest similarity to the idealized situation presumed by these law. In other words, they try to eliminate every influence of the so-called disturbing factors. We all remember the consequences of this elimination very well: in school we confirmed Galileo's laws by playing with heavy marbles falling over small distances to avoid the disturbing resistance of the air. What is disturbing in these experimental contexts? The answer is always deduced from the mathematically formulated S-model and "disturbing" means here: not being in accordance with the idealized mathematical model.

(c) The task of the designing engineer, finally, is contrary to that of the experimentalist. His *Fingerspitzengefühl* informs him about the relevant disturbing factors which influence the factual path of the projectile. Therefore, in the context of technology these factors loose their meaning of *disturbing* in the strictest sense: they could hinder the target from being hit but they could also assist in hitting it. This means: the engineer *uses* the so-called 'disturbing' factors to achieve his aim and the experimentalist *eliminates* these factors to achieve his aim.

Galileo was convinced that scientific analyses of disturbing factors are impossible. From a modern point of view he was right in a certain sense. The formulation of his argument is, however, unclear to us. The form of the projectile, he said, can vary infinitely. And scientific analyses of infinities, he thought, are impossible. This does not convince us. In set theory we calculate with infinite sets! And many physical models presuppose infinities!

However from the methodological point of view, modern ballistics still follows the same way as Galileo did. Even the fundamental laws are similar. Figure 3 shows that Leonardo da Vinci was already convinced of the parabola law and that he had already discovered the geometrical forms similar to those of the projectiles described in modern ballistics on projectiles used during the Gulf war [fig. 4].

Many things have changed since then. Modern ballistics works with computer models nowadays. Calculations are much better now than in previous centuries. But, we can still find the famous 45° which were prescribed before - first in the Middle Ages and later by Galileo - as the best angle of projection. I often meet scientists who joke about the S-model of the Middle Ages [fig. 5]. But they should not forget that corrections of the S-model of the 45° are only required, if we use super guns (like the *Dicke Berta*) to shoot projectiles into higher spheres where air resistance is significantly less.[6] If the distance is more than 30 kilometres, then again new kinds of 'disturbing' factors have to be considered in the computer model: the Coriolis forces. These are apparent forces resulting from the rotation of the earth.

Why are these 'disturbing' factors still the subject of the engineer's *Fingerspitzengefühl*? After Galileo, Newton tried, in vain, to determine these factors scientifically. He formulated the general law which says that air resistance is equal to projectile velocity. Other factors, however, are of more importance. The local mass density and viscosity of the air influence the path of the projectile. This means that local circumstances cannot be neglected: the gunner has to take the weather-forecast into consideration. The diameter and the velocity of the rotation of projectiles are important too. This is especially the case when gas bombs are used as Iraq perhaps intended to do during the Gulf war. Finally, we have the Mach number which is important if a projectile breaks the sound barrier. The Mach number is experience-based information and is not a result of mathematical deductions concerning sound waves.

All these circumstances show how disciplines of experience-based technology function: the judgement about the relevancy of the disturbing factors is based on the *Fingerspitzengefühl* developed by experience; this judgement is not based on the ballistic S-model which has an empirical character. In Breda and in Den Helder the professors in ballistics do not use quantum mechanics. They do not even use Newtonian mechanics. Why do they wear blinkers? These blinkers are justified by *the divergence between science and technology* [fig. 6]. In ballistics they help to avoid overcomplexities of the computer models. Mathematical deductions concerning the sound waves could be added to the computer model. This kind of an experiment is interesting from a scientific point of view. But the technological prognoses would not necessary be a better one. Quantum mechanical deductions could be added too. But they would not help the gunner to negotiate with weather prognoses. This shows another consequence of overcomplexity: it undermines the engineer's *Fingerspitzengefühl* dealing with the complexity of the concrete situations. These situations cannot be dominated by mathematical deductions alone. This was the reason why we distinguished the S-factors from the Te-factors. The necessity to avoid overcomplexities justifies the engineer's blinkers.

1.3. NATURAL BLINKERS?

The first meaning of *blinkers* is neutral. The blinkers of an horse are a pair of flat pieces of leather fixed beside its eyes to prevent it seeing objects on either side; in other words: they prevent the horse's paying attention at disturbing phenomena. The word has also a derogatory meaning: blinkered people are unable to understand or to accept anything beyond their own familiar insights or ideas. My use of the metaphor presupposes the following negation of the pejorative aspect: engineers are at least partly able to understand but they do not use insights beyond the demarcation of what is relevant to a successful design.

We can distinguish three steps in this negation. And the training of our students in engineering contains (or should contain) these three steps. They learn, first of all, the formal laws dominating the field in which they like to become specialists. They learn, second, to deal with the disturbing factors; these are left out of consideration during the idealization preceding the mathematical formulation of the formal laws. Therefore, as long as mathematically formulated laws form an essential part of the training, insights blinkered by idealizations cannot be avoided. And as long as engineering is oriented at the solving of practical problems, the looking for relevant disturbing factors behind the blinkers forms an essential part of the training too. Galileo discovered this second step as an essential part of engineering as we have seen in 1.1.

The third step is the fore mentioned *divergence between science and technology*. Science strives to gain the most sophisticate explanations of phenomena. Therefore, relativistic (Einsteinian) mechanics is better than classical (Newtonian) mechanics from a scientific point of view. From a technological point of view, however, it is not useful to prefer relativistic mechanics in all circumstances. Explanations and predictions based on the classical approach are often sufficiently precise. In relativistic mechanics mass depends on velocity whereas it is a constant in classical mechanics. Calculations based on the relativistic mass concept and applied on the behaviour of an airplane lead to results deviating from those of classical calculations. In this case, however, the difference is not more than (as Casimir expressed it once) a little perspiration drop of the pilot. The application of relativistic mechanics does not lead to new information relevant to airplane designs. That is the reason why this application does not belong to the competency of airplane constructors. The same can be said about other technologies: neither relativistic nor quantum mechanics forms a crucial part the scientific education of architects, of mechanical engineers or of electrical engineers.

Not only engineers but also practically minded scientists are familiar with this third step or aspect of my blinkers doctrine (methodology, however, did not pay attention to this crucial aspect of applied science). None of us will ever use the Van der Waals' equation of state as long as our predictions based on Boyle's and Gay-Lussac's laws are precise enough for our practical needs in given situations. It is true that the boundaries are fuzzy: our *Fingerspitzengefühl* (and not an exact demarcation) informs us where these classical gas laws can be used and where the application of the equation of state is required. This use to take it easy is common not only in technology but also in science. In technology, however, this use has the aforementioned special aim: it helps to avoid overcomplexity and to concentrate on the manifold aspects of the concrete situations in which practical problems are immersed. In other words: it helps the engineer to dominate competently the field in which s/he is active.

This is the reason why I like the in 1.2 mentioned example of the professors in ballistics. Their know-how illustrate precisely the methodological meaning of blinkers. They are well trained physicists; they know not only classical mechanics but also relativistic mechanics; and they know very well how far classical mechanics is relevant for their practice of the problem solving. Galileo was not familiar yet with this third restrictive aspect of engineering. Of course he was not. But this aspect is an essential part of the training in engineering: engineers know or should know why modern *scientific* theories are of none relevancy or are only partly relevant to their practice. And it is the aim of the 'STeMPJE'-approach that engineers know or should know to what extend modern economic and other insights in social events are relevant to their practice too. The blinkers should function as filters: all knowledge has to be filtered and judged to guarantee successful designs. This judging and filtering is or has to become a natural feature of the engineer's competency. Only a multifactorial approach can assure that an invention will lead to an innovation as the concerns require.

1.4. 'EXPERIENCE-BASED KNOW-HOW', WHAT DOES THAT MEAN?

Experience based know-how, in the meaning as I used this expression up to now, is partly scientific, partly extra-scientific and partly prescientific.

Aristotle's lever law, Euclides' and Ptolemeus' optical laws of reflection and refraction and Galileo's laws represent science based knowledge. The way of applying these laws in technology has practical aspects and is, therefore, a subject of an extra-scientific know-how. Finally, the know-how of technicians has often been a source of scientific knowledge. The lever law was known to the Babylonian and Egyptian technicians long before Aristotle was born. The parabola law was already known to Leonardo. Sailors applied insights which later became a formal subject of aerodynamics.

The experienced craftsman knows his trade, not by the book merely, but by long practice under many different circumstances. "Experience" denotes the capacity to do something, learned in the habit of doing it and guided rather by rule-of-thumb precept than by theoretical understanding. Usually, the technologies of house-, bridgeconstruction and mining are merely based on experience. The *Fingerspitzengefühl*, based on experience, does not exclude that we expect the engineers working in these experience-based technologies to have been trained to apply formal laws in their fields. We also expect them to be familiar with the materials of these fields. Let us describe 'experience based technologies' in a negative way: the axiomatic deductive structure of these laws does not play an important part in these technologies as it does in macroscience-based technologies. The meaning of this expression is the subject of the next paragraph.

2. Macroscience-based designs

2.1. S-MODELS OF MACROTECHNOLOGY

Blinkers play a part in experience-based technology, as we have seen. Since Newton we have had a second kind of blinkers: those of the macrotechnologies. In the context of these technologies more general theories are applied, but the behaviour of microscopic particles is not considered [fig. 7]. Macrotechnological S-models are based on macroscientific theories: on classical mechanics, on classical thermodynamics or on the classical theory of electromagnetism including the macroscopic theory of optics. Many technologies are based on these theories.

Experience-based knowledge only justifies prognoses of isolated phenomena whereas macroscientific theories claim completeness. For the first time this claim was presented by Newton in his *Philosophiae Naturalis Principia Mathematica* of 1687. This aspect of the *Principia* already becomes clear from the table of contents [fig. 8]. After the presentation of the mathematical principles in the introduction the deductions of the first book show how the description of idealized mechanical motions can be deducted from those principles. There is a similarity between the task of Galilei's physicist and the aim of this first book: the mathematical description of idealized motions. However, there is also an important difference. Galileo's laws were based on empirical contexts. The laws of the first book were deduced mathematically from the principles introduced in the beginning of the *Principia.*

In the second book the difference between Galilei's experience-based approach and Newton's macroscientific approach becomes greater because Newton used physical laws to predict the behaviour of disturbing factors or - as he called them - resisting mediums. With pendulum and other experiments he tried to justify the acceptance of these laws. Galileo, on the contrary, had been convinced of the fact that these factors are inaccessible to scientific analyses, as we have seen.

The pendulum experiments of the second book also serve another aim.[7] Newton used the results of these experiments to introduce a well based demarcation between mechanics and the microscientific conjectures which still lacked any empirical foundation at his time. Varying the construction of the bobs Newton tried to demonstrate that little particles, like cartesian ether particles, did not play any part in the mechanical motions. These considerations clarify the ultimate claim expressed by the title of the third book: the complete domination of the whole mechanical world. There is no macroscopic mechanical motion which cannot be described by mathematical principles or laws.

Newton's mechanics needed many corrections and additions. In the 19th century physicists were still dealing with this task. Newton's approach, however, became an example of how a macroscientific theory has to be constructed. His example became paradigmatic for the work on thermodynamics during the 19th century. Newton's axiomatic method had 'synthesized' the mechanical laws of Galileo, of Kepler, of Huygens, and of others. With a similar intention Carnot introduced his principles in 1824: he was the first to 'synthesize' the insights in heat phenomena which had become comprehensible in the contexts of steam and hot air engine construction. Many others contributed to the growth in this field during the 19th century. Nernst's equation of 1906 is usually considered as the last contribution to this process. Since then, classical thermodynamics, like Newtonian mechanics, are considered a 'finished' theory.[8] This means that physicists are convinced of the fact that all macroscopic heat phenomena can in general be predicted with descriptions deduced from basic equations; the details are left to special disciplines or to the engineer's *Fingerspitzengefühl*. These methodological principles are still in practice today.

Newton's approach also became paradigmatic example for the work on electromagnetism during the 19th century. Maxwell's equations can be compared with the mathematical principles introduced at the beginning of Newton's *Principia*. These equations create a unity in the different experience based insights into magnetic, electric, electromagnetic and optic phenomena just as Newton's principles did in the field of mechanics. It is true that Maxwell tried to add a microscopical foundation of his equations. He had concrete fantasies about an S-model of atoms and molecules turning around in a complex system of wheels. These ideas formed a starting point of many discussions about a universal ether during the 19th century. These microscopic ideas, however, do not play any part in our technological applications of the Maxwell equations at present, as we will see [at the end of 2.2].

2.2. Blinkers also in the context of macrotechnological S-models and Te-systems?

Blinkers seem to be in contradiction to the claim of completeness with regard to the macroscientific theories. The table of contents of Newton's *Principia*, however, mentions the crucial point: the second book introduces experience-based laws. And in a similar way, the modern special disciplines on classical mechanics, on thermodynamics and on electromagnetism introduce experience-based insights. Moreover, we also have to consider that the system behaviour of steam engines and other engines depends very heavily on the material properties.

Laymen in technology think that modern science can explain everything. They believe that modern science can satisfy the quest for absolute certainty: in technology, nothing should be left to coincidence. In a certain sense, they are right: science tries to explain everything *theoretically*. But the jump from theory to the practice of engineering means the transition from nicely idealized worlds to the 'dirty' complexity of the concrete world.

Why should we not try to eliminate all uncertainties, for instance, by applying the modern microscientific theories on the behaviour of 'disturbing' factors in macrotechnological systems? The answer is that the concrete world is too complex for these microscientific theories. It is, indeed, true that the research in solid state and surface physics has led to *ab initio* calculations, which make it possible to predict the behaviour of materials.[9] But these prognoses can only be applied if the materials have an extremely high degree of purity. The production of these materials - for instance in the context of chip-technology - requires a kind of production of small samples in situations which are very similar to those of the laboratories.

Such kinds of materials are not used in the context of mechanical, thermodynamical or electrical engineering. Metals, used in these technologies, are impure from the microscientific point of view: they do not have a nice crystal structure; their chemical elements or their chemical compounds are mixed with odd atoms of other elements or of elements which do not belong to the compound; many of their atoms are not ideal because they miss the prescribed number of electrons. In this sense the material of concrete bridge constructions are impure too. No physicist would ever think of applying quantum mechanics to predict the behaviour of the concrete used for a bridge construction.

Nor is quantum mechanics applied in the context of construction engineering: explosions or implosions of macroscopic barrels are the results of disturbing factors which are, for a great deal, the field of the experience-based *Fingerspitzengefühl.*

The aerodynamic construction of airplanes forms another example of macro-technological systems of which several Te-factors belong to the field of experience-based insights. The most important technical defect causing airplane accidents is faulty metal. Sometimes this phenomenon is described in terms of lack of cohesion or adhesion of little particles. These microscientific descriptions should, however, not mislead us. They don't allow exact prognoses. Impurities play an important part in the fatigue. The impurities are the result of the special kinds of ores and of the special modes of production used in the blast-furnaces of the different countries. That is why the metallurgy applied in macrotechnologies is experience-based knowledge and is not deducted mathematically from the formalisms of a microtheory.

This can be said about the metallurgical properties of metals used by electrotechnical engineers as well: the degrees of isolation and conduction depend, for a great deal, on the impurities of the materials. Another interesting example in the context of electrotechnical engineering is the behaviour of the ionosphere. *Ionosphere* suggests that the behaviour of little particles is microscientifically examined. The opposite is true. Electrical engineers deal with the reflections of electromagnetical waves in a macroscientific way; electromagnetic equations, together with experience based insights, form the guides in antenna technology [fig. 9].

3. Blinkers even in microtechnologies?

Electrical engineering belongs to the macrotechnologies. Electronics is based on a know-how using theories on the behaviour of little particles.

In microscience the ideal of the physicists to explain, to predict and to dominate all kinds of behaviours is approached. In the context of technological developments, however, this ideal cannot be realized. It is true that, for instance, the invention of the transistor was based on microphysical insights. But the improvement of the production of transistors and of IC's is still for a great deal the field of *Fingerspitzengefühl and experience-based know-how*. Nor should we forget that the R&D leading to the invention of the transistor would have been impossible, if metallurgical assistants had not helped to produce the required instruments and materials.

At the beginning of the transistor technology (before 1947) the metallurgical aspect was left entirely to experience-based know-how. This becomes clear from the following historical anecdote. In 1947 Gordon Teal was working at the Bell Labs doing research in chemistry. His subject did not interest him very much. Although the discovery of the first transistor had to remain top-secret, he was informed about several details of the design. He knew, for instance, that germanium had played an important part in the discovery. And he was a specialist in this field. As soon as he was informed about the mechanism of the solid state amplifier, Teal developed the essential concept for transistor production: the single crystal technique of making p-n and n-p junctions to achieve the required conductivities. If these ideas had not been worked out, the commercial production of transistors and IC's would have remained impossible. However, Shockley was not convinced of that in 1948. In that time he was not able to believe that ideal crystals could be of any relevance to the production. Therefore, he did not allow Teal to work on this topic.[10] Later on, Shockley had to give up his blinkers.[11] Many aspects of transistor technology are, however, still based on intuitions. The reduction of drops out during the production of transistors and IC's is, even nowadays, still a field which is dominated by 'alchemistic' thoughts and *Fingerspitzengefühl.*

Blinkers are justified to come to specialisations based on S-models. Without S-models science-based designs are impossible. If we pay attention to the factors hidden by a special kind of blinkers, then we can reduce the dangers of these blinkers. This is illustrated by the anecdote about Teal and by the other examples mentioned before.

In conclusion we must say something about high energy physics. Particles, created during experiments in this field, often exist so very briefly that computers are required to signalize these phenomena. Extremely high energies are needed for the generation of these particles. Because of these aspects the industrial relevancy of high energy physics is still doubted. Therefore, people who work in the context of industrial microtechnology are usually not trained in this field of physics. Here we meet blinkers again.

4. Interactions between the three different fields of technology

I do not intent to justify the explained distinction between the three fields of technology: this distinction is deduced from practice. Students in electrical engineering are usually not trained in the technological application of quantum mechanics. Maxwell equations do usually not belong to scientific curriculum of students in architecture. This is true in all countries.

The three fields, however, do not have isolated evolutions. Experience-based assistance is a prerequisite for macrotechnological designs as we have seen. And we have also seen that microtechnological inventions are impossible without the help of experience- and macroscience-based know-how. On the other hand, microtechnological innovations (like the transistor and the IC) are able to cause profound revolutions in the other fields of technology. These interactions do not affect the distinction: CAD has changed the activities of the designers; an architect using a plotter has, however, not become a specialist in microtechnology; after all his scientific education is not aimed at the application of microscientific theories. The same can be said about electrical engineers; they introduce transistors in their designs; they lack, however, the scientific training to design new kinds of transistors; they deal with these devises as if these would be black boxes. We will meet some other examples of these technological interactions afterwards.

Theoretical considerations of the interactions have often led to the priority question: what factors are pushing or pulling technology decisively? Many aspects of this question lack practical relevance: it is true that the designers combine adequate reactions on the given factors, but they are seldom forced to analyze the interactions between the factors in detail or to judge about priorities. On the other hand, the 'STeMPJE' approach prescribes to reflect on the profitable production of designs. Therefore, designers have to consider the prerequisites for the production and the introduction of their results. This also implies a concentration on the evolution of the factors. In this sense, designing is strongly con-

nected with the planning of the combination of the different kinds of know-how.
We meet seldom insoluble priority questions if the factors are considered in the *concrete* contexts of technology. This will be clarified in the next sections. Only if we wish to find the factor which is decisive for *every* technological progress then we look for a satisfying answer in vain.

B. 'STeMPJE' DYNAMICS

5. Insoluble priority questions

5.1. THE 'CHAOTIC' DESIGN REALITY

One of the often discussed questions, turning out to be insoluble, concerns *the priority of the S- and Te-factors:* exact scientists time and again show how technical effects can be derived from fundamental principles; designers, on the other hand, usually think that their rich fantasy, as the resource of many inventions, is so complex that it inaccessible to exact analyses. In a certain sense the exact scientists are right: technological effects can be described in scientific terms. On the other hand it is in conflict with experiences, that technological inventions are derived from fundamental theories without design problems. Reference books suggest this. But the complex practice of designing shows the contrary. It is true that the aforedefined concepts imply that the S-model of a technical effect can be deduced mathematically from general principles in theoretical analyses. But these analyses hide behind blinkers the other required kinds of technological know-how and the competent treatment of the other factors. To illustrate this I am going to pay attention at one macrotechnological design activity (Huygens' work on his clock) in § 6 and at two microtechnological design activities (that of the transistor and that of the Plumbicon) in § 7 and § 8. We will see that the priority questions loose their apparent relevancy in the context of these concrete engineering activities: a sample of a pendulum clock worked effectively (in 1658) whereas the required S-model was still unknown to Huygens; a sample of the transistor functioned effectively (in 1947) whereas its scientific explanation was not yet satisfactory at that time; the tv pickup tubes, at contrary, belonged to the field of the vacuum tubes and the electron theory could explain the relevant phenomena from the beginning (since 1933); but many technical improvements and other scientific understandings were necessary to achieve the required design. Daarom is er moeilijk een algemeen geldend schema te vinden dat elke ontwerpsituatie stuurt vanuit bekende S-factoren naar onbekende Te-factoren of omgekeerd.
It was the intention of the economical push-pull discussion to indicate the decisive factors in innovations univocally: it would be either the S- and T-factors together or the M-factors. What determines tthe technological innovation: techological science or the market? A decision about this 'exclusive or' appeared

to be difficult. The reason for this will be clarified by examples. Sometimes the market asks for a completely new design and this was the case with the start of the transistor research in 1929 (see 7.1). When the first transistors functioned in 1947/48 it appeared to be a hard job to introduce the new device in the various market sectors (see 7.2). With television the development went quite differently: the technology already had a high level of sophistication (see 8.1) before the economical progression and the market were ripe for the introduction of this technology (see 8.2). Sometimes S- and T-fators influence the M-factors and sometimes vice versa.

Also with the P- and J-factors in environmental issues it is difficult to make reliable predictions with respect to the chronological order in solving the problems. Sometimes the S- and T-factors are controllable and the M-factors form a barrier. Sometimes the M-factors are favourable for the introduction of environmental friendly products and the S- and T-factors cause problems. Sometimes the P-factors stimulate technological-scientific research. In other cases a new controllability causes new legislation.

E-factors play an important role in architecture. Architects that design from a certain style will initiate ideas from about those factors and then search for solutions for questions related to S-, T- and M-factors. On the other hand E-factors usually form the finishing touch in the design of electronical devices. Also with regard to these factors it is not possible to derive a nice scheme from the design practice. It is hard to find a model to which the design activities in all technological disciplines will obey.

Is the design practice 'chaotic' or is our expectation that the design practice will follow simple schemes irrealistic? I will first pay some attention to this question. This is necessary to create clarity about the question which insights can be derived from the examples of design developments in the sections 6-8.

5.2. PROTOTYPES OF THE 'STEMPJE' APPROACH VERSUS PREMATURE GENERALIZATIONS

In discussion on general models I think design methodology can learn from physics. Physicists base their models on carefully chosen idealizations and on mathematically formulated laws that have been tested empirically. And yet they are very careful in the matter of general applicability. This appears for example when Casimir talks about the 'wide range of phenomena' that can be described satisfactory with classical mechanics.

"But this domain of validity has limits. That is not surprising. What is surprising is that a theory originally suggested by a limited number of observations is found to be valid for a much wider domain . . . Classical mechanics breaks down for very high velocities; there the theory of relativity replaces it . . . Classical mechanics also breaks down for very small dimensions; there quantum mechanics holds sway"[12].

When physics must be very careful with generalizations, than this holds even stronger for design methodology, because that does not even (yet) rely on carefully chosen idealizations and mathematically formulated laws, that have been tested empirically[13]. As an escape for this problem I have chosen for the 'STeMPJE' approach with the following characteristics:

(a) It is concrete: as a starting point a concrete existing design is chosen. When searching for an optimal solution designers always take a test design[14]. Therefore it is possible to speak of a series of designs before the final solution is found. This will become evident from my examples.

(b) It is practical, just like all design oriented disciplines are practical. The analysis of concrete design situations do not aim for a generally valid model but for an instruction that illustrates how students in their later engineer's practice will have to work.

(c) It is empirically critical: a satisfying analysis of an example of a design can serve as a prototype for further analyses. But this function of protoypes has it limits, like the 'domain of validity' of physical theories has limits.

(d) It is multifactorial and follows the route of the piecemeal search for an optimum: a design is a provisional response to many given factors. No general procedure guarantees an instantanious optimal reaction to the various factors. The 'STeMPJE' approach does not imply a multidisciplinary decision taking that offers the final solution to all problems via a burocratical way. It shows how via trial and error an optimum is found.

The 'STeMPJE' approach, though, is tolerant to methods that lead to a balanced survey of data. With some of our projects 'STeMPJE' is combined with qualitative function analysis (like in the design department of Fokker, where one tries to match the design of aircraft doors to the desires of airline companies). In other projects 'STeMPJE' appears to be combinable with value engineering. I will deal with these projects in section 11.

6. Huygens' clock designing and the interaction of the STeMPJE factors

In courses of physics the behaviour of pendulums is described with Galileo's and Huygens' pendulum laws which belong to the corpus of classical mechanics. The description presumes also the distinction between mass and gravity. In this context Huygens' design of the pendulum clock seems to be the result of an application of insights which can be deducted from the mechanical principles in an axiomatic deductive way. In fact Huygens achieved his inventions in a different way.

In 1637 the States General of Holland were in a difficult situation. Designers had been invited for a clock to determine the longitudes at sea. Nobody had been able to present the desired solution. Only Galileo had communicated in his letter of 1636 that he could solve the problem with a pendulum clock. Therefore, the States General wanted to negotiate with him. The Pope, however, dissuaded Galileo from accepting offers of the Calvinistic republic.

Twenty years later Huygens patented his design of a pendulum clock and in 1658 his *Horologium* appeared, describing the invention. From historical documents it seems to be probable that Huygens had used one of Galileo's sketches. In the spring of 1659 he was accused of plagiarism by, among others, Prince Leopold de' Medici.[15] Huygens pointed out that Galileo's pendulum was a solid arm mounted directly onto the shaft and would, therefore, be susceptible to distortions and even to stopping, as Huygens himself knew by experiment. In the same year his new experiments, concerning the period of a pendulum, led to the discovery that it was independent of its amplitude only in the case of small oscillations. He conjectured that the circular path of the pendulum bob was not more than an approximation and that it should be corrected in order to get a real isochronous path. His experiments led to a new curved path: that of the cycloid with vertical axis equal to half the length of the pendulum [fig. 10]. It took him fourteen years to prove that this insight can be deducted mathematically from Galileo's laws of fall and Descartes' principle of inertia. The deduction was presented in *Horologium Oscillatorium* (1673). It is striking that this study does not contain any consideration of the cartesian particles although Huygens was at that time and also remained afterwards a convinced defender of a cartesian approach in mechanics. The reason why he suppressed his predilection for this approach was that he did not like to mix the two discussions: that on the priority and that on the theoretical explanations of mechanical phenomena. Newton was impressed by Huygens' purely mathematical deduction. The *Horologium Oscillatorium* became a paradigmatic example to Newton: it showed him how a mathematical mechanical theory had to be constructed. In § 2 we have seen how Newton realized this aim.

In the meantime Huygens' search for a marine clock had a spin-off effect. In 1658 churches at Scheveningen and Utrecht were provided with pendulum clocks. Clockmaker Samuel Coster guaranteed that the clock of the church in Utrecht would not deviate more than eight minutes in a week. Huygens invention became also in use as a scientific instrument. Mechanical clocks had been constructed before, but Galileo and others had preferred a water clock for their experiments. Huygens became one of the first members of the *Académie des Science* founded by Colbert in name of king Louis XIV in 1666. Six years later the astronomer Jean Richer was send by the *Académie* to Cayenne, an island near to French Gyuana. He had to test the behaviour of Huygens' pendulum clock. Richer discovered that the pendulum length had to be 2.8 mm shorter on Cayenne than in Paris if the oscillation time had to be the same, one second, on both places. Huygens ascribed the deviation to the rotation of the earth: the centrifugal force on a body is, he thought, stronger at the equator than near the poles. Later, Newton mentioned many other discoveries of the same phenomenon in the second book of his *Principia*. He concluded that weights of bodies do not remain the same "in different regions of the earth". In 1742, finally, Jean Bernoulli introduced the distinction between mass and weight. The surprising facts are that Newton did not know this distinction yet and that Huygens' pendulum clock is not the result of an application of this distinction. At the contrary, this distinction is based on phenomena discovered by using this clock. The same can be said about the principle of the cycloidal bob motion.

My historical sketch shows Huygens' way of designing and, especially, *his dealing with the S-factors*. After his discovery of the principle of the cycloid motion, Huygens had to work out his S-model based on the laws of fall and on the principle of inertia. In our days engineers, active in classical mechanics, will also have to work out their S-model if they do not get the explanations or predictions required in the special field in which they are working. They possibly can deduce insights from the general principles; other insights will be empirical laws. My historical sketch shows, furthermore, that Huygens was a sinner: he offended against the design methodological rule prescribing the (scientific) analysis has to precede the synthesis of the design. He, on the contrary, started with a technical correction of one of Galileo's sketches. Then he constructed his design with a new pendulum. He patented his idea in 1657. The experiments leading to the new empirical insight about the cycloidal motion were done two years later. Fourteen years again were needed for the mathematical deduction of this new insight. A design had given rise to this scientific analysis. Should not writers of 'modern' text books mention this scandal in design methodology? Or should they give up their rule that the (scientific) analysis always has to precede the synthesis? Should not they accept that designing is a practical activity and that designers like to deal with concrete problems shown by provisional constructions?

Now, let us pass to *the relation between the S-factors and the other Te-factors*. With his adapted S-model Huygens was able to explain and to predict the bob behaviour of his design of 1673. The form of the two cycloidal jaws, guaranteeing the cycloid bob motion, also formed a part of his formal treatment [fig. 11]. He was, however, not able to deduct mathematically all details of this clock from his S-model. Many technical and not mathematical problems had to be solved: his clock had a horizontal escapement-wheel whose teeth alternately act upon pallets of a horizontal axis connected with the pendulum; the escapement-wheel transmitted the bob motion to two wheels to move the seconds-hand and the hours-hand; and so on, and so on. The details of all these problems could not be solved scientifically and were left to Huygens' experience-based *Fingerspitzengefühl*.

With regard to the M-factors we have to say that many different parts of the market were interested in the improvements of his invention stimulated Huygens to work on it during his whole life: in 1657 he presented his first design when he was 27 years old; the last one was a marine clock in 1694, that was one year before he died. We have to admit that he did not work for his subsistence. He never had to, because he was the son of a rich family. This latter aspect is, however, of no importance with the regard to these factors.

With regard to the P- and the J-factors we can say that the political situation encouraged him. The States General were favourable. The intervention of Leopold de' Medici could have led to a juridical process about the priority. In fact this intervention stimulated Huygens to improve his design, to illustrate that his invention was the better one, and to work on *Horologium Oscillatorium*.

The pendulum clock was not Huygens' final aim. The *Horologium Oscillatorium* already contained a design for a marine clock: special weights on the cord of the pendulum and different hinges had to avoid the influence of the swells. In 1675, however, he could master all these difficulties in a different way after his invention of the balance-spring for clocks and watches. At the same time he paid attention on *the E-factors*: from an esthetic point of view we can say that his drawing of the marine clock was intended to be a good looking instrument fitting very well in the luxurious lay-out of a captain's cabin. Esthetic aspects of instruments played a much more important role at that time as they do nowadays.

Was Huygens' balance-spring clock the result of an unordered trial-and-error process? Or was his design activity steered by a plan containing well defined steps in the direction the final aim of that clock? If we go through Huygens' *Oeuvres completes* then we meet the aforementioned series of his designs, beginning in 1657 and ending in 1694. Every design represents an elimination of shortcomings of its predecessor and can, therefore, be seen as a step of a piecemeal rationality; the whole series is steered by an idea of the possibility to achieve a marine clock. An idea, like this, is an intuitive concept; its content is not only cognitive; it also includes the wish to realize a new instrument which changes the daily life praxis.

The marine clock was the idea of one individual. How is it possible that a group, like the employees of a business corporation, supports and realizes the idea of one design program? This question has two aspects: the problem of the identification of different individuals with one intuitive concept and the problem of the historical identity of an idea. We will see that design ideas are fuzzy and allow, therefore, the identification of different groups. Furthermore, we will see that the 'STeMPJE' factors can change a design idea qualitatively. This is often the case in the contexts of microtechnological programs like that of the transistor program and that of the Plumbicon, as we will see.

7. Transistor designing

7.1. THE STRUGGLING WITH S- AND TE-FACTORS

The first transistor, invented at the Bell Lab in 1947, was a point-contact transistor (PCT). The name of this device indicated that a polystyrene *point* pressed closely to each other the connectors of the two circuits (input and output) on one of the three layers in which the amplification took place. Research efforts had been made at several places. The continuity of the program, however, had been the strongest at the Bell Labs.[16] These laboratories were founded by the AT&T and the WEC, being both active on the telephone market, in 1925. The program was stimulated by Marvin Kelly since 1928 when he became head of the Vacuum Tube Department; in 1934 he became general research director; and in 1944 he reorganized several departments for an effective realization of the program.

In the late 1920s a combination of *M- and Te-factors* formed the occasion to think about the transistor program. The telephone market had been expanded tremendously in the 1920s and the increase of telephone connections was also expected for the future. This implied that the telephone systems had to meet highly technological requirements. Relays and vacuum tubes formed the crucial parts of the systems. Through their switching ability, relays made feasible complex interconnecting networks. For rectification the diode tube was used. Amplification was realized by the triode tube or by one of its successors (for instance the pentode tube). The vacuum tubes enabled the extension of telephony to great distances. The devices, however, had several limitations. The relays were slow. The tubes, thought fast, had other disadvantages too: they were made of glass and, therefore, fragile; because of the filament, their number of hours they function was limited. They were expensive, unreliable, bulky, and wasteful of energy.

In the late 1920s other Te-factors, than those just mentioned, and special S-factors were in favour of the program. A new type of a solid state rectifier (the copper-oxide rectifier of Grondahl and Geiger) was invented. By experiments Russell Ohl of the AT&T's Radio Division established the frequencies at which the new device could be used. Its physics was unknown. In 1925-30, however, quantum mechanics had been born; physicists expected that this theory could lead to new insights in the behaviour of the electrons in solids. The first tasks of the program were, therefore, the physical understanding and the improvements of the new diode rectifier. For that reason Bell researchers attended lectures on the new theory at the universities. The substitution of the triode tube by a solid state device formed the more advanced task of the program. No sample indicated that this substitution belonged to the real possibilities. The motivating - idea, however, was taken from history and was based on an argument by analogy. In 1907 Lee de Forest had invented the triode tube by introducing a third electrode between the anode and the cathode of Fleming's diode tube; this third electrode, the grid, controlled the electron stream from the cathode to the anode. At the beginning, the physics of the tubes had also been a mystery. Studies in electron theory, however, had led to a better understanding and to improvements of both tubes. From the transistor program, a similar evolution was expected. In 1929 Walter Brattain was hired by the Bell Labs. In a later interview he formulated in the following way the general idea of the program:

"Anybody in the art was aware of the analogy between a copper-oxide rectifier and a diode vacuum tube and many people had the idea of how do we put in a grid, a third electrode, to make an amplifier"[17]

In the 1930s this idea led to a series of designs [fig. 12]. None of these designs led to the expected amplification. Two factors failed to achieve the desired success: a S-factor and a Te-factor.

The lack of physical insights, though quantum mechanics, remained an obstacle: the explanation of the special behaviour of electrons in different solids turned out to be difficult. At the universities people tried to surmount this task: John Slater at the MIT (where Shockley studied), Eugene Wigner at Princeton (where Bardeen became his training), Nevill Mott at Bristol, Walter Schottky in Rohstock, and Robert Pohl at Göttingen. A new discipline, solid state physics, was born. Kelly discovered that the designers's introductory training in quantum mechanics did not guarantee the required knowledge. Therefore, he hired Bill Shockley and other graduates of the new discipline. Since 1936 Kelly was convinced that theoretical research had to accompany the designers's efforts. Solid state physics, however, was still in its infancy. And, especially, a satisfying explanation of rectification was not yet available: why was the flow of charge carriers (electrons, holes) easier in one direction than in the other? Wilson's theory of 1931 explained this phenomenon with quantum mechanical concepts like *tunneling*. This theory turned out to be false after its application on the copper-oxide rectifier: it predicted the easier flow in the wrong direction. Later on, the theory of Mott and Schottky led to a more satisfying description. The Bell researchers used the Mott/Schottky-theory as the starting-point for their designs (till Bardeens discoveries of 1945 which will be mentioned later on).

The desired success was also hindered by the lack of a Te-factor: combinations of specially adapted semiconductors were required to achieve the transistor effect; the metallurgical knowledge and semiconductor technology of the 1930s was not yet able to perform this task. In this field the aforementioned Ohl, and also Southworth, Grisdale, Scaff, and Theuerer were active (later on, after the invention of the first transistor in 1947, Gordon Teal developed his single crystal technique, as we have seen in § 3). In the thirties these metallurgists dealt with the cat's whisker; this solid state rectifier had been used in the radio's before its replacement by the diode tube. An old sample was found while rummaging in a secondhand radio market in Manhattan. After experiments this cat's whisker turned out to be more sensitive at high frequencies than the diode tube. This excited Ohl's inquisitiveness. He tested more than hundred materials and he, finally, found out that the optimal functioning of the cat's whisker was achieved if the whisker point was in contact with silicon (galena was one of the materials used before). He also found out that the device sometimes rectified in one direction, sometimes in the other, and sometimes not at all. This discovery of 1939 formed the starting point of many experiments, finally leading to the insight in the n-type conducting silicon and the p-type conducting silicon. In the 1940s it was discovered (by Karl Lark-Horovitz and others) that electrons are the majority current carriers in the first type and that the holes have this function in the second type. *It is a surprising fact, that this insight could have been already deducted from solid state physics in the early 1930s!* This crucial discovery of the different types of conducting was, however, not a result of mathematical deductions. It was achieved by experiments with devices and by trials to improve them.[18]

We can conclude that the research was done on three levels or lines. *Experimental physicists*, like Brattain, tried to find new solid state designs of the diode and of the triode. Since 1936 *solid state physicists*, like Shockley, were analyzing the theoretical possibilities leading to several pencil designs: a pencil design is for me a science-based sketch without its materialization. *Metallurgy and semiconductor technology* formed the third line. A strong organizational coordination between these three lines lacked. Furthermore, the importance of the third line was ignored. In 1979 John Bardeen remarked on this topic:

> "In retrospect, there was a large gap between the scientific and development efforts in that no one was responsible for the development of semiconductor technology independent of any particular device."[19]

His judgement was even more sarcastic when left the Labs in 1951 (and we will have to consider this judgement in its context later on in 7.2):

> "Many of the things that the Bell Laboratories are proudest of now were done in spite of the management"[20]

The three lines crossed only accidentally each other. This was the case in 1939. Experiments and analyses, concerning Wilson's theory and the Mott/Schottky-theory, had led to the following insight with regard to the solid state diode: the effect could be realized, when the distance between the 'electrodes' was about 10^{-4} cm. This result contradicted Wilson's theory, presupposing tunneling with the maximum width of 10^{-6}. This result implied also a difficulty with regard to the design principles. *The layer principle*, as I call it, was based on the belief that in solid materials, similar to the triode tube, a grid layer had to be insert into the distance between the two other electrodes. Brattain had been designing in this way during the 1930s. In 1939 he was doubting about the possibility of inserting a control layer into a space of less than 10^{-4} cm. He asked Shockley for advice. Shortly afterwards, on 29 december 1939, Shockley had developed the *field effect transistor principle (FET principle)* and presented his first pencil design, based on this principle: instead of a grid layer, a field had to control the flow of charge carriers between the 'electrodes'. Since then, Brattain tried to realize a transistor in accordance to this new principle. Other designers, however, continued working on the layer principle. This situation confirmed what I have said before [at the end of 5.2]: design ideas, like that of the solid state amplifier, are fuzzy and allow an identification of many groups looking for different solutions of design problems.

During World War II the research efforts were stopped. Later on, this interruption turned out to be fruitful. The radar program had led to better insights in the behaviour of silicon and of germanium and in the p-type and n-type conductions. And, second, Kelly's reorganization guaranteed a more effective collaboration between two of the three aforementioned research lines: John Bardeen and other well trained theoretical physicists were hired, and they worked intensively together with designers like Brattain.

The research was continued where it was broken off before the war; in April 1945 Shockley presented a new pencil design of a FET, based on the new insights in the behaviour of semiconductors: an applied field had to change the conductance of three very thin semiconductor layers [fig. 13]. Brattain, Ohl and others tried to construct this amplifier. But the success was bound to come. Bardeen was asked for advice. After several months of searching he believed to have found the failure: at the interfaces, between the semiconductor layers and the plates of the input circuit, electrons recombined with holes and became trapped in surface states; these trapped electrons formed a shield reducing the influence of the field to 1/1500 of the effect predicted by the Mott/Schottky-theory. This hypothesis of Bardeen is usually considered to be the birth of surface science.

At the Bell Labs, the surface states hypothesis led to new kinds of experiments and to a new series of FET designs. The immediate predecessor of the point-contact transistor was also a FET design. On 11 december 1947 Bardeen's and Brattain's used an oxide layer to avoid the trapping at the interface between the connector of the input circuit and the p-type layer [fig. 14]. And now, finally, amplification was realized. The effect, however, was opposed to what was expected. If the design had been a FET then the positive bias of the input circuit would drive away the holes from the surface and out of the p-type layer; the result of this would be the decreasing of the output circuit. But the design did not work in this way: the output circuit increased as the input circuit increased. Different experiments led to the conclusion that the oxide layer did not isolate. This discovery led to the insight that holes were injected into the p-type layer and that they flow into the field set up by the output circuit. The discovery also indicated that the output circuit injected holes into the n-type layer, where they became minority charge carriers before entering the p-type layer. Bardeen suggested that greater amplification could be obtained by placing the two contacts closer to each other. *The point-contact transistor (PCT) was born by Brattain's realization of this proposal.*

The new insights in the behaviour of the minority charge carriers (electrons in p-type layers, holes in n-type layers) was of great importance for the later designs in the transistor technology. These insights, however, could have been deduced already from B. Davydov's theory of 1938. This theory was, however, unknown to the Bell researchers. Two reasons are discussed in the literature to explain this special lack of knowledge.[21] Few attention was paid at Russian physics in Western countries at that time. Furthermore, the theory was formulated with a sophisticated formalism; its mathematically physical aims seemed to be too far away from the practical problems of transistor designing.

In conclusion we must say that M-factors did not influence directly the problems which had to be solved by the designers: since 1929 these factors were favourable of investments in research on a new amplifier. However, these factors did not prescribe yet detailed requirements, which the designers had to keep in mind. With regard to this aspect the program changed qualitatively after 1947 when the first transistor worked effectively. I will pay attention at this point in 7.2.

With regard to the planning we can say that the design idea did not guarantee the steering of one well planned process, beginning in 1929 and ending with the desired success in 1947. What we have seen was a step by step process: every design formed a new starting point in the evolution, revealing often new S- and T-factors, which had to be studied more in detail. To achieve the final aim, every unsuccessful design led to a new mobilisation of different kinds of know-how for eliminating the shortcomings. In this sense, the progress was a piecemeal one. The insight in the required assistance remained also limited. Kelly's 1944 reorganisation, for example, did not guarantee the collaboration of all required kinds of know-how. The aforementioned gap between R&D and semiconductor technology did not even disappear immediately after the 1947 invention. The motivating aspect of the design idea, however, became strongly reinforced by this invention: series of new designs formed the consequence. The influence of the M-factors on these designs changed qualitatively: the efficaciously working transistor stimulated the quest for more efficiently working designs. In this new situation the before neglected third line of the R&D (metallurgy and semiconductor technology) became, slowly but surely, recognized as the most important source of know-how to improve the designs. This will be discussed in detail now.

7.2. The Strength of a Design Idea: The Struggling with M-Factors in the Context of Transistor Designing

Shockley was visiting laboratories in Europe when the crucial experiments of december 1947 led to the invention of the PCT. After his return, at the end of the month, the success of the others provoked conflicted emotions in him. He admitted later:

> "The birth of the point-contact transistor was a magnificent Christmas present for the group as a whole. I shared in the rejoicing. But my emotions were somewhat conflicted. My elation with the group's success was tempered by not being one of the inventors. I experienced some frustration ... . In response to this frustration, for the next five years, I did my best to put the Labs - and myself - in the lead of transistor patents." [22]

Frustration stimulated his efforts. Even on New Years Eve he was working and searching for new semiconductor amplification principles applying the new insights in the minority charge carriers. His calculations led to the invention of two new pencil designs: the p/n/p and the n/p/n junction transistor [fig. 15]. *They can be considered as concretizations of the aforementioned layer principle.* Shockley himself used the term 'sandwich structure' to indicate that the middle layer (in the first case the n-type layer, in the second case the p-type layer) had to be very thin; the reason is that holes are the charge carriers of the p/n/p transistor and they would recombine in a large n-type layer; electrons are the charge carriers of the n/p/n transistor and they would be trapped by holes in a large p-type layer.

Shockley's refusing to support Teal's idea of the single crystal technique (as we have seen in § 3) was in a certain sense dramatic because, later on, this technique turned out to be necessary to construct junction transistors. It was not less dramatic that Shockley's own designs got few support at the beginning: the majority of the Bell researchers preferred to work on the development of the PCT.

From the commercial point of view this was a reasonable desire: Bell's primacy had to be established by the effective production of the PCT. From the technological point of view this desire was reasonable too: detailed experiences of difficulties, surmounting during the development of the production system, had to be gathered. These aims were achieved in october 1951, when four companies had started the commercial production of PCTs. In 1952, however, an article in *The Bell System Technical Review* was published by Jack Morton who was charged with Bell's production system for the PCT. He admitted the bad reproducibility of the PCT, its poor reliability, and also its bad designability.

However, Bell organised a lucrative symposium on transistors in 1952. The fees were $ 25.000; a reduction was granted to participants who bought the licence to produce the transistors. In 1956 a similar symposium was held (its topic will be mentioned later on); it was not less profitable. Was Bell selling anything more than an idea? In the 1950s and in the early 1960s transistors were much more expensive than tubes. They were more temperature-sensitive and they made much more noise. Their frequency was more restricted too. The junction transistors, available since 1952, were less noisy but their frequency was even more limited. At that time British physicists believed that the transistor was nothing more than a publicity stunt of their colleagues at Bell.

How can we explain the final success of the transistor whereas its entrance on the market showed many shortcomings? The answer requires, I think, a precise insight in the dynamic interaction between the S-, Te- and M-factors of that time. Can such an insight be deduced from the following concept of Braun and MacDonald about transistor evolution of that time?

> "The integrated circuit was a commercial innovation developed by scientists working in a technological industry. Thus, it contrasts with the transistor, which was a scientific invention discovered by scientists who had little connection with industry."[23]

This concept is clear but not realistic: why should companies or consumers accept new designs or products which have more technical shortcomings than traditional products? The predicate *new scientific invention* can excite the curiosity of some individuals; but it guarantees only a small market for a short time. Furthermore, Braun's and MacDonald's judgement contradicts the facts. In 1953 not scientists but fifteen companies believed to be making transistors commercially. They can be subdivided in two groups. Nine of them (General Electric, RCA, WEC, Westinghouse, Sylvania, Raytheon, Philco, CBS, Tung-Sol) were established tube producers. It is true that transistors were regarded as of interest and possibly of long term to these nine companies: they were not considered to be the main concern. But the remaining five companies were new; and in 1956, again eleven new companies were active in the field of this production. Their foundation was based exclusively on the idea of a profitable transistor production. One of them was the Texas Instruments (TI). It was founded in 1953. At the Laboratories of this new company the first CI was invented in 1958. Therefore, the TI and many other companies were already active on the transistor market long before the CI came into existence.

The transistor program was not, as Braun and MacDonald express, developed by scientists without any commercial objective alone. We will achieve a more sophistical insight by our multifactorial analysis and by paying attention at *the dynamic character* of the interactions: changes of the Te-factors led to changes of the S-factors, changes of the Te-factors led to changes of the M-factors, the believe in the transistor idea was reinforced by explanations of the S-factors, and so on. Some of these interactions are already mentioned before.

Kelly, restarting the program after the war, had already a commercial aim in mind: this was Bell's quest for the primacy in solid state devices. It was partly realized when the PCT and the junction transistors were patented by 1948. The next step, the demonstration of the producibility of the new device, seemed to be achieved when the aforementioned companies were making PCTs since 1951. Morton admitted the technological weakness of the PCTs in 1952. In the same article, however, he tried to reinforce the belief in the transistor idea by announcing the development of the new germanium junction transistors:[24]

> "With respect to reproducibility and interchangeability, transistors now under development appear to be equal of commercial tubes."

From a 'scientific' point of view, this way of introducing a new technology was not done very well: Bardeen's aforementioned negative judgement was characteristic for a theoretical physicist, who had an other way of planning in mind; before starting the development the theoretical research should have led to a better physical understanding, the designs of the junction transistors (as alternative for the PCT) should also have been worked out, and Teal's ideas should have been realized too. It is true that none of these tasks was performed when the first transistors were in production. From *his* point of view, Bardeen was right. The insights in the S-factors of the PCT and in those of the junction transistors were, indeed, very restricted at the beginning. This can be illustrated by the anecdote on Shockley presenting his paper on his invention of the p/n/p transistor to the *Physical Review* in 1948. The editorial board refused to publish it because of the theoretical weakness of the quantum mechanical explanation. Shockley was forced to publish it in Bell's own Technical Review sixteen month later. This example illustrates that there was still a gap between the practical insights and the scientific understanding at that time. This gap was not bridged before 1951, when Shockley published his book on *Electrons and holes in semiconductors with applications to transistor electronics*. Shockley's 1952 analysis of the FET showed that an other new solid state devices could be realized *if* the required materials could be produced. *In the meantime technological experiences had been growing by the trials to develop and to improve the PCT. Bardeen and Shockley failed, however, to recognize the crucial importance of this latter aspect to transistor designing.*

Braun and MacDonald are only in a certain sense right: Bardeen and Shockley were those "scientists who had little connection with industry". Bardeen, being such a scientist, was happy to leave Bell and to become a professor in Illinois in 1951; he never did industrial research again. More than Bardeen, Shockley had some commercial objectives in mind. He had good contacts with the Department of Defence which bought nearly all 90.000 PCTs produced by the WEC in 1952. He organised Bell's lucrative symposium in 1952. In 1954 he started his own laboratories following the same formula as Bell had followed before: getting patents by doing basic research and earning money by permitting productions in licence. At that time, however, the Bell Labs had given up their old formula: the fundamental aspects of transistors were not considered to be their main concern any more; the 1956 symposium was dedicated to production techniques and not to fundamental questions. Shockley had overseen this necessity to react in a new way at the technological situation; his commercial career was short. In 1959 he sold his firm. He had failed to see that the dynamic interactions between the S-, Te- and M-factors had changed the situation.

In this new situation Teal's single crystal technique had a higher relevancy than Shockley's basic research efforts. In 1952 Teal was hired as research director by the TI. Two years later, he made silicon junction transistors commercially available. This was an important success to his method. In 1958 an other impressive advance was made in his laboratories: Jack Kilby's invention of the first effectively working IC, employing a concept that made possible the implementation of many functions on a single chip of single-crystal silicon. [25] The availability of silicon is easy: sand contains silicon. Because of its high melting point, it is less temperature-sensitive than germanium. This implies, on the other hand, a lower mobility of the charge carriers (its energy gap is more than that of germanium), an inferior frequency performance, and designing problems because of a smaller space between the electrodes. An other disadvantage is its stronger tendency to absorb unintended impurities. Therefore, other chemical elements were tried out. The efforts remained unsuccessful. And Teal's production technique remained without an effective competition in the 1950s. For his company this was a comfortable situation: "TI's sales rose almost vertically; the company was suddenly in the big leagues".[26] In the 1950s, however, General Electric, Fairchield, Philco, Bell, and other companies were already searching for new production techniques, like planar process technique, jet-etching, diffusion technique. Planar technique turned out to be the most efficacious. In spite of this, it remains true that Teal was the first who discovered the high relevancy of the semiconductor technology to transistor designing.

An other specific feature of the transistor innovation was the necessity to search for its applications in the 1950s. During the 1930s and 1940s, its primarily intended application had been the telephone systems. But the introduction of the transistor in these systems turned out to require too drastic changes of the systems [one of the reasons will be discussed in § 10]. Therefore, the first use of the transistor was tried out in smaller technical systems: in hearing aids. Bell stimulated this use by renouncing their fees for this production in licence. The hearing-aids market was, however, not without risks. Raytheon, producing more transistors than any other company in 1953, lost its dominant position in the 1960s because it failed to extend its transistors to other applications. The profitable production of transistors exclusively for this market was guaranteed (by a special interaction between the S-, Te- and M-factors) only for a short time. Other small system applications were those in pocket calculators and pocket watches. In 1954 TI started the production of transistor radios. The following year transistors were constituent parts of the computers introduced by IBM. The result was impressive; its size was small and the energy consumption was reduced by 95 %. In the early 1960s one-sixth of all transistors were used in computers whereas the application in telephony was still problematic at that time. The automobile industry was considered as one of the greatest potential markets for solid state components from the beginning. But it was difficult to adapt the components to the conditions into the car engine. Social obstacles had to be removed too: designers were not ready to accept and incorporate semiconductor devices in their arrangements of the car engine; the introduction of the new technology also needed special training of the local auto mechanics, by which cars must be serviced and repaired. It is symptomatic that only 1 % of the Ford cars produced for the American market contained microprocessors in 1978. These examples show that there does not exist *one* scheme characterizing the use of a new technology into many different fields. The successful introduction always depends on special combinations of S-, Te- and M-factors. In conclusion we can say that progress on the following levels was needed for the transistor introduction since 1947:

- a better physical understanding,
- alternative designs for the PCT (junction transistors, FET) and more sophisticated designs (ICs),
- detailed experiences of the difficulties, surmounting during the development of the production system, had to be gathered,
- progress in semiconductor technology and
- tentative trials to introduce the new devices in different products and production systems; the success of these trials depended on concrete combinations of the factors and lacked a general scheme; a successful introduction did not even guarantee a long term success.

The efforts on the two first levels was preferred "by scientists who had little connection with industry" (Bardeen and Shockley). From the technological and commercial point of view others were right by preferring to work on the other levels. In this sense the development since 1947 was an other confirmation of what I have said before [at the end of § 6 and what was confirmed in an other sense by the different transistor design principles about which 7.1 informed]: the design idea of a technological program is fuzzy; it is strong if it is realistic; and then it allows an identification of many groups looking for problem solvings on different levels. The progress in designing and product introduction is usually a piecemeal but not a peaceful process.

8. The designing of tv tubes

8.1. Battles on P- and J-factors and Philips' Plumbicon

Different design programs have to be compared with each other to avoid unjustified generalizations and to achieve unambiguous descriptions, explanations or predictions in the STeMPJE approach. Therefore, I will pay attention at the Plumbicon [referring to other publications for detailed analyses[27]].

The Plumbicon is a television camera tube developed at the Philips Labs (the Philips Research Laboratories in Eindhoven). It was presented in 1962, and it turned out to be very useful for colour television, requiring the broadcasting of three colours. In the Philips camera PC-70 of 1965, three small Plumbicons together with small prisms replaced the older cameras's bulky assemblage of tubes and mirrors. There are similarities between the transistor and the Publicon:

(a) Both inventions required applications of semiconductor technology.
(b) Both were intended to be used as devices in lager systems of other products.

With regard to the R&D, however, there were deep differences.

(a) The transistor program was an independent activity. The Plumbicon project, at the contrary, was the final period of intensive working on the evolution of pickup tubes; and this evolution was a part of a more comprehensive program: the introduction of television. The reason of this difference is clear: the television technology was still problematic in the 1930s.
(b) Nationalistic preferences led to P- and J-factors influencing directly the designing of pickup tubes. Such an influence lacked during the transistor research.
(c) Furthermore, the transistor designs did not work before 1947. In the 1920s, at the contrary, three systems could be considered as first realizations of the television idea: Baird's system (UK by 1924), Karolus' system (Germany by 1924) and Jenkins' system (USA by 1925). It is a bit misleading to use the same word for these systems because they had few in common with the television in its present meaning:

only the broadcasting of static pictures was made possible. However, different objectives stimulated the realization of the idea. Applications in the context of advertisement were intended in the USA, the Russians thought about military uses, the transmission of texts by telephone was realized with Karolus's system in Germany, and again other objectives were regarded to be possible in Great-Britain.

Since the invention of the electronic scanning in the 1930s, the P-factors of the standards influenced directly the designing. To understand this phenomenon we have to go back to the original invention of Paul G. Nipkow.

Nipkow was a 24 years old student when he got his television idea in 1884. According to this idea, a picture had to be illuminated via a rotating disk with holes in it. By this way of illuminating the picture was divided in elements (pixels) and, because of the rotation of the disk, these elements described arc lines. The intensity of the light reflection of every pixel varied with the parts of the picture. A selenium cell transformed this optical signal into an electrical signal: lighter parts caused higher current voltages and dark parts lower voltages. The picture had to be reproduced by using a similar disk and a lamp whose brightness was controlled by the signal voltage of the current.

In the 1930s this mechanical/electrical scanning with a disk and a lamp could be replaced by a scanning beam of electrons in a vacuum tube. This improvement had been made possible by the developments of the vacuum tubes during the 1920s. In the 1930s several countries got their 'own' tubes: the Americans had the iconoscope and the orthicon, in the UK the emitron was developed, and the Frenchmen had also their own tube (a predecessor of the later emitron of 1947). Governments or other national authorities protected the designs of their own industrial companies by prescribing special lines systems. The differences dominated the whole history of camera tube designing. They even influenced the application of technological methods, like for instance the ways of dealing with the electrons in the tubes. In 1951 three Philips Labs researchers (Schagen, Bruining and Francken) describe the consequences in the following way:

> "Pickup tubes are divided as follows: (1) pickup tubes with low electron velocities ... , in which the target is stabilized on the cathode potential ... , and (2) pickup tubes with high electron velocity ... , in which the target is stabilized on the potential of the collector ... . To the first group belongs the image orthicon which is mainly used in the USA, to the second belongs the image iconoscope, which is preferred in Europe. One of the reasons for this preference has to do with the great number of lines for which was decided on the Western continent (625, in France 819). With high electron velocity one can better fulfil the high requirements the definition of such a great number of lines poses to the focusing of the scanning beam".[28]

At that time a 405 lines system was used in the UK, and a 525 lines system in the USA. In case of the USA, this standard had been established in 1941 after a great battle between companies and committees during many years. At the 1950

CCIR conference in Geneva several countries had accepted the 625 lines system. Among them were Belgium, Denmark, Italy, the Netherlands, Sweden and Switzerland. No agreement was achieved an agreement with France and Great-Britain. Similar battles dominated the later developments. In the 1950s, colour television led to a competition between the NTSC of the USA, Telefunken's PAL and the French SCAM. Nowadays such a competition exists between different kinds of HDTV.

As the quoted words of the Philips Lab researchers express, two main traditions dominated the series of camera tube designs: the American tradition with the image orthicon as its representative and the European tradition with its image iconoscope. Some engineers tried to eliminate the gap between the traditions by establishing the *economic optimum*: the increasing of this number of lines would increase the costs without improving the image quality, and the reducing of this number would worsen the image quality without saving much cost. However, no foundation of an optimum led to a political consensus. The activity of other engineers was based on a more realistic thought: they designed *line translators* for broadcasting between countries using different lines systems.

How came this situation into existence? To answer this question we have to go back to 1933. In this year RCA's director Vladimir K. Zworykin presented his *iconoscope*. It was the first tube based on electronic scanning. The images passed an optical lens and became projected onto a mosaic layer of the target [fig. 16]. This layer was formed by a mosaic of isolated cesium grains which became positively charged, depending on the light intensity. The scanning beam of electrons discharged the grains one after the other. A disadvantage was the disturbing influence of the so-called secondary electrons, not belonging to the scanning beam. Some of them were collected by an anode, others fell back on the cesium layer and spotted the final image.

This shortcoming was, for a deal, reduced in the *image iconoscope* by using a photocathode. This cathode transformed the optical images in images of photoelectrons. At high speed these photoelectrons hit the target surface, formed by a mica layer and not - as it was the case of the iconoscope - by a layer of cesium grains. This was the second type of tubes, described by the three Philips researchers: because of the high velocity of the photoelectrons, parts of the mica layer became more or less positively charged (depending of the light intensity): the photoelectrons emitted secondary electrons from the mica layer. These secondary electrons became captured by a high voltage anode. The image iconoscope, presented in 1934, was a result of a collaboration between Zworykin and RCA's licensee Telefunken. The image iconoscope was considered as a typically 'European' tube although its predecessor was invented in the USA. In 1935 the Reichspost started the public broadcastings using this tube and applying a 180 lines system. In the next year the fascist government had its great publicity stunt: in Berlin and Leipzig the Olympic Games could be observed in seventeen public television rooms. At that time it seemed to be that the Reichspost had won the competition with the BBC. On 11 December 1936, however, the BBC had its own stunt: using a British tube, the emitron, and applying a 405 lines system, it broadcasted the coronation of King George VI.

The principle *improvement by simplification* led to RCA's *orthicon* of 1937. The double channel tube was characteristic for the iconoscopes [fig. 16]: the image reproduction followed a horizontal line, the scanning came obliquely from below. The orthicon had to be simpler; the Greek *orthos* means: perpendicular; by producing the optical image on one side of a transparent target and by scanning it from the other side, a perpendicular form of the tube could be realized. To control better the secondary electrons the velocity of the scanning beam electrons had to be low. This was achieved in the following way: the gun cathode potential of the scanning beam was the same as the target potential; and the secondary electrons were attracted by a large potential difference between the target and the anode. Therefore, the orthicon was called *cathode stabilized*, as we can see in the quoted sentences of the three Philips researchers. The orthicon is in a stronger sense an American tube than the iconoscope in which there was a large potential difference between the gun cathode and the target and in which target and anode had the same voltage.

The *image* orthicon was the successor of the orthicon; its name was based on an analogy: a photocathode was added to the previous design like it had been the case with the image iconoscope succeeding the iconoscope.

None of the aforementioned designs led to a satisfying solution of the problem of the secondary electrons spoiling the image signals. The semiconductor technology, developed in the 1940s, enabled RCA people to try out an other solution by applying photoconduction: the conductivity of some semiconductors increases as they become illuminated. RCA's *vidicon* of 1950 was based on this principle [fig. 17]. Its target consisted of two plates: a transparent and conductive signal plate and a photoconductive semiconductor plate. The conductor was connected to the earth, so that, there was a potential difference between conductor and semiconductor in the dark: the semiconductor was isolated because no electrons could pass the voltage difference. The semiconductor became conductive after being hit by a light image which had passed through the signal plate: the electrons went from the semiconductor to the signal plate, whereby the voltage of the semiconductor became locally higher in correspondence with the light intensity of the image parts. When the scanning beam hit the illuminated place with its higher voltage, the original potential was restored, which caused a small current; this was send, as a video signal, from the target to the amplifier. However, the vidicon was not a great success. The used materials reacted too slowly on changes. By this the reproduction of rapidly moving objects became vague. Nevertheless, in 1964 three vidicons were applied in RCA's TK-42. This camera was nearly a "technical disaster". The vidicons exhibited serious colour smearing if operated in a high-sensitivity mode. Without success, RCA researchers tried to find a solution and, finally, they accepted this phenomenon as an inherent feature of photoconductivity.

They were flabbergasted by Philips' demonstration of its new camera with three Plumbicon tubes, based on the same photoconductivity principle, at the 1965 NAB convention in the USA. General Electric, Marconi and other companies were enthusiastic and decided to incorporate Plumbicons in the new designs of their cameras. RCA tried to copy the Plumbicon. In spite of its know-how and experience, it was never able to produce a sample. Finally, it considered the manufacture as a "black power". In its research and production, Philips turned out to have developed a know-how and *Fingerspitzengefühl*, especially with regard to lead oxide and other materials easily absorbing X-rays.

Plumbicon is written with a capital because this is the name of a special vidicon. Its target is made from lead monoxide. This material can be doped, so that it becomes a pin-diode, consisting of a strong p-type layer, a intrinsic semi-conductor i-type layer (which is, in fact, a weaker p-type layer) and an n-type layer. In the dark the interfaces of the layers block the current: holes do not enter into the i-type layer from the p-type layer and electrons do not enter into it from the n-type layer. This isolation is due to the recombinations at surfaces, as we have seen in the context of the transistors. The lack of any dark current means an essential improvement. The i-type layer extracts, if it is illuminated, electrons from the n-type layer and holes from the p-type layer. The p-type layer is scanned by the beam. Because of the high efficiency of its target, the Plumbicon could fulfil the American and the European standards.

Was the new tube "developed in great secrecy in Philips's laboratories"? This judgement of Inglis is not true.[29] In 1954, the predecessor of the Plumbicon was already described by Heijne, Schagen and Bruining in their article 'an experimental pickup tube for television with applications of photoconduction'. The RCA Labs did, probably, not follow the tv research of the Philips Labs. Therefore, it is more interesting to deal with the question: how could the tube, - bridging the two design traditions, be invented and developed by Philips, which had not yet been considered as a serious competitor on this market? To answer this question we have to go back to the 1920s.

8.2. HOLST'S STRUGGLING WITH M-FACTORS

In the late 1920s Philips was active in television technology only for a short time. On 4 October 1927 a television system, similar to those of Karolus and Baird, was patented in the name of Gillis Holst, who was the director of the Philips Labs since their foundation in 1914. In December 1928 the system was demonstrated during a session of the Dutch Physics Association. The *Radio-expres* reacted with contempt for the exposed technology: experiments of a good amateur would lead to the same results! Holst himself could have written this critics because two reasons. First of all, he had demonstrated that the Philips Labs was able to deal competently with the new technology; and this was recognized by the Radio-expres too. The second reason was that he did not like to work for the very small market of amateurs enjoying the broadcasting of static or almost static pictures. On this topic he disagreed with the comity of the company's vice-directors who wanted a scientific and technological competition with Telefunken and RCA. The demonstration in 1928 was in favour of Holst's opinion: television was not yet ripe for serious R&D.

In 1935 Holst and the top management of Philips were still in doubt about the chances of a commercial mass production. On 25 November managers (Anton Philips, Otten and others) and researchers (Holst, Van der Pol, Oosterhuis) were together at a meeting and they came to the following conclusions:

> "As far as we can see at the present time television will always remain very costly and therefore available only to a few people. The definition still falls far behind that of mediocre film. To get a better TV image it will be necessary to use shorter wave lengths resulting in a smaller range of the transmitters. With the wavelength used at present reception is possible within a radius of 40 km from an aerial 100 metres high. With a transmitter located on the Brocken Mountain (1600 m) it was impossible to get satisfactory reception in Berlin at a distance of 100 km. For this reason TV will be limited to large population centres. In addition the cost of a receiver will probably not fall below 1000 guilders."

From other documents we know that these were Holst's arguments. He was also aware of the specific requirements for TV programs, compared to radio programs. After his visit of RCA he had written from New York on 2 February 1932:

> "They have not solved the problem of making television programs interesting, everything is still in the laboratory stage."[30]

On the other hand he was aware of the dynamic character of the interaction between S-, Te- and M-factors: this interaction could change completely the situation. In 1936, the year of the Olympic games and of George's VI coronation, three research groups were formed: one dealt with everything relating the studio, including the cameras and the generation of the signals; an other studied the modulation of video and audio signals; and a third worked on the receivers. To the first group belonged Herre Rinia, a man with 'golden fingers'. At every stage of the development, his excellent designs demonstrated clearly the level of television broadcasting based on the known S- and Te-factors. Van der Pol and Van der Mark belonged to the second group, and Haantjes to the third. A document of 28 August 1940 shows that at that time about thirty other excellent scientists (like Posthumus and Tellegen) were active in the field. A group of technical assistants of high quality (Van Vlerken, De Vries, Venis, Hepp) supported the research. During the war the program was continued. For colour television a frame sequential system was developed and tried out, applying alternate pictures in red, green and blue.

However, Holst was not yet convinced a large-scale tv market in the future. In 1940 he expressed his doubts to Haantjes. These doubts formed the reason, why he did not stop the alternative program based on the home cineac idea: every day, the news would be delivered to subscribers by the way of a film. The group, working on this program, had calculated in 1937 that a home cineac set would cost less than a tv set. In june 1944 the members were still convinced that home cineac was, compared with television, something "superior in many aspects".

Holst's doubts were justified and his way of directing the laboratories was not wrong. On one side technological improvements, market changes and social reorganisations were necessary to realize television as a mass communication mean; the predictions about the developments of these conditions were unsure. On the other hand, the Philips Labs was prepared for the situation in which these conditions would become more favourable: since the 1930s competent research groups were formed. Holst's doubts stimulated critical studies. We can conclude, that the tv research tradition, which became successful later on, was started at the time when Holst was the director.

In 1946 he retired. Rinia, one of his successors, became responsible for television. In Europe few people felt the need for television broadcasting at that time. However, like it had been already the use before 1940, again after the war Philips managers and researchers often visited the USA to observe the production methods and markets and to come into contact with the research laboratories. In 1948 one of them, Bouman, was impressed by the television market which had been made possible by the growing income rates in the USA. Such a evolution could be expected for Europe too. In his telegram, dated March 5, he urged to eliminate all doubts and to mobilise all forces for the sake of television [fig. 18]. Since then the research activities became intensified and the production group of the electric tubes was persuaded to collaborate. In the Labs H. Bruining was made leader of the television group. Together with Heijne and Schagen he invented the first Philips photoconducting tube in 1954, as we have seen before. But this tube did not yet imply a clear and definitive line for the research program. In 1955 a new kind of the image iconoscope, the scenioscope, was designed by Schagen and others. This confirms the theses which we have met before: the design idea of a technological program is fuzzy so that it allows the identification of groups looking for different problem solvings. Every design can form a new starting point for the planning of a new design. With regard to microtechnological programs the planning does not only need a collaboration of physicists and chemists:

> "Pickup tubes are masterpieces of glassblowing skill and because of the motivation of very skilled glassblowers the most unorthodox constructions have been realised. Mr Schampers was the man with the golden hands, who began as a glassblower, but by self study rose to the level of scientific assistant, and had contributed significantly in all stages of Plumbicon development."[31]

This confirms again, what became already clear in § 4: microtechnological inventions are impossible without new experience-based know-how.

9. Casimir on logistics in the context of microtechnological designing

Three design developments have been discussed here: that of the marine clock, that of the transistor and that of the pickup tubes. There seems to be a strong similarity: every design was stimulating the search for the scientific or technological insights, required for its improvements. Is not this conclusion undermining the necessity to distinguish microtechnology from microtechnology?

I have to admit that I got the idea to develop my views on these technologies and on this distinction after my studying of Hendrik Casimir's publications and after my fruitful discussions with him. He is not only a famous physicist.[32] He also was, together with Rinia and Verwey, one of Holst's successors as director in 1946. Since 1956, when he became member of the General Board of the company, he was responsible for the R&D, all over the world, till he retired in 1972. In the fifties and sixties, he developed his so-called spiral doctrine to express in a systematic way his views on researchmanagement and on the relations between modern science and technology.

Spiral means, like in *wage-price spiral*, in this context: a continuous upward or downward movement caused by strongly interdependent factors. In the most strict sense, this interdependence between modern science and technology is realized in high energy physics: its theoreticians need the support of experimentalists and of technical 'assistants' who themselves are scientists of a very high level. Indeed, not only the theoreticians Glashow, Salam and Weinberg got the Nobel price because of their contributions to this discipline (in 1979) but also the experimentalists like Rubbia and technical physicists like Van der Meer received this honour (in 1984). The interdependence in high energy physics implies a barrier against societal influences and steerings too: the aims and the relevance of this kind of science and technology can not be, and therefore should not be the subject of public discussions.

In a less strict sense this interdependence forms the organizing principle for the research of those companies which intend to be active in the fields of microtechnology. On the one side, the barrier between the research and the existing production methods is more transparent in this industrial context: at least a fuzzy relation between the phenomena of research and the intended market should be guaranteed. On the other hand, fruitful industrial research requires also a barrier protecting against the intentions of special production groups. Examples from the Philips research confirm the rationality of this quest for protection and they illustrate at the same time that, even at the level of the fundamental research, applications and concrete designs are intended:

> "We [Philips researchers] originally studied ferromagnetic ferrites for a special application in carrier telephony: many years later we found that the bulk of our production goes into television receivers. From a long-range point of view our work on ferromagnetism has been successful and profitable but if our nascent carrier telephony department had had to bear the initial expenses it would have been killed right at the start."[33]

Bell's transistor research confirms this view too: it was originally started for the application of a device in telephony and its first results were used in radio's and computers long before the telephone systems could be adapted for the use of the new device, as we have seen. The protection of the research does not imply an undermining of the necessity of cooperations. Casimir for example in his Holst lecture at the Eindhoven University in 1981 states:

> "In principle a university should be the ideal place to establish cooperation between various disciplines. In practice not much of that was and is realised. The Philips Labs with respect to that could serve as an example".

A barrier must protect not only the microscientific research but also the research of electromagnetic phenomena. Casimir's argument is that these phenomena led to impressive innovations and they would never have been discovered in the context of industrial production:[34]

One could question whether the electrical energy and the electrical transmission of force would not have been found by manufacturers of steam engines, who found that increasingly longer driving-belts were annoying way of transferring the power of the steam engine to looms and other equipment. Well, it did not happen that way and I can hardly imagine that it would have happened. At best they would have invented the V-string in stead of the flat driving-belts and in the end they did. Electricity, on the other hand, came forth out of research into natural phenomena. Researchers like Faraday and Maxwell found the laws of electromagnetic induction, and afterwards the dynamo and electromotor came (. . .) One could question whether radio waves would not have been invented by directors of post offices who thought the speed of (postkoetsen) should be increased and the crippling of horses laid severe barriers to the transport of letters. This is not the way electromagnetic waves were found. (. . .) This went via the theoretical insights of Maxwell, via speculations that light could be a vibration in electrical and magnetic fields. From that the idea came forth that such vibrations would also exist in lower frequencies and Herz proved this by experiments. Ten, twenty years later these electromagnetic waves played a role in the transmission of data, first of morse signals, later also of language, images (television) etcetera.

A prerogative for the industrial research is the University work on theories like electron theory and quantum mechanics. Casimir is convinced that transistor research confirms his thesis in the following way:

> "Now first of all let me say that the transistor is not only an extremely important and useful device ..., but that the research that led to the transistor is a very beautiful piece of work indeed, combining well planned and yet imaginative experiments with a penetrating and elegant analysis. ... And yet I think the transistor is a good example for my thesis. [After the foundation of the electron theory by J.J. Thomson and after the formulations of the quantum theory by Sommerfeld and other university researchers] the problems became less fundamental ... . Now the one new principle that was added by the Bell

people to existing notions was the idea that electrons can persist for some time in p-type germanium ... ; similarly injected holes can persist for some time in n-type germanium. This is both an important experimental fact and an interesting theoretical idea, but from a philosophical point of view it is certainly not on a par with the great new ideas of quantum mechanics.[35] The work on transistors is essentially a brilliant and novel application of known principles, that would be anyway, but happens to be of great technical consequence too. Once more, the fundamental notions that are required are coming from the universities."[36]

In this account we meet several prerogatives for successful industrial research. The university research is leading and has to lead to the fundamental notions which can be applied in industrial research; and the autonomy of the research has to be protected in companies which intend to be active on the market of electronics and of electrical devices. Translated in a practical methodology, these postulates imply that the designing and the production can only lead successfully to innovations in these contexts if the social environment of those companies guarantee the required education and if the companies are financially strong enough to protect its fundamental research. From this it becomes clear that *the domination of the S- and the other Te-factors depends on market factors and on social factors*.

One could discuss about the question why the spiral doctrine defends the special quest for a protecting barrier in favour of the research applying the aforementioned physical theories. The age of the theory of electromagnetism is almost one and a half century. Electron theory, quantum theory and quantum mechanics are more than sixty years old. This is true. However the microtechnological applications of quantum mechanics in electronics or photonics (laser technology) requires a special theoretical training, a special equipment, a set of special instruments and a support of well trained assistants. Therefore, we meet a university like spirit if we enter the laboratories of microtechnological industries. And we smell a technological air if we enter the laboratories of the chemical industry, of the blast-furnace companies or of the airplane manufacturers. If we apply the same word 'research' to the work of these different laboratories then it becomes ambiguous. In a differentiated meaning it remains true that *the domination of the S- and the other Te-factors of these fields depends on special market and social factors*.

10. Progress in designing by analogies

10.1. S- AND TE-ANALOGIES

In the philosophy of science there are many disagreements. Usually, however, there seems to exist one unanimity among scientists and philosophers: the final methodological aim of science is the so-called theory of everything. The formalism of this theory will exclude all ambiguities. A subject, not fitting completely into the rigorous framework of constants and variables of this formalism, will be idealized. If disturbing factors occur, then they will have to be explained by other parts of the formalism. In this sense we all are still Newtonians because Newton was the first, who formulated this ideal.

It is not necessary to discuss the details of this view. Only two points are important for us. If *scientific* is taken in its Anglo-Saxon meaning then a scientific theory excludes ambiguities. And such a theory is better than its predecessor if it closer to the theory of everything: if it is more universal. Design methodology lacks these two clear criteria.

Designers do not always prefer the most universal or the most sophisticated theory. Discussing the experience-based and macroscience-based technologies, we met examples of that. The account on transistor research delivered other examples: the designers preferred the more classical Mott/Schottky-theory to Wilson's theory (using quantum mechanical concepts). They did no pay any attention at the more sophisticated theory of Davydov till the recombinations, discovered in Shockley's FET design, had shown its relevancy. The preference of a theory is deduced from design problems in the context of engineering.

Design methodology lacks also the second criterion: the rigorous exclusion of any kind of ambiguity. Progress starts with the analysis of a given design: it is compared with an other better design. Sometimes this second is nothing more than a sketch or it exists only in the designer's imagination. The comparing implies analogies: similar technical functions have to be realized; the second design has to meet also new requirements and the shortcomings of the first design must be eliminated or at least reduced.

We can say that engineers are active in fields which are two-piece: the unambiguous application of the formalism of scientific theories goes together with the quest for better designs than the previous ones. This second aspect, the reasoning by analogies, becomes more dominantly present and is more explicitly present in the minds of the engineers if the required scientific theories with their formalisms lack. This becomes clear if we compare the pickup tube research with that of the transistor. When the tv research restarted in the 1930s, the S- and the other Te-factors were well known from the electron theory and the tube technology. Therefore, the analogies are not often mentioned explicitly in the history of camera tube designing. There is one exception: the similarity between the expressions *image iconoscope* and *image orthicon* indicates the application of the same technical principle in an analogical way, as we have seen in 7.2. Of course, there were many other analogies. There are similarities between the aforementioned targets. And a similarity can also be discovered between Nipkow's concept of scanning and Zworykin's concept of it. Historical descriptions use often expressions belonging to the discipline of analogies: 'similar', 'dissimilar', and so on. However, the explicit use of the term *analogy* is seldom in this context. The descriptions of transistor research, at the contrary, often use this term which was already introduced by Britain, Shockley, Teal, Bardeen and others, as we have seen. The reason for this difference is clear: the idea of the transistor had to be formulated without the knowledge of the S-factors and without the know-how about the Te-factors needed to realize this idea. In the thirties the surface science did not exist, and the solid state science could not yet solve the relevant problems, as we have seen. In the texts of that time we can distinguish at least two kinds of analogies:

(a) the first one is the function analogy: in the telephone network one wanted to fulfil the function of the tube triodes in a different way;

(b) the second one is the geometrical analogy: in the designing of the solid state triode, the chosen starting point was a form that resembled geometrically the already successfully completed design of the solid state diode, of which the triode deviated by the third electrode.

Analogies are often regarded to be vague concepts. But in these R&D developments one perceives conceptions with concrete and exact aspects. This had already been the case with the diode: the solid state version of that had been found by starting from the technical function, that had been realised by means of a physical effect, that is described in an exact way with the mathematical equation of the characteristic of a diode. In a similar way the desired technical function of the solid state amplifier could be deduced from the tube triode. Seen in the light of this assignment each step is assessed during the R&D period.

The steps themselves were influenced by the geometrical analogies. They were started from the structuring of the effects of the element that was to be replaced: the grid [fig. 12]. One could compare the design by Glaser, Koch and Voigt in 1939 with this tube triode. Then it becomes clear in what way the researchers tried to stick to the same structure of effects: one first tried to insert the grid between the anode and the cathode in the existing solid state diode. The imitation of the geometrical form, i.e. of the spacial succession of the electrodes in the tube, was characteristic for this kind of analogy, stimulating the transistor designs in the 1930s. In 1939 there was a split-up in the series of designs, as we have seen. But the difference between the layer principle and the FET principle did not lead to the negation of the geometrical analogy. A manipulation of the charge carriers, similar to that of the grid of the tube triode, had to be realized by a field or by an other process in a layer between the two other electrodes which names still were *anode* and *cathode* in the thirties.

Was the geometrical analogy still maintained in the predecessor of the PCT [fig. 14]? One could defend that the input circuit of the PCT amplifies the output circuit *between* its connecting points. In that case the geometrical analogy is saved. Bardeen, however, considered the design in an other way when he described the amplification of this design. For him the golden plate (for which he used the letter "A") functioned as a cathode and the tungsten point (for which he used "B") as a anode. With the "(sic)" in his notebook he expressed his surprising that the functional analogy was realized in a 'wrong' way:

"The explanation is believed to be as follows. When A is positive, holes are emitted into the semi-conductor. These spread out into the thin p-type layer.

Those which come in the vicinity of B are attracted and enter the electrode. Thus A acts as cathode and B as a plate in the *analogous* vacuum tube circuit. The ground corresponds to the grid, so the action is *similar* to that of the grounded grid tube. The signal is introduced between A (the cathode) and ground (grounded grid). The output is between B (the plate) and ground. *The signs of the potentials are reversed from the (sic) those in a vacuum tube* because conduction is by holes (positive charge) rather than by electrons (negative charge). The *analogy* was suggested by W. Shockley".[37]

After the invention of the PCT, the terminology of 'anode' and 'cathode' was replaced by 'emitter' and 'collector' because in solids not only electrons but also holes could function as the charge carriers. This implied that the function analogy was realized in a 'wrong' way and this again implied one of the difficulties to replace the tubes by transistors in the existing telephone networks. However, these analogies - together with the growing insights in the S- and the other Te-factors - led via the series of designs to the successful invention.

The analysis of analogies is not elaborated further here. For more thorough logical and methodological analysis I refer to other publications.[38] To avoid misunderstandings two remarks are necessary:

(a) The two types of analogies that were mentioned in practice - it appears - are confused easily. Figure 19 gives an example of this. Therefore it is crucial to distinguish the analogies from a *logical* point of view, as I have done elsewhere.

(b) Besides the two aforementioned kinds of analogies at least two other kinds have to be distinguished: the familiar analogy (whose similarities are weak from a logical point of view) and the isomorphy (whose similarities are strong from a logical point of view).

(c) The geometrical analogy played an important role in the architectural designing of Leonardo da Vinci, Le Corbusier and Van der Laan. The problems of this application can not be worked out here.

10.2. M-, P- AND J-ANALOGIES

Decisions about the M-factors are often dominated by arguments by analogy. It is difficult for a company to enter a market where it does not have yet stabilized sales channels or where it meets new protected markets. One of the arguments to stop the aforementioned home cineac project was that Philips did not deliver *similar* products which presupposed contacts with branches like that of the newspapers and of the films. If a company or a division of a company is active on extremely different markets then the organizing splitting up of the responsibilities becomes necessary: too weak analogies between the markets are not tolerated.

Also in the juridical contexts arguments by analogy play an important role. Sometimes some of our young researchers establish that there does not exist any juridical rule or jurisprudence with regard to the new product whose chances they are studying. But they forget that those rules can be applied by analogy. An interesting example is formed by the juridical aspects of computer science. In the Netherlands the entering into a computer system becomes considered as the entering of a house; this analogy leads to the possibility to apply all the laws against the disturbance of domestic peace on the entering into computer systems.

The logical and methodological analysis of these kinds of analogies and norms is important; but it was not yet successful up to now.

11. In practice, STeMPJE reduces complexity

Perhaps the preceding introduction gives the impression that it is not easy to apply the 'STeMPJE' approach in practice. I have to admit that I was hesitative when we started our first projects. Later on this appeared to be unneccesary; several projects have been finished successfully and the ongoing projects proceed well. When applied to concrete subjects STeMPJE reduces the complexity of their situation for clients and the analysis is easier to carry out for the researchers than the preceding introduction suggests. I will limit my illustrations to one example.

From the very beginning I expected that 'STeMPJE' analyses of Stirling engines would yield interesting results. Scientifically and technologically they are interesting: they work economically, quiet and clean. I wondered to what extent M-, P- and J- factors had limited the development of those engines. Marc de Vries has now finished a historical analysis to clarify the concepts[39]. A project with a practical aim has also been finished now. This project had ben set up with Stirling Cryogenics and Refridgeration, a small company with about 150 co-workers. In first contacts I was constantly corrected by the director, because I kept speaking about Stirling engines, while the company did not deal with car engines, but refridgerating machines. The purpose of those corrections became clear to me later on: the director wanted to avoid that his merchant's practices would be related to the S-, Te- and M-factors of Stirling engines. The negotiations lead to a simple formulation of the assigmnment of the project: what is the chance of success for the development of predesigns for the company. The project has been carried out by Marc de Vries and a young researcher.

The more technological-scientific aspect of the roject consisted of a systematic comparison of the knowledge and know-how of the company with competitors: the state-of-affairs with respect to S- and Te-factors and the possibilities of their development. J-factors were also included: the favourable or unfavourable circumstances because of patents. Then the various market sectors were studied: home refridgeration, commercial refridgeration, refridgeration with food processing, air conditioning, etc. In the analysis of the environment legislation special attention was paid to regulations that had been announced in Brusels with respect to CFC's that affect the ozone layer, and with respect to successors of CFC's that stimulate the greenhouse effect. This study yielded a clear list of requirements. This intermediate result was compared with other results and the comparison led to the following result: the company would beat all competitors because in the potential of developing one specific design, that scored highly with respect to requirements of those market sectors that ask for cooling at temperatures of lower than -30 °C. Usually the situation in design departments is complex, because the S- and Te-factors allow the development of several predesigns. The 'STeMPJE' approach reduced this complexity by answering the question which predesign was most relevant with respect to the other factors. The formulation of the final outcome was not unconditioned: a condition for the chance of success remained the political-juridical development with respect to the environmental legislation. The company can follow these developments; the project has made clear which information sources are relevant. Another condition was formulated as a question with respect to the sales channels and the capture of a reputation: is cooperation with another company that directs itself toward other market sectors necessary to acquire this?

This way STeMPJE leads to a systematic analysis and reduction of complexities. In her application this approach is contrary to holistic approaches that require responsabilities for the design to be taken by individual members of an multidisciplinary team of experts. The problems in designing that ask for a intelligent and creative solution are then shifted from the shop-floor to a forum with burocratical decision making.

References

Introduced abbreviations:

PAD = Philips Archive Document
PTR = Philips Technical Review

References

Atherton, W.A. (1984): *From compass to computer*, San Francisco: S.F. Press.

Bardeen, J. (1979): 'The early days of the transistor', in: Raju, G.V.S., ed.: *Proceedings 1979 Stocker symposium*, pp. 3-10.

Bos, H.J.M. (1986): 'Introduction', in: C. Huygens, pp. 11-29.

Brattain, W.H. (1968): 'Genesis of the transistor', *The Physics Teacher* (March), pp. 109-14.

Braun, E. & MacDonald, S (1982): *Revolution in miniature; the history of semiconductor electronics*, Cambridge: UP.

Bull, G.V. & Murphy, C.H. (1988): *Paris Kanonen - the Paris guns (Wilhelmgeschütze) and project HARP*, Herford: Mittler.

Casimir, H.B.G. (1966): 'Science and industry', in: Casimir & Gradstein, pp. 75-80

Casimir, H.B.G. (1970): 'De maatschappij in de maalstroom van de wetenschap', *Wending* 25, pp. 499-510.

Casimir, H.B.G. (1979): *De kringloop van natuurkunde en techniek in de 20e eeuw*, Haarlem: Hollandsche Maatschappij der Wetenschappen.

Casimir, H.B.G. (1983): *Haphazard reality; half a century of science*, New York: Harper.

Casimir, H.B.G. & Gradstein, S., eds. (1966): *An anthology of Philips research*, Eindhoven: Philips.

De Haan, E.F. (1962): 'The Plumbicon, a new television camera tube', *PTR* 24, pp. 57-58.

Demuth, J.E. (1990): 'Revolution in surface science: from reciprocal space to real space', in Sarlemijn & Kroes, pp. 57-73.

De Vries, M.J. (1993): 'The Philips Stirling engine development: a historical-methodological case study into design dynamics', *Methodology and Science* (in print).

Durbin, P.T. (1988): *Technology and contemporary life*, Dordrecht: Reidel.

Galilei, G. (1638): *Dialogues concerning two new sciences*, New York: Dover, 1954.

Gerrits, G.C. (1948): *Grote Nederlanders bij de opbouw der natuurwetenschappen*, Leiden: Brill.

Heijne, L. (1960): *Photoconductive properties of lead-oxide layers* (thesis), Amsterdam: Free University.

Heijne, L. (1991): Letter to A. Sarlemijn and M.J. de Vries.

Heijne, L., Schagen, P & Bruining, H. (1954): 'An experimental pickup tube for television with application of photoconduction', *PTR* 16, pp. 43-54.

Heisenberg, W. (1948): 'Der Begriff der "abgeschlossene Theorie" in der modernen Naturwissenschaft'; reprinted in: Heisenberg, W. (1984): *Schritte über Grenzen*, München: Piper, pp. 73-80.

Hoddeson, L. (1977): ' The roots of solid-state research at Bell Labs', *Physics Today*, 30 (March), pp. 23-30.

Hoddeson, L. (1981): 'The discovery of the point-contact transistor', *Historical Studies in the Physical Sciences*, 12 (1), pp. 41-76.

Hoddeson, L. (1981a): 'The emergence of basic research in the Bell Telephone System, 1875-1915', *Technology & Culture*, 22 (July), pp. 512-44.

Hoddeson, L. (1990): 'Innovation and basic research in the industrial laboratory: the repeater, transistor and Bell Telephone System', in: Sarlemijn & Kroes, pp. 181-214.

Hogarth, C.A. (1973): 'The transistor - its invention and its current prospects', *Physics in Technology*, 4, pp. 173-186.

Huygens, C. (1934): *L'horloge à pendule ou à balancier de 1666 à 1695; anecdota* (= Oeuvres complètes de C. Huygens, tome XVIII), La Haye: Nijhoff.

Huygens, C. (1986): *The pendulum clock or geometrical demonstration concerning*

the motion of pendula as applied to clocks, Ames: Iowa State UP.
Inglis, A.F. (1990): *Behind the tube, a history of broadcasting technology and business*, Boston: Focal Press.
Johnson, V.A. (1969): *Karl Lark-Horovitz, pioneer in solid state physics*, Oxford: Pergamon.
Kelly, M.J. (1950): 'The Bell Telephone laboratories - an example of an institute of creative technology', *Proceedings of the Royal Society*, 203-A, pp. 287-301.
Kilby, J.S. (1976): 'Invention of the integrated circuit', *IEEE transactions on electron devices* ed-23 (7), July, pp. 648-657.
Krafft, F. (1970): *Dynamische und statische Betrachtungsweise in der antieken Mechanik*, Wiesbaden: Steiner.
Krafft, F. (1982): *Das Selbstverständnis der Physik im Wandel der Zeit*, Weinheim: Physikverlag & Chemieverlag.
Kroes, P. & Bakker, M., eds. (1993): *Technological development and science in the industrial age*, Dordrecht: Kluwer.
Morton, J.A. (1953): 'Present state of transistor development', *Bell System Technical Journal*, 31 (3), pp. 411-42.
PAD: The story of the Philips laboratory at Eindhoven, 1914-1946.
Pearson, G.L. & Brattain, W.H. (1952): 'History of semiconductor research', *Proceedings IRE*, 50, pp. 1794-1806.
Petritz, R.L. (1962): 'Contributions of materials technology to semiconductor devices', *Proceedings IRE*, pp. 1025-1038.
Petry, M.J., ed. (1993): *Hegel and Newtonianism (Archives Internationales d'Histoire des Idées)*, Dordrecht: Kluwer.
Rinia, H. (1938): 'Television with Disc according to Nipkow and interlinear scanning', *PTR* 3, pp. 289-95.
Rinia, H. & Dorsman, C. (1937): 'Television installation with disc according to Nipkow', *PTR* 2, pp. 72-76.
Rose, A. & Iams, H. (1939): 'The orthicon, a television pickup tube', *RCA Review* 4, pp. 186-199.
Sarlemijn, A. (1987): 'Mechanica van 'const' tot wetenschap', in: Sarlemijn, A.: *Van natuurfilosofie tot technische natuurkunde*, Eindhoven: EUT.
Sarlemijn, A. (1987a): 'Analogy analysis and transistor research', *Methodology and Science* 20 (3), pp. 40-61.
Sarlemijn, A. (1988): 'Newton's pendulum experiment and specific characteristics of his scientific method in physics', in: Scheurer & DeBrock, pp. 113-31.
Sarlemijn, A. (1990): *Het ontwerp, spil van de techniekkultuur* (inaugural lecture), Eindhoven: EUT.
Sarlemijn, A. (1990a): 'Science and technology, present relations in historical perspective', in: Sarlemijn & Kroes, pp. 3-23.
Sarlemijn, A. (1992): *Technische fysica in ontwikkeling*, Eindhoven TU.
Sarlemijn, A. (1993): 'Pendula in Newtonian mechanics', in: Petry (forthcoming).
Sarlemijn, A. & Kroes, P. (1988): 'Technological analogies and their logical nature', in: Durbin, pp. 237-257.

Sarlemijn, A. & Sparnaay, M.J., ed. (1989): *Physics in the making, Essays on developments in 20th century physics*, Amsterdam: North-Holland.

Sarlemijn, A. & Kroes, P., eds. (1990): *Between science and technology*, Amsterdam: North Holland.

Sarlemijn, A. & De Vries, M.J. (1992): 'The piecemeal rationality of application-oriented research, an analysis of the R&D history leading to the invention of the Plumbicon in the Philips Research Laboratories', in: Kroes & Bakker, pp. 99-131.

Schagen, P. & Boerman, J.R. & Maartens, J.H.J. & Van Rijssel, T. (1955): 'The "scienioscope", a new pickup tube for television', *PTR* 17, pp. 173-182.

Scheurer, P.B. & DeBrock, G. (1988): *Newton's scientific and philosophical legacy*, Dordrecht: Kluwer.

Schuurmans, M.F.H. (1990): 'Predicting properties of materials: dream or reality?', in Sarlemijn & Kroes, pp. 33-56.

Shockley, W. (1949): 'The theory of p-n junctions in semiconductors and p-n junction transistors', *Bell System Technical Journal*, 27, pp. 435-89.

Shockley, W. (1950): *Electrons and holes in semiconductors with applications to transistor electronics*, Princeton NY: Van Nostrand.

Shockley, W. (1952): 'Transistor electronics: imperfections, unipolar and analog transistors', *Proceedings IRE*, pp. 1289-1313.

Shockley, W. (1952a): 'A unipolar "Field Effect" transistor', *Proceedings IRE*, pp. 1365-1376.

Shockley, W. (1973): 'The invention of the transistor, an example of creative-failure methodology', *Proceedings of the conference on the public need and the role of the inventor*, June 11-14, Monterey, Calif (National bureau of standards special publication 388), pp. 47-89.

Shockley, W. (1973a): 'Creative failure methodology', *Electronics & Power*, February 22, p. 59.

Shockley, W. (1976): 'The path of the conception of the junction transistor', *IEEE Transactions on electron device*, ED-23 (7), July, pp. 597-620.

Snow, C.P. (1969): *The two cultures, and a second look; an expanded version of the two cultures and the scientific revolution*, Cambridge: UP.

Teal, G.K. & Sparks, M. & Buehler, E. (1952): 'Single crystal germanium', *Proceedings IRE*, pp. 906-9.

Teal, G.K. (1976): 'Single crystal of germanium and silicon - basic to the transistor and integrated circuit', *IEEE Transactions on electron devices*, ed-23 (7), July, pp. 621-639.

Weiner, C. (1973): 'How the transistor emerged; a technical invention was aided by social inventions which translated basic physics into practice', *IEEE Spectrum*, January, pp. 24-33.

Yoder, J.G. (1988): *Unrolling time, Christian Huygens and the mathematization of nature*, Cambridge: UP.

Zworykin, V.K. (1935): 'The iconoscope', *Proc. IRE* 22 (January), pp. 16-32.

Zworykin, V.K. & Morton, G.A. (1954): *Television, the electronics of image transmission in color and monochrome*, New York: Wiley.

Notes

1. I thank dr. Marc de Vries for his invitation to present a new version of Sarlemijn (1990).

2. For the meanings of 'science and 'art' in the Middle Ages see Sarlemijn (1990a).

3. Snow (1969) is its 14th edition.

4. For the discussions on this topic, I have to thank prof.ir. M.A.W. Scheffelaar (of the RMA) and prof.dr.ir. J.H. Hendriks (of the RAN).

5. Galilei (1638), pp. 251-55; Sarlemijn (1987) and (1992).

6. Bull, G.V. & Murphy, C.H. (1988).

7. I dealt with this subject in Sarlemijn (1988) and (1993).

8. This expression is used in Casimir (1983), pp. 26-38 and Heisenberg (1948).

9. Schuurmans (1990).

10. Shockley (1973), pp. 82-3: "My position at that time was that we could do adequate scientific research by cutting speciments from the relatively large crystals that appeared naturally in the polycristalline ingots resulting from solidified melts". Teal "reminisces about fears impareted to him by management that his obstinacy might cause him to loose his job."

11. In 1952, after mentioning Teal (and Little), he recognized: "For the last few years, practically all advances at Bell Telephone Laboratories in transistor electronics and transistor physics have been based on the availability of single-cristal material", Shockley (1952), p. 1291.

12.Casimir (1983), p. 32.

13.Only very specific subjects can be dealt with mathematically; this holds for the discipline of the design analogies; see section 10.

14.Bos (1986), Gerrits (1948), pp. 92-129 and Yoder (1988), esp. 130-147.

15. Bos (1986), Gerrits (1948), pp. 92-129 and Yoder (1988), esp. 130-147.

16. Hoddeson (1977), Hoddeson (1981), Hoddeson (1981a), Hoddeson (1990).

17. Q.b. Braun & MacDonald (1982), p. 37.

18. Pearson & Brattain (1955), Hoddeson (1981).

19. Bardeen (1979), p. 7.

20. Q.b. Braun & MacDonald (1982), p. 42.

21. Hoddeson (1981), p. 75, Atherton (1984), p. 241, Hogart (1973), p. 173-86.

22. Shockley (1976), 612.

23. Braun & MacDonald (1982), p. 88. This judgement is not in agreement with their more sophisticated account of the three stages of the transistor introduction on pp. 182-3.

24. Morton (1952), p. 412.

25. Kilby (1976), Teal (1976).

26. *Fortune*, Nov. 1961, p. 226; q.b. Teal (1976), p. 635.

27. Sarlemijn & De Vries (1992), Sarlemijn (1992).

28. Schagen, Bruining, Francken (1951), p. 73.

29. Inglis (1990), pp 286-89.

30. (PAD), p. 95 and p. 94.

31. Heijne (1991).

32. For the future of physics, his name is perpatuated by expressions like 'Casimir Operator', 'Casimir Effect', 'Onsager-Casimir Relations', and so on. Sarlemijn & Sparnaay (1989) deals with developments in the 20th century physics and with the role which Casimir played in them.

33. Casimir (1966), p. 86.

34. Casimir (1970).

35. Can this opinion be considered as a confirmation of what we established in 5.1: a technical effect, achieved after many strenuous efforts and clever thoughts, time and again can be derived from fundamental principles in the mathematically elegant deductions of theoretical physicists? In a certain sense it is, but Casimir is not neglecting the engineers's efforts and know-how.

36. Casimir (1966), p. 88.

37. Q.b. Hoddeson (1981), p. 75, italics added.

38. Sarlemijn & Kroes (1988), Sarlemijn (1987a), Sarlemijn 1993a).

39.De Vries (1993).

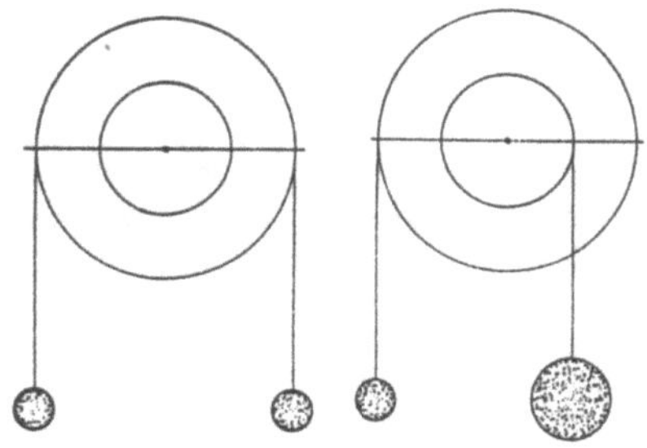

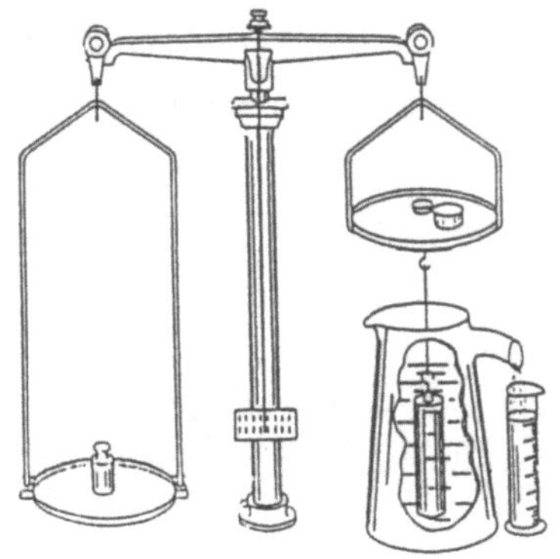

left
pulleys

right
scales

Figure 1 & 2: Aristotle was the first to formulate the law of levers. The application he thought of were the pulleys that are shown left. Finding the law was based on an idealzation: the friction at the point of action was left out. Archimedes made a relation between Aristotle's statics and hydrostatics by means of the test shown right.

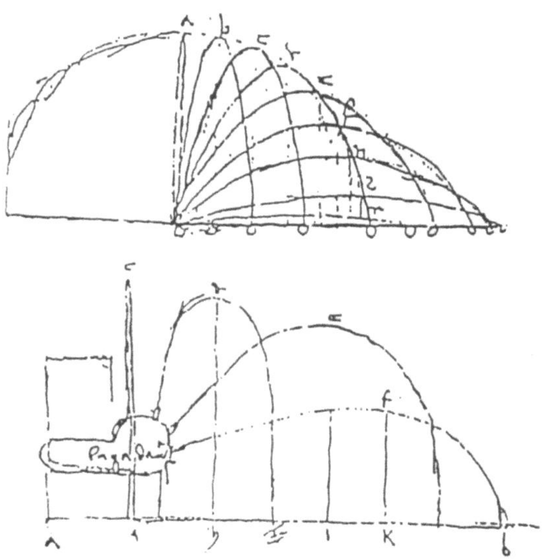

Fig. 3: Leonardo da Vinci already drew parabola shaped projectile trajectories. The lower drawing shows an experimental demonstration: water is pressed out of the pipes of a water bag; the water streams according to parabolas. The drawings are printed in retrospect because that was the way Da Vinci made his drawings.

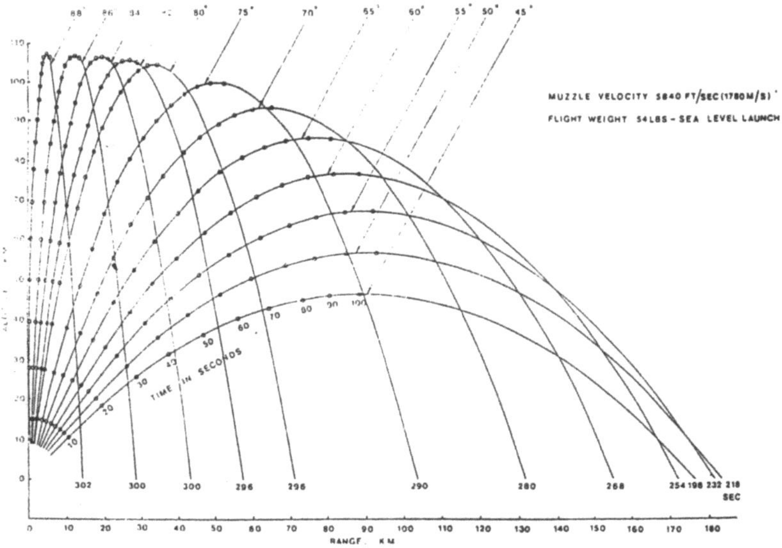

Figure 4. Trajectories of 20th century missiles (with permission derived from Bull & Murphy (1988)).

Figure 5. The Medieval 'mathematical' model for buller trajectories. This drawing has been derived from D. Santbech [1561]: *Problematum astronomicorum et geometricorum sectiones septem*, Basel.

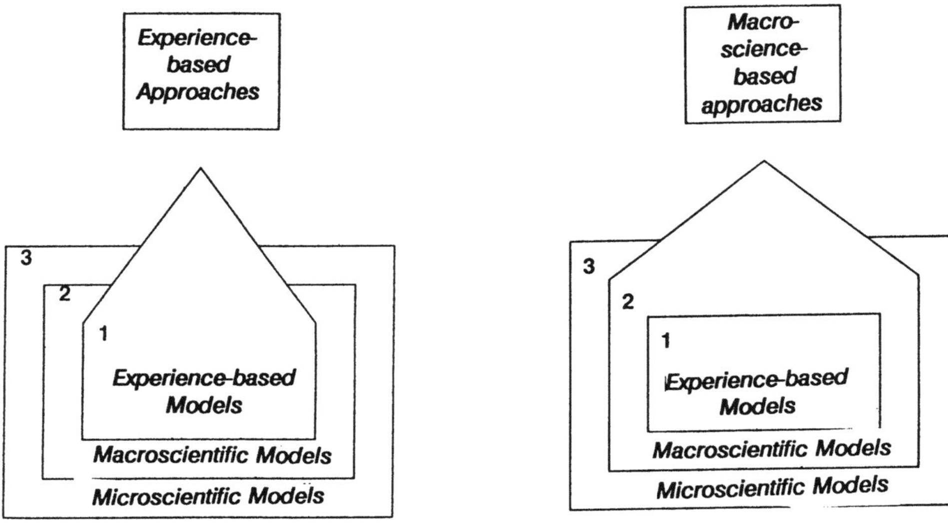

Figure 6 & 7: Divergence between science and experience-based technology (left) and the divergence between science and macrotechnology (right).

Axioms, or laws of motion

Book I: The motions of bodies

Book II: The motions of bodies
(in resisting mediums)

Book III: The system of the world
(in mathematical treatment)

Figure 8: The main parts of Newton's *Principia*

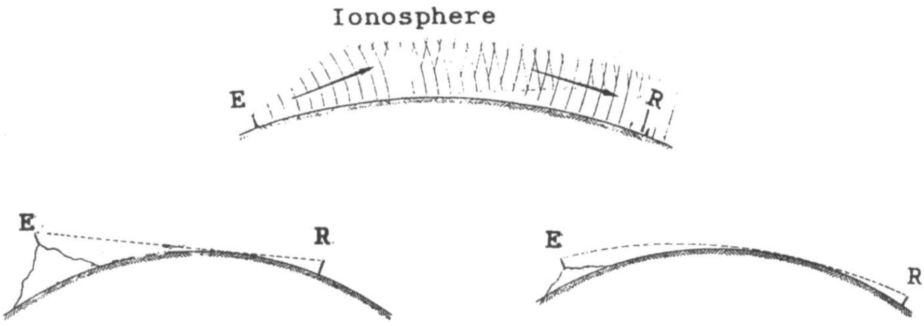

Figure 9. The radio waves are reflected by the ionosphere, as shown in the upper drawing. The range of the shorter television waves between transmitter E and receiver R is determined by the optical horizon (lower left) and the refraction of the electromagnetic waves (right).

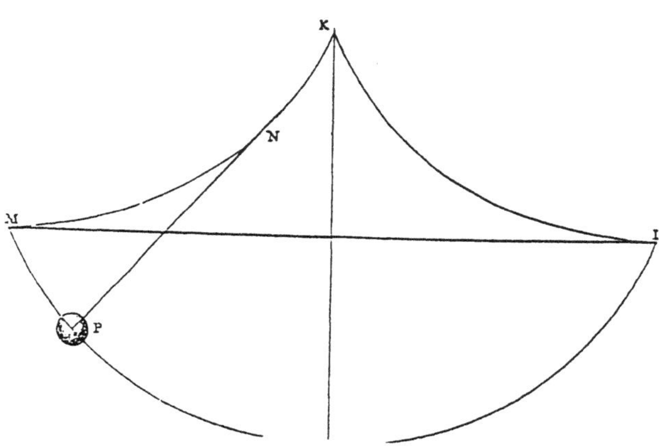

Figure 10. The cycloid path of a bob according to Huygens's drawing.

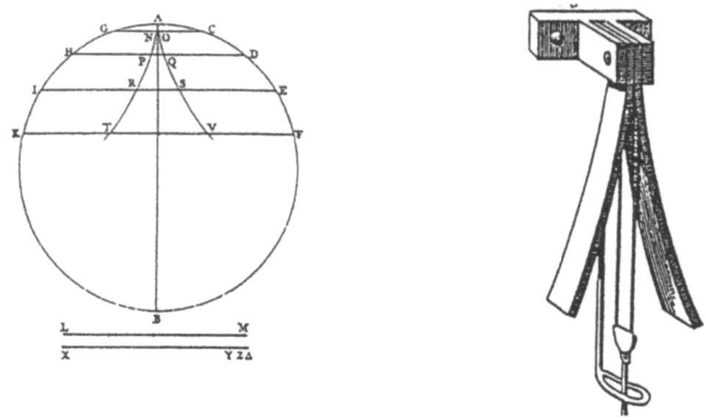

Figure 11. pendulum clock Huygenbs derived the function of the cycloid curves for the pendulum (drawing right) from a geometrical proof for which purpose he used the drawing shown left (drawings derived from Huygens' Oevres Completes).

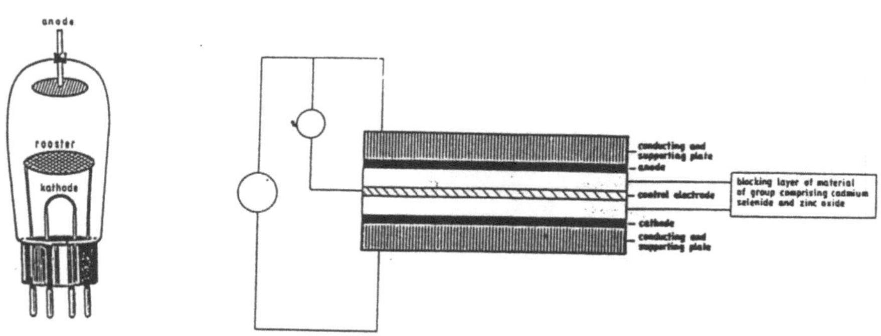

Figure 12. On the left side the troide tube; on the right the layer transistor design by Glaser, Koch and Voigt in 1939. Clearly recognizable is the geometrical analogy on which the layer principle was based: in the solid state three layers replace the cathode, grid and anode in the same spatial order as in the vacuum tube.

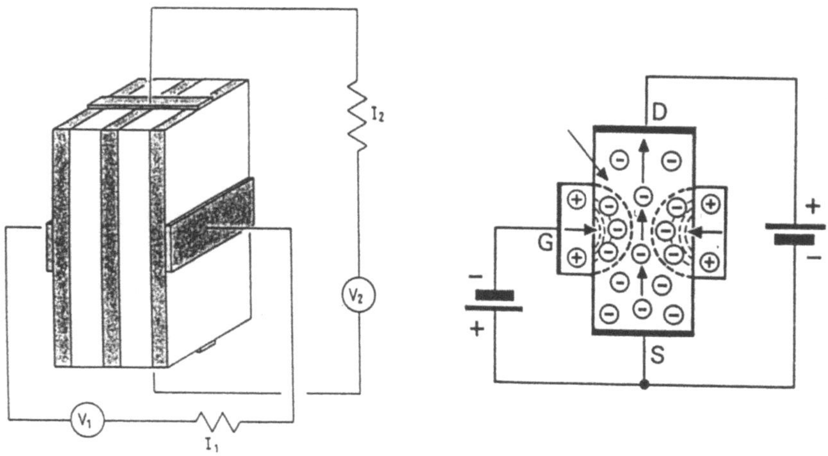

Figure 13. On the left Shockley's FET design in 1939. The illustration on the right shows the principle of a modern FET, in which the channel is formed by p-type silicium. Electrons flow from sourse S to drain D. The input circuit influences the electron flow via gate G, which is p-type conducting. In his explanation for the chosen terminology Shockley (1952), p. 1368 points to the analogy: "The choice selected is 'sourse' for the electrode through which the cariers flow into the channel, 'drain' for the electrode into which the carriers flow out of the channel, and 'gate' for the control electrodes that modulate the channel. One reason for selecting 'gate' . . . is that the subscript 'g' reminiscent of 'grid' and the analogy is close between the two".

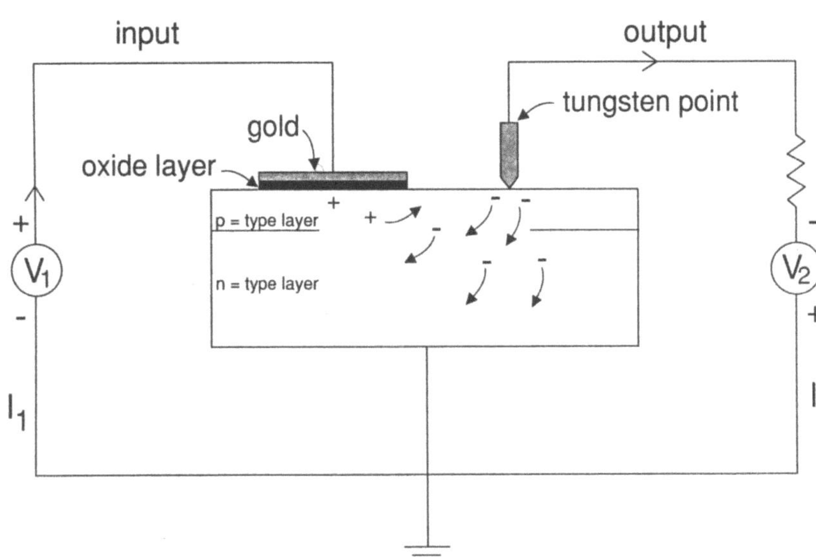

Figure 14. Bardeen and Brattain designed this predecessor of the point-contact transistor on December 11, 1947.

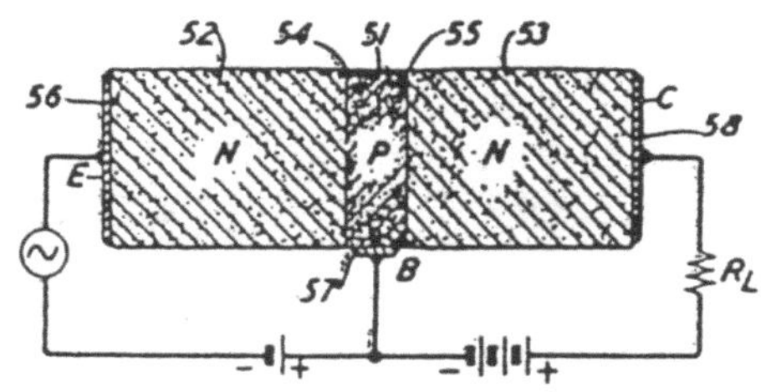

Figure 15. Drawing of the n/p/n junction transistor derived from the patent application. Shockley started to work on the layer principle of the junction transistor at December 31, 1947.

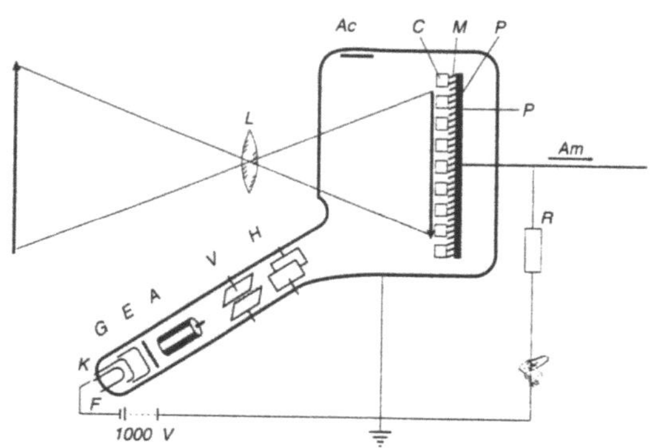

Figure 16. Zworykin's iconoscope: according to a line system plates V and H send an electron beam to the cesium grains C at the isolating mica plate M. A_c is the anode.

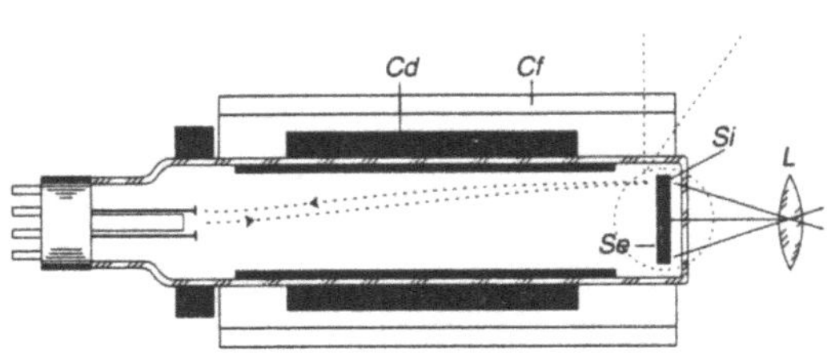

Figure 17. The vidicon; is signal plate Si is a conductor, its second plate Se is a semiconductor.

<u>C.O.B.-GRAM</u>

U R G E N T

Date: 5/3/48
Time: 9.15

Lopes Cardoso - Try to stop developmentwork broadcast-
receivers and concentrate all efforts on television stop
Television is our biggest chance stop
Protelgram is allright but we have to work on followup like
hell! stop
Write on all doors and walls and blackboards <u>TELEVISION</u> stop
Make everyone televisioncrazy stop
We have enough people to do the job but most of them work on the
wrong items stop
There really is only one item: <u>TELEVISION</u> stop
The only actual televisionfront we have at the moment is right
here in U.S.A. stop
We are able to force it if we are ready to fight AND TO KEEP
FIGHTING! stop
Mobilise Eindhoven please stop
No time to lose TELEVISION IS MARCHING ON HERE AND FROM HERE OVER
THE WHOLE WORLD stop
The only question is: WHO MARCHES ON WITH TELEVISION, PHILIPS OR
THE OTHERS? stop
THE OTHERS <u>ARE</u> ALLREADY MARCHING! P H I L I P S E I N D H O V E N,
T A K E T H E L E A D ! fullstop

Bouman

Figure 18. Bouman's telegram on March 15, 1948.

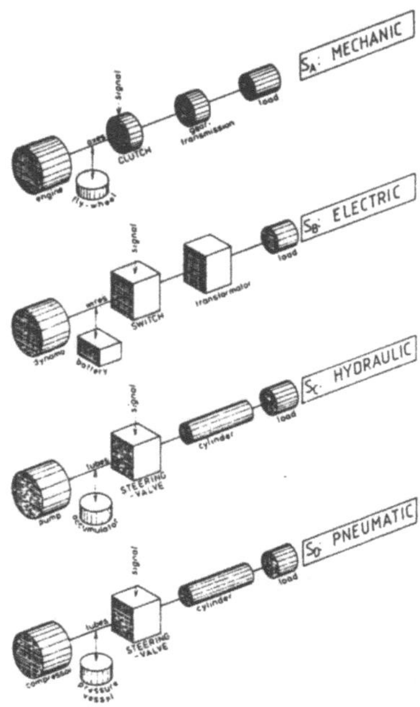

Figure 19. Adapted figure of four systems, with permission derived from a course 'motor car engineering' of a Dutch teacher training. The drawing suggests a similarity between the following elements: source of energy, control signal, transmission, load, energy storage and energy transport. The course also points out the mathematical analogy between the physical laws that apply to the systems. A research revealed that the figure causes confusion with the students. They know that valve and switch are not necessarily positioned the same way. Indeed there is only a functional analogy and not - as the drawing suggests - a geometrical analogy. Less than 50% of the students knows that the validity of the mathematically simlar physical laws are not based on the geometrical analogy between the systems. After a one hour explanation 85% of the students understood that the functional analogy must be distinguished from the fysical laws. From this it became evident that the analogies need to be distinguished clearly.

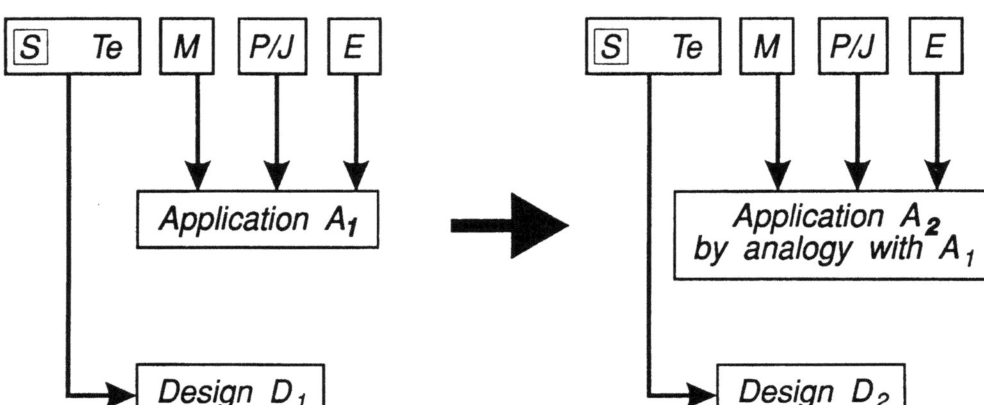

Figure 20. The 'STeMPJE' approach.

METHODS FOR MADNESS: FORMALIZATION AND AUTOMATION OF GENERATIVE PROCESSES IN 'CLASS 1' CREATIVE DESIGN

M.D. ECKERSLEY
University of Minnesota
USA

ABSTRACT. Creative design is a non-deterministic but goal-directed process of search through a broad problem space of hypothetical artifacts. Creative design is an understudied topic in design science and computer science. The paucity of knowledge about creative design contributes to certain fundamental limitations in formal design models. Procedural computing principles offer a powerful means to facilitate discovery and innovation in knowledge-based design. Issues surrounding the development of procedural design systems are elaborated.

1. Introduction

Increasingly, the "special purview" of design science is tied to research and advancements within computer science. The field of artificial intelligence research in design stands, by virtue of its growth and strategic vigor, in a position to dictate if not envelope the future of design research generally. But to some degree this has long been the case, at least since the early engineering and architecture computing applications of the 1960's. This paper will argue that design science suffers a certain shortsightedness resulting from the way in which computing theory and technology developed historically. Because a radical sea change in computer science is now occurring in the form of massive parallelism, design science now may reassess its long held biases toward qualities of non-determinism and creativity in design. The nature of contemporary change in computer science will be reviewed, and advance indication of massive parallelism's impending effect on design computing will be discussed.

1.1. BACKGROUND: COMPUTING AND DESIGN

After 1950 computer science conformed to the implementation architectures of von Neuman and others for the calculation of, for instance, ballistic trajectories

M. J. de Vries et al. (eds.), Design Methodology and Relationships with Science, 249–266.

and hydrodynamics. During this period, electronic computing was defined by long established methods of manual calculation and well known algorithmic procedures. James Bailey (1992) writes that, "The first electronic computers of the 1940s succeeded so quickly because they copied the sequential architecture of human computers. In so doing, they inherited all the sequential ways of expressing and formulating science that had developed over twenty-five hundred years, a period in which computers (human) shaped science far more than science shaped computers" (P. 67). Thus, from Aristotle to Decartes and to the present, the premium placed on order and the sequential has been remarkable and largely dictated by the limited short-term memory faculties of the human mind.

When called upon to model important features of physical phenomena and human problem solving, AI programs were developed around the solution of partial differential equations. The simple yet powerful conceptual device known as the Turing machine (Turing, 1937) led to the complex and arguably intelligent implementations currently in use. However, it was not until the late 1970s when the physical limitations of conventional computing architecture became apparent. Realization that von Neuman-type serial processing is not the necessarily the optimal implementation of the primal Turing machine, especially for modeling complex non-linear phenomena, came as a shock to many people. Nevertheless, it is increasingly clear that the overwhelming emphasis on linear, sequential symbol processing effectively blinded computer science from the prospect of potentially better computing models (Hillis, 1985).

Design science came into being during the 1960s, a period of significant expansion in computer science. For this reason it is hard to imagine how design science might have developed independent of the electronic computer's profound influence in science and engineering. The literature of design science has long evidenced a strong engineering perspective on design research. The somewhat shorter history of artificial intelligence (AI) research in design--a distinguishably separate movement drawing researchers mainly from computer science and engineering fields--has shown a similarly circumscribed engineering focus on knowledge based design. The result of which has been a disproportionate emphasis on a particular class of design activity over other classes in design science generally. A recent classification of design problems by Brown and Chandrasekaran (1989) points up the emphasis placed by design science on computability as a prerequisite for the scientific study of design. The classification scheme paraphrased in the following is useful and will be employed throughout.

Class 1 Design: Comprising open-ended 'creative' design wherein methods for effective problem decomposition are not readily available, requiring development of original routines usually leading to innovative results. Brown and Chandrasekaran explain that "The average designer in industry will rarely, if ever, do Class 1 design, as we consider this to lead to a major invention or completely new products... This then is extremely innovative behavior, and we suspect that very little design activity is in this class" (P33).

Class 2 Design: Comprising problems in which general and sometimes specific methods for effective decomposition exist and may be applied, but selective aspects of the problem may require original design plans or substantial modification from an existing state.

Class 3 Design: Comprising routine design problems wherein effective methods to deal with potential failures are known and directly applicable, nevertheless requiring plan selection and often complex backtracking.

Traditionally among computer scientists studying design and design scientists developing intelligent computer applications, the sole emphasis has been on Class 3 problems because of the perceived computability of such problems, their deterministic nature, and their tendency to yield concrete, predictable outcomes. They exhibit, for instance, greater specificity, a smaller data set, and therefore a smaller array of variables to symbolically represent. It can also be argued that because Class 3 and some Class 2 design is also more attractive from a direct commercial standpoint, its study represents for AI researchers a greater return on resource investment. Thus, the strong tendency in AI has been to focus on Class 3 design for practical reasons, and effectively ignore less tractable forms of design behavior.

1.2. PROBLEM REPRESENTATION AND NEW ECONOMIES OF SCALE

Problem space is a well known concept suggesting the context of all relevant information bearing on a problem. Borrowing on Newell's (1980) conception of problem space, Brown and Chandrasekaran (ibid.) define the design process as "formally a search...in a very large space for objects that satisfy multiple constraints" (P. 20). Eastman operationally defined problem space under a strict information processing paradigm of human cognition, offering that "the processing that cognition and problem solving involves can be modeled as a series of transformations generating a sequence of information. The total number of states generated by applying all permutations of applicable information to all information states defines the total problem space" (P. 669-670). Problem space is represented operationally in the extent of computing resources required to perspicuously represent intrinsic problem variables. The heuristic of "going specific" evidenced in AI's early shift to intelligent domain applications was motivated far more by pragmatic economic concerns (i.e., the limitation of computing and funding resources) to manage problems space than the inherent "interestingness" of practical applications. Likewise in the area of design-applied AI, the vast majority of resources are devoted to practical problems in which variables (e.g., walls, circuits, accessibility) and procedures (e.g., weighted optimization, backtracking) are well known but not particularly interesting in a conceptual sense. But given the economic considerations involved, it is understandable that few resources in AI are devoted to Class 1 design, which undoubtedly would require costly search of a less determinate problem space over an indeterminate period, having outcomes that are far less predictable.

Until now. The overriding heuristic to converge quickly on an essential problem description is no longer necessarily warranted given the introduction of massive parallel computers. Having memory that is both capacious and cheap, parallel computers, are efficiently adaptive systems that compute through the interaction of great quantities of simple identical processing/memory cells. The concurrent processing capability of such systems improves processing rates exponentially, and system performance qualitatively. These developments have great portent for problem space representation.

Hillis (1992a) writes, "Data parallelism is ... the greatest source of economy of scale in computing. For example, a massively parallel computer may have sixty-four thousand processors that can add together sixty-four thousand numbers in one step..."

The beautiful thing about data parallelism is that the opportunity for doing more than one thing at once increases with the amount of data being processed. The power of the solution scales with the size of the problem. A conventional computer takes twice as much time to process twice as much data. A massively parallel computer can often process twice as much data in the same amount of time by applying twice as many processors (Hillis, 1992a, 4-5).

In other words, the long standing trade-off of program conciseness for processing efficiency is no longer necessary given parallel processing. "If an algorithm is twice as bulky but gets the answer twice as fast, that is a wonderfully good trade-off," writes James Bailey (ibid.). "The essential characteristic of a parallel computer is that it keeps thousands of intermediate results active at once... In particular, algorithms that operate on all the data values at once are now very plausible" (P. 71). Parallel processing has tremendous potential for the "navigation" of complex problem space in search of better designs.

1.3. PROBLEM SPACE AND ITS EXPLORATION

Complementing the above linear, sequential characterization of problem space by Eastman is the classic depiction of problem space by Charles Eames (Neuhart, Neuhart, Eames, 1989). Eames' conception of problem space was defined by three separate though intersecting parameter sets: *client/sponsor* concerns; *user/society* concerns; and *designer* concerns. The intersected space comprises the *critical set* of common issues to which Eames considered a design must be addressed. The value of the Eames' diagram is its simplification of a dynamic process of creative imagination and negotiation occurring under time and resource constraints.

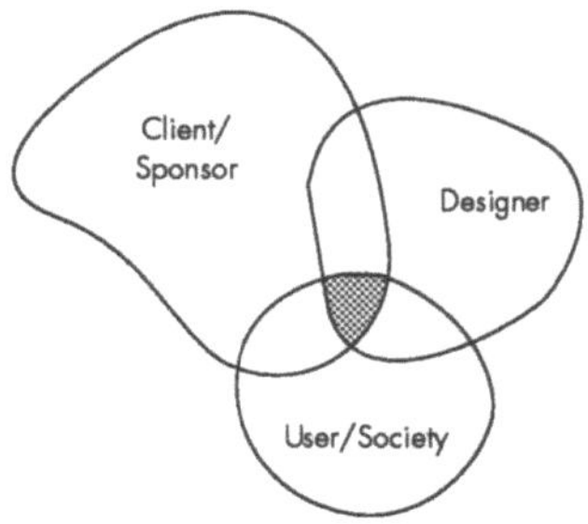

Figure 1. Problem Space Diagram
(after Charles Eames)

Enhanced dimensionally (Eckersley, 1992), Eames' diagram can be visualized as a cluster of interpenetrating volumes--the volumes being formed by nodes, each corresponding to a variable or parameter defined within a given problem. The aggregate dimensions of the cluster represent the conceptual size of the total problem space. The cluster constitutes a conceptual space containing an unlimited set of hypothetical artifacts having both formal/syntactic and semantic qualities. The critical core space shared by the three volumes contains the (still) unlimited set of conceptual artifacts satisfactory to all parties involved. Articulated problems and resulting artifacts inevitably bear the imprint (e.g., knowledge, myths, needs, intentions, motivations, physical structure) of the actors to a design process. But the most important dimensional enhancement of Eames diagram is the introduction of *time*, and with it the inevitable change, growth, contraction, and bifurcation of factors in problem space. How much more meaningful is the concept of design space when view spatio-temporally rather than linearly and sequentially as in Eastman?

Problem representation in design can be defined as the ability to recognize relevant elements in a global problem gestalt. Given the above cellular metaphor, design visualization is process of imaging the set of virtual artifacts (occurring within the critically important core space) that promises problem satisfaction. Design visualization--just as natural visual perception--is enhanced by feedback to input under dynamic conditions (e.g., movement, reaction,

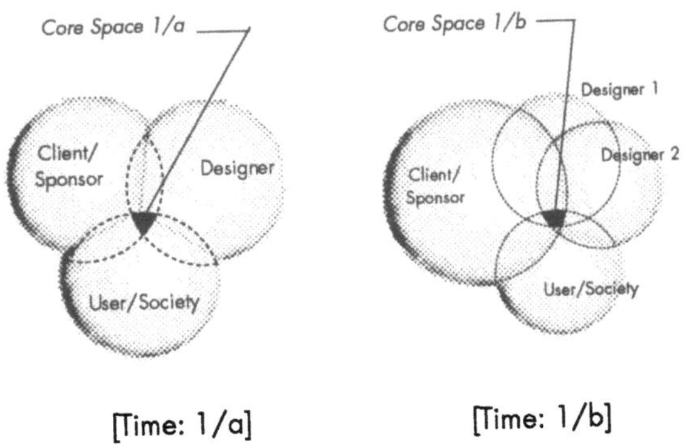

Figure 2. Multi-Dimensional Problem Space

reiteration, alteration, mutation, evolution). Information that enhances the visualization of artifacts in design is highly valuable as it leverages the evolved capacity of the mind/visual system to rapidly test informal hypotheses. Application of this principle is discussed below in section 3.

Visualization of virtual artifacts in problem space can be enhanced significantly by computational means (Eckersley, 1990; 1991). Sims (1991) calls the process *procedural structure creation,* and he approaches it using the metaphor of biological evolution and the creation of artificial "life" forms. Exploiting the tremendous power of *The Connection Machine*, Sims effectively places the user inside problem space, not unlike a video game display with a joy stick for the user to steer by. Thus, problem space has become more concrete than ever anticipated. Procedural design computing, suggested by this and other applications below, extends far beyond conventional forms of knowledge-based design. Its generative, non-linear, and collaborative nature offers the prospect of meaningful computational support for Class 2 and Class 1 design.

2. Creative Design

Early forays into knowledge-based design (Negroponte, 1970) generated interesting results, but ultimately disappointing. However, more recent engineering and planning applications have shown promise in selective instances (e.g., Tommelmein, Johnson & Hayes-Roth, 1987; Maher, 1988), suggesting that conventional AI techniques appear adequate to address central aspects of Class 3 design in which components and their relations can be confidently specified. However, the same techniques are ill-matched for modelling less formalized Class 1 and Class 2 design. A lingering challenge to AI is the possibility that Class 1 design is not wholly *rational*, or that its rationality is bounded in ways that are, at least, unconventional (see Bazjanac, (1975) and Bucciarelli, (1988)). A position commonly held in computer science is that determinism is prerequisite to computation; that belief has served to exclude many apparently non-deterministic activities from serious study.

Creativity is one such behavior that has evaded explication through the lens of formal modeling. Nevertheless the importance ascribed to creativity in the marketplace of ideas, and the way in which innovation tends to change things--thereby enlarging the axiom set and uncovering new predicates--makes creativity an attractive though illusive target for research. Recognition of *creation* as an issue central to the advancement of knowledge in design AI is indicated by the 1992 IFIPS-sponsored conference on "Computational Models of Creative Design".

2.1. THE QUESTION OF ORIGINALITY

Creativity has been variously defined over the years (Stein, 1974; Newell, Shaw, and Simon, 1964) with definitions focusing on one or more of the following: the creative personality, the creative process, and the creative product. Kim (1990) posits:

"*Originality* relates to the novelty of the implementation. If the end result of a design activity is already part of the stock of human knowledge, we may call it a *conventional* implementation, and the associated process of discovery a *self-learning* phenomenon. A design that adds to the human knowledge base, on the other hand, is a *novel* product, and the process *original* (pp. 181-182).

In the sense of Kim's definition, the process of conventional Class 3 design design is different *in kind* from original Class 1 design. That is a direct challenge to Simon's (1973) assumption that structural differences between problem classes are merely *of degree*, and therefore amenable to similar analytical and theoretical treatments. Certainly twenty years of AI literature have not substantiated Simon's claim. Nor has AI offered much insight into creative problem solving generally--suggesting that if the differences were merely of degree, then evidence of greater mutual interconnectivity would have long since appeared.

A judgment of basic structural differences between Class 1 and Class 3 design leads to divergent models for intelligent design systems, and different procedures for operation. For instance, in Class 1 design the best heuristic is not to necessarily to rapidly collapse the problem space to its smallest possible extent, but rather to explore parameters until a promising problem definition or solution path is discovered.

For instance, A postmodern model of design might also borrow on the vehicle of game theory (von Neuman & Morganstern, 1953). Jean-Francois Lyotard (1984) invokes the body of literature on game theory in asserting that there are two different kinds of 'progress' in knowledge: one corresponding to a *new move* within the established rules; the other, to the invention of *new rules*, thus establishing a new game. Like Hillis (1992b), Lyotard references biological systems in asserting that a problem too narrowly defined yields a limited range of solutions, and that a perfectly controlled system is a weak system. By contrast, a problem defined broadly will, by virtue of sheer combinatorial probability, yield a diversity of unanticipated artifacts, most of which will be redundant and unusable. But found within generated "waste" is a small ratio of potentially useful "genetic material", the appropriation of which is vital to transforming and improving the system itself. Constraining a system to produce only appropriate artifacts is inimical to the creative design process. Like a biological system suffering the effects of inbreeding, a shape grammar with limited access to raw random influences becomes functionally bounded.

Game theory offers a scenario of non-linear design--a process of evolving rules and generated outcomes of limitless variety. Viewed as such, design is a dynamic game of induction in which the *means* are known (or invented on the fly) and the *ends* are specifically unknown but recognized for their potential value when seen. This is a process leading from general principles to particular instances. It involves decision-making under conditions of uncertainty from an often wide set of parameters. Chance is a dummy player to the game (i.e., it makes choices but receives no payoffs). The objective of the game is to formulate, by informed trial and error, optimal rule sets (i.e., strategies) that, when played out, yield form of maximal significance and utility to the user(s). The players can be thought of as *meta-designers*. The ultimate opponent is system entropy. The process is focused not on procedural details but on aggregate qualities and the attainment of global objectives.

Such a non-linear system was anticipated in Lenat's (1977) AM and Eurisko systems (1984). The purpose of AM was to discover interesting concepts and then develop conjectures upon them. Criteria of interestingness and heuristics for handling the search were developed which led to a marginally autonomous system that "discovered" for itself a set of fairly interesting mathematical facts. However, unlike mathematics, design-related knowledge is not always explicit and available a priori. For instance, interestingness in procedural design might be defined loosely as the judged promise of hypothetical forms/concepts inferred from existing forms/concepts. For these reasons it can be argued that computer supported Class 1 design problem-solving is, for the time being, better suited to a strategy involving an adaptive learning framework (see Holland, 1992) capable of flexible interaction with the expert use--this in order to learn how the creative designer: (1) operates under conditions of uncertainty, (2) assigns credit differentially to phenomena and (3) discovers rules that advance the design process, exploiting the computer's power and graphics capabilities to visualize progress toward solution. This line of reasoning suggests the attractiveness of a more integrated model of human-computer interaction in design. Such a model centers the task of contemporaneous analysis of graphical feedback and synthesis on the human designer who is privy to the rich universe of semantic relations so fundamental to creative fields of design.

2.2. SYNTAX OVER SEMANTICS

Issues of syntax have long been dominant over semantics in design science and AI. This tendency is typified in Sutherland (1975) and Stiny (1975), and is symptomatic of the course taken by computer science related above in section 1.1. Emphasis on syntactic relations traces back to Chomsky's (1965) operational theory of grammar (of which Minsky [1992] has been highly critical for its failure to address semantic issues). Natural language, like visual language, is a *generative* phenomenon, the complexity of which is connected--but not exclusively--to complexity in the environment. The subtlety of natural semantic associations can be appreciated in Minsky's (1974; 1985) concept of *frames* as a fundamental representational architecture of knowledge, as well as in the concept of *mental models* articulated by Holland, Holyoke, Nisbett, & Thagard (1986).

A common assertion in AI is that semantics is an *emergent quality* of syntactic structure, and therefore indistinguishable from syntax (i.e., a non issue). A comparable argument in Modern design is that form, in fact, *is its own content*, or at least, form offers a semantics all its own. Nevertheless, it appears intuitively obvious that the richness of natural semantic associations can only be formalized crudely in a predicate logic, and therefore, something key is lost between the symbol and its referent (Fodor, 1986; Pylyshyn, 1984).

Building meaning into the computational equation is less important in the context of Class 3 design as constraints are tight and virtually definitive. But the debate over structure and meanings attributed to it in Class 2 and Class 1 design is unsettled as it derives from basic theoretical limitations concerning knowledge. The advancement of knowledge-based design is contingent on some sort of *reconciliation* of this dichotomous issue (Stiny, 1990). Prospects for reconciliation can be found in alternative computing models and implementations, like those offered by Hillis (1985) and Holland (1990), Wolfram (1986), and in Friedhoff (1989), the strategy of which is *not* the (effectively futile) reconstruction of natural human intelligence and perception from scratch, but rather the substantial *augmentation* of such by the development of critically relevant theory and technology. A plausible strategy is to integrate computing more intuitively into the design process, thereby to function as a natural mind extension of the designer for dealing with the functional complexities of design problems (Owen, 1989). This implies greater interaction between the user and system, with the system learning designer preferences in-process and responsively prompting the user to consider alternative paths and patterns.

3. Concretion Of The Conceptual: A Model Of Procedural Design

Creative design is a learning process (Cross, 1983) requiring motivation and persistence on the part of the designer, as well as relative mastery over pertinent rules, concepts, and discrimination skills (Stahl & Webster, 1976). Unknowns are initially plentiful as a problem has yet to be negotiated; the values and criteria for judging the outcome(s) have yet to be clarified. Alexander (1970) describes the process as the reconciliation of two intangibles: a form which is not yet designed, and a context which is not yet properly describable. The overall goal is to derive an artifact informed by previous knowledge but evidencing new knowledge born of a unique process. Visualization of spatial relationships, and contemporaneous feedback to alteration and alternative options are key. Incorporating a *procedural* model into the design process shifts the locus to the graphics generating computer, a mutual and active soundboard for the generation of design propositions. Such a model stands in stark contrast to Stiny and March's (1981) concept of a design machine.

Viewed in this way, design is essentially nothing more (and nothing less) than an iterative process of rule (algorithm) formulation and reformulation when existing rules fail to yield desired results. The aggregate design process can be seen as a cybernetic machine (Weiner, 1948) in which the rules define the capabilities of the machine, selective randomness provides the input, and the design proposition is the output. Once the basics of the machine have been established, the *meta-designer* organizes the machine in such a way as to get satisfactory results. The designer's creativity can be traced to the intrinsic clarity, sophistication, and inventiveness of the rules he/she conceives and implements.

3.1. EXAMPLES OF CLASS 2 AND CLASS 1 PROCEDURAL DESIGN

3.1.1. Class 2 Design. Procedural design principles have wide potential application in Class 2 design in which practicality dictates that a problem be classified relatively early on in the process; deadlines are often looming and general problem parameters are well known. For instance, in product design or graphic design the general form and function of the artifact (e.g., electric shaver, brochure) may be well understood, but *specifics* are unknown and thus amenable to procedural techniques. Time/resource constraints commonly preclude the use of potentially advantageous, but otherwise time consuming search of design alternatives. However, procedural techniques make possible a broad but efficient search of problem space in a compressed time frame, thus diminishing the tradeoff of depth for breadth and yielding more knowledge to the user of pertinent problem parameters, unforeseen solution paths, and formal options that can be exploited in design. Unpromising possibilities are effectively selected out from contention, almost in a Darwinian sense, to ensure the "survivability" of scripts yielding more promising outcomes.

Examples of Class 2 procedural design are increasingly common in the literature and in practice. For instance, Akin's (1992) HeGeL program for the generation of architectural layouts demonstrates aspects of limited procedural design. The defined parameters of Akin's experimental problem were comparable to those of almost any office layout problem, but incorporated into the program were rules derived and later formalized from analysis of designer problem solving protocols. HeGeL succeeded in generating multiple unique layouts comparable those generated intuitively by the human subjects. Larry Keeley (1992) and associates at the Doblin Group in Chicago have developed a proprietary publication layout production system that operates on limited procedural principles. The system orders text/copy and other graphic elements according to selected parameters (e.g., quantity, timeliness, nature of source material, production quality, usage pattern), yielding a suggested layout strategy that is locally unpredictable, but which may inform the eventual design. Such examples grow out of what was originally theoretical and ex post facto research by Koenig & Eizenberg (1981), Kirsch (1986), and others. Indications are that the future looks bright for procedural techniques of rule-based design in the arena of Class 2 problem.

3.1.2. Class 1 Design. Class 1 procedural design has less obvious commercial application, perhaps, than it does theoretical application. Suspicion about open-ended design activity is perhaps nowhere more intrenched traditionally than in the design professions. Closer akin to art-making--where constraints are largely self-imposed--than to engineering design, Class 1 design is far more process-intuitive than formalistic in nature. However, artifacts derived from such a process can offer compellingly original insights and ancillary applications.

As mentioned above, Sims (1991, 1990) and his contemporaries (Dawkins, 1986) approach procedural design from a framework of simulated biological evolution, finding computational counterparts to such concepts as genotype, phenotype, selection, fitness, reproduction, mutation, and sexual combination. The "phenotypes" (artifacts) generated from the process are visually impressive simulations of natural physical and biological manifestations. But of even more implicit importance than the generated images is the powerful concept of using symbolic expressions themselves as genotypes which, in turn reproduce new genotypes. Sims writes: "For evolution to progress there must be variation or mutations in new genotypes with some frequency. Mutations are usually probabilistic as opposed to deterministic." What this means is that the parallel processing Thinking Machine system used by Sims has the means to *learn*--to perpetuate and optimize those symbolic instructions generating preferable outcomes and alternately dampen the effect of those instructions generating less preferable outcomes. Sims writes further :

Consider an array of knobs, each controlling one parameter, that can be experimentally turned to adjust the results. As more options are added to the procedure for more variation of results, the number of input parameters grows and it can become increasingly difficult for a user to predict the effects of adjusting particular parameters and combinations of parameters, and to adjust the knobs effectively by hand.

An alternative approach is to sample randomly in the neighborhood of a currently existing parameter set by making random alterations to a parameter or several parameters, then inspect and select the best sample or samples of those presented. This allows exploration through the parameter space in incremental arbitrary directions without requiring knowledge of the specific effects of each parameter. This is artificial evolution in which the genotype is the parameter set, and the phenotype is the resulting structure. Selection is performed by the user picking preferred phenotypes from groups of samples, and as long as the samples can be generated and displayed quickly enough, it can be a useful technique. (P.3)

The approach taken by the author (Eckersley, 1988, 1990) to Class 1 procedural design is remarkably similar to that of Sims, albeit originating from a different contextual source: artistic illustrative imagery. The model in question was implemented in a conventional production rule system and reflected a similar perceptual selection mechanism to that of Sims, but without the important learning function. The process is guided perceptually by the user in a global sense, but the graphical results may still reflect considerable local unpredictability. An unanticipated aspect of the process, however, has been the occasionally evocative *content* of yielded artifacts. In other words, though the process of procedural design is exclusively syntactic in nature, the user will favor those genotypes and phenotypes that evoke *semantic* associations (e.g., zoomorphism, biomorphic). This phenomenon, not unlike the construing of cloud formations to resemblance of animals, might be called a *semantics of association*. A similar *semantics of perceived utility* has been observed in which the user, guided by the task of generating, for instance, illustrative imagery for T.S. Eliot's poem, *The Wasteland*, will almost certainly be able to construct from the raw graphical data, meaningful imagery. Examples of procedurally developed artifacts generated in the author's software application, Apprentice, are seen in Table 1 and Table 2. Images in Table 1 are edited down from a large series generated for use in a poster composition. Raw graphic elements and selected digitized photographs were manipulated procedurally and later pieced together to form the ultimate composition. Table 2 is a series of selected discrete images some of which were generated completely stochastically, others were pieced together from random chunks of graphical data.

4. Conclusion

The foregoing discussion has attempted to contextualize a gamut of issues: computing theory, process modeling, problem classification, interactive procedural principles, creative processes, and semantics. Such connections are not trivial, but rather have become central to the future of rule-based design computing. The question of meaning should not forever be neglected in models of design computing as meaning occupies a central position in contemporary design thought (Rheinfrank and Welker, 1990). Meaning is individually and collectively attributed to, and negotiated with artifacts in the environment. That process can now begin in the virtual design stage, and should be built into computing models.

"Cookie cutter", design has long had limited emotional appeal to humans. Application of procedural principles to product design, graphic design, and architecture is an answer to sterile and unimaginative modularity in contemporary design. The architectural application of procedural techniques in Seaside, Florida by Plater-Zyberk and Duany testifies to the practicality and utility of such principles.

Design is viewed popularly by professionals and lay persons alike as an inherently creative form-giving activity, a process engaged in singly or collectively to develop original artifacts of value. Unfortunately, functionalism has been so narrowly defined as to exclude the consideration of *values* driving a design, and ways in which artifacts may be more representative of their human makers and users. Design science has shown an unimaginative orthodoxy in focusing almost exclusively on Class 3 and Class 2 problems in domains of product design, engineering, planning, and architecture. Whatever academic respectability the field gains in delimiting itself so, it loses in professional relevance. Simon's (1981) rationale for a science of design lacks consideration of application. Perhaps the general lack of interest evidenced by the professions for design science today (Jacques, 1992) is attributable to the field's early trade-off of priorities. Design science's historical omission of Class 1 design as a serious topic of study, has isolated the field from participation in design disciplines that deal in issues of communication and semantics, novel expression in form and content. If, as the above discussion suggests, design science was formed by the tools it used, perhaps imminent new tools will reform it meaningfully.

5. Acknowledgements

The author wishes to thank Larry Keeley (Doblin Group), Karl Sims (Thinking Machines Corp.), and David Litman (George Mason University) for their assisstance.

References

Akin, O. (1992) 'Heuristic generation of layouts (HeGeL): based on a paradigm for problem structuring', Environment and Planning B: Planning and Design 19, 33-59

Alexander, C. (1964) Notes on the Synthesis of Form, McGraw Hill, New York.

Baily, J. (1992) 'First we reshape our computers, then our computers reshape us: the broader intellectual impact of parallelism', Daedalus 121(1) 67-86.

Bazjanac, T. (1975) 'The promises and the disappointments of computer-aided design', in N. Negroponte (ed.), Reflections on Computer Aids to Design and Architecture, Petrocelli/Charter, New York.

Brown, D. C. and Chandrasekaran, B. (1989) Design Problem Solving: Knowledge Structures and Control Strategies, Research Notes in Artificial Intelligence Series, Pitman, London.

Bucciarelli, L. (1988) 'An ethnographic perspective on engineering design', Design Studies 9 (3), 159-168.

Chomsky, N. (1965) Aspects of the Theory of Syntax, MIT Press, Cambridge, MA.

Cross, N. (1983) 'The relevance of cognitive styles in design education', Design Methods and Theories 17(1), 37-49.

Dawkins, R. (1986) The Blind Watchmaker, Harlow Logman.

Eastman, C. M. (1969) 'Cognitive processes and ill-defined problems: A case study from design', Proceedings of the first Joint International Conference on Artificial Intelligence, Washington, DC. 669-675.

Eckersley, M. (1988) 'Computer graphic analysis of random number arrays: a source of abundant and diverse forms', in CDE'88: Lectures and Writings from Computers Design Education. University of Industrial Arts, Helsinki.

Eckersley, M. (1990) Thoughts on a computer-based design apprentice', Computers & Graphics: Chaos & Graphics 14 (3/4), 511-517.

Eckersley, M. (1991) Factors of meaning, chance, and utility in shape production systems', Journal of Environment & Planning B: Planning and Design 18, 301-314.

Eckersley, M. (1992- in press) 'Generative design systems and the navigation of problem space, Journal of Formal Languages for Word, Image and Sound.

Fodor, J.A. (1988) Psychosemantics: The Problem of Meaning in the Philosophy of Mind, MIT Press, Cambridge, Mass.

Friedhoff, R. M. (1989) Visualization: The Second Computer Revolution, Abrams, New York, NY.

Hillis, W. D. (1992a) 'What is massively parallel computing, and why is it important?', Daedalus 121(1), 1-16.

Hillis, W. D. (1992b) 'Computers as creativity amplifiers', Proceedings of the 42nd International Design Conference (ICDA), June 16, 1992, Aspen, Colorado.

Hillis, W.D. (1985) The Connection Machine, MIT Press, Cambridge, MA.

Holland, J.H. (1992) 'Complex adaptive systems', Daedalus 121(1) 17-30.
Holland, J.H. (1990) 'Emergent models', in A. Scott (ed.), Frontiers in Science, Blackwell, Cambridge.
Holland, J.H., Holyoak, K.J., Nisbett, R.E., and Thagard, P.R. (1986) Induction: Processes of Inference, Learning, and Discovery, MIT Press, Cambridge, Mass.
Jacques, R. (1992) 'Chairman's report', Design Research Newsletter of the Design Research Society, March.
Keeley, L. (1992) Personal communication, September 23.
Kim, S. (1990) Designing Intelligence, Oxford Univ. Press, New York.
Kirsch, J.L. and Kirsch, R.A. (1986) 'The structure of paintings: formal grammar and design', Environment and Planning B 13, 163-176.
Koenig, H. and Eizenberg, J. (1981) 'The language of the prairie: Frank Lloyd Wright's prairie houses', Environment and Planning B 8, 295-323.
Lenat, D. B. (1977) 'Automated theory formations in mathematics', Proceedings of the Fifth International Joint Conference on Aritficial Intelligence, pp. 833-842.
Lenat, D.B. and Brown, J.S. (1984) 'Why AM and EURISKO appear to work,' Artificial Intelligence 23, 269-294.
Lyotard, J.F. (1984) The Postmodern Condition: A Report on Knowledge, Minneapolis: University of Minnesota Press.
Maher, M. (1988) 'HI-RISE: An expert system for preliminary structural design', in Rychener, M. (ed.), Expert Systems for Engineering Design, 37-52.
Minsky, M. (1974) 'A framework for representing knowledge', MIT Artificial Intelligence Laboratory Memo 396, Cambridge.
Minsky, M. (1985) Society of Mind, Simon & Schuster, New York.
Minsky, M. (1992) 'What's the difference', Proceedings of the 42nd International Design Conference Aspen (IDCA), June 15, Aspen, Colorado.
Negroponte, N. (1970) The Architecture Machine, MIT Press, Cambridge, Mass.
Neuhart, J., Newhart, M. and Eames, R. (1989) Eames Design, Abrams, NY.
Newell, A. (1980) Reasoning, problem solving and decision process: the problem space as a fundamental category, Attention and Performance, VIII., L. Erlbaum, 693 - 718.
Newell, A., Shaw, J.C. and Simon, H.A. (1964) 'The process of creativity thinking', in H.E. Gruber (ed.), Contemporary Approaches to Creative Thinking, Atherton, New York.
Owen, C. (1989) 'Design education in an information age', Design Process Newsletter, Design Process Newsletter 4(3), Illinois Institute of Technology.
Pylyshyn, Z. (1984) Computation and Cognition: Toward a Foundation for Cognitive Science, MIT Press, Cambridge, Mass.
Reinfrank, J.J. and Welker, K.A. (1990) 'Meaning', Design Journal 1, American Center for Design, 54-61.
Simon, H. A. (1973) 'The structure of ill-structured problems', Artificial Intelligence 4, 181-201.

Simon, H. A. (1981) The Sciences of the Artificial, MIT Press, Cambridge, (first edition 1969).

Sims, K. (1990) 'Panspermia', Siggraph Video Review.

Sims, K. (1991) 'Artificial evolution for computer graphics', Computer Graphics 25(4) 319-328, ACM Press.

Stahl, R.J. and Webster, N.C. (1976) 'Instructional activities to develop student concept learning and problem solving skills in art education: a proposed model', ERIC Document Reproduction Service, No. ED 138 509.

Stein, M. (1974) Stimulating Creativity, Academic Press, New York.

Stiny, G. (1975) Pictorial and Formal Aspects of Shape and Shape Grammars, Birkhauser, Basel.

Stiny, G. (1990) What designers do that computers should', in M. McCullough, W.J. Mitchell, P. Purcell (eds.), The Electronic Design Studio, MIT Press, Cambridge 17-30.

Stiny, G. and March, L. (1981) 'Design machines', Environment and Planning B 8, 245-255.

Sutherland, I. (1975) 'Structure in drawings and the hidden-surface problem', in N. Negroponte (ed.), Reflections on Computer Aids to Design and Architecture, Petrocelli/Charter, New York.

Tommelein, I.D., Johnson, M.V., Hayes-Roth, B. and Levitt,R.E. (1987) 'SIGHTPLAN: a blackboard expert system for construction site layout', J.S. Gero (ed.), Expert Systems in Computer-Aided Design, Elsevier Science Publishers, IFIP: Amsterdam.

Turing, A.M. (1937) 'On computable numbers, with an application to the entscheidungsproblem', Proceedings of the London Mathematical Society 42, 230-265.

von Neumann, J. and Morganstern, O. (1953) The Theory of Games and Economic Behavior, Princeton, New Jersey.

Weiner, N. (1948) Cybernetics, MIT Press, Cambridge, Mass.

Wolfram, S. (1986) Theory and Applications of Cellular Automata, World Scientific Publishing Co. Pte Ltd., Singapore.

Table 1

Series:
sketches generated procedurally from Design Apprentice for use in development of poster: James Thorpe, poster designer.

Routines & Variables:

> digitized photos: 7,10
> cut & move routine
> type/text routine
> distortion/skew routine
> color inversion routine
> shape series routine
> rotation routine

looping iterations:
12/200
odds: variable, contingent

Table 2

Misc. stochastic 2-D sketches generated procedurally from Design Apprentice

RHETORIC AND THE PRODUCTIVE SCIENCES: TOWARDS A NEW PROGRAM FOR RESEARCH IN DESIGN

R. BUCHANAN
Carnegie Mellon University
USA

ABSTRACT. One of the central problems of design studies is the diversity of views about the subject matter, methods, and principles of design practice. These views stem from different philosophic assumptions and are part of the nature of design thinking. A new program for research in design should include an effort to clarify the diverse meanings of central concepts and the assumptions upon which they are based. This can provide a platform for further research into the peculiar indeterminacy of subject matter in design. This, in turn, could lead to a new type of rhetorical inquiry into the integrative disciplines of design thinking which give direction and purpose to the particular techniques and methods revealed in the study of designing. Investigation of themes such as communication, construction, strategic planning, and systemic integration--each a form of rhetoric that enables the cultivation of abilities such as invention, judgment, decision-making, and evaluation--is a critical task in the development of design theory. But the rhetoric of design thinking should be balanced by a new type of inquiry that is directed towards kinds of products and their related processes of making which constrain designers. This type of inquiry could be called "productive science," in contrast to the forethought that characterizes design.

1. Introduction

An international conference on design and science quickly reveals one of the central problems of design studies that stands as a potential roadblock to productive discussion. The participants in such a meeting hold exceptionally diverse views about what constitutes the proper subject matter for attention, the most suitable method or methods for investigating that subject, and the principles that best serve to test and ground various claims about the subject in objective reality. This is fundamentally a philosophic problem, because different views of design and the relationship between design and science are often based on different philosophic assumptions. Yet, philosophic assumptions are seldom addressed directly and openly in a setting where one might reasonably expect

M. J. de Vries et al. (eds.), Design Methodology and Relationships with Science, 267–275.

such differences to be clarified and alternative points of view brought to bear on common problems of inquiry.

One reason that fundamental beliefs are not discussed is that some participants are unaware of the philosophic dimension of claims about design and are unable, as a consequence, to articulate their own beliefs or perceive the practical significance of beliefs held by others. For these individuals, philosophic assumptions remain tacit and unexamined. In contrast, other participants are, indeed, aware of the philosophic assumptions that guide their work and the work of others, but they regard many of the assumptions made by colleagues as errors to be fought and corrected. However, the strategy is not to confront and consider alternative assumptions on their own grounds. This could "confuse" a group, since the open exploration of alternative ideas and methods sometimes reveals the sense in which a position *is* reasonable and defensible. Furthermore, such discussion may call into question assumptions that one would prefer to leave unchallenged. Rather, the strategy is to overcome other beliefs indirectly, by attacking specific claims about design. If claims that are a consequence of certain philosophic assumptions can be made to seem foolish or misguided, then the assumptions themselves may be foolish or misguided as well. Finally, there are participants who adhere to the philosophy of anti-philosophy. Some colleagues (scholars of design as well as designers) regard talk about philosophy as irrelevant to the practical problems faced by designers. All forms of thought and knowledge should be judged by their practical applications, and the meaningful questions of practice may be distinguished from the utopianism, dogmatism, and formalism of philosophic discussion. The "practical," in this case, means what can be accomplished by persuasion, and persuasion, in the subtle steps of a determined person, degenerates into a kind of force. The practical is what can be *made* to work, without regard to truth or tests of independent validity or ultimate questions about what is better or what is good. Such a view is sometimes defended as "empirical," as if deliberate attention to philosophical questions is not, itself, fundamentally empirical, leading away from, rather than towards, the concrete texture of human experience. Evidently, the meaning of "empirical" is not empirically determined. The scope of the term expands or contracts to include or exclude different subjects according to philosophic assumptions.

How is design studies possible in a situation where individuals hold strikingly different views about the subject matter, methods, and principles of design, yet are unable or unwilling to make their deepest assumptions explicit and enter into a discussion that deliberately seeks to connect philosophic beliefs with practical problems?

2. Design and Science

Little effort is required to sketch the alternative relationships between design and science that have been explored in the twentieth century. The strategies are timeless in their simplicity, echoing similar strategies for considering the relationships between art and science or between science and the practical that continue to be explored in ongoing debate in other fields. Design is reduced to science; design is viewed as essentially different in kind from science; science is reduced to design; or science and design are viewed as an inseparable union of theory and practice carried into productive action and the life of the community. These four types of relationship, expressed even in such terse form, provide the thematic directions that make design studies a pluralistic enterprise of great complexity.

First, design may be regarded as an **intuitive, instinctive activity for meeting human needs** that stands, itself, in need of the application of scientific laws and scientific knowledge to improve its efficiency and effectiveness. The goal is to discover a science of design, recognizing that some part of design may forever remain beyond scientific analysis because of aesthetic factors and the irrational, ever-changing tastes and preferences of human beings.

Second, design may be regarded as an **intellectual discipline of forethought and planning** that is different from science but draws on knowledge gained by the sciences to achieve ends of utility, pleasure, and justice in the everyday world. The goal is not to reduce design to science, as in the first case, but to discover a discipline of design thinking that may use scientific knowledge for practical purposes while also resolving the artistic and moral dimensions of a specific practical problem.

Third, design may be regarded as the **art of operations and performance, the art of the practical and the possible, the art of making things work** to achieve any purpose, serve any interest, satisfy any desire. Within such an art, knowledge is power, and power comes from a clearer understanding of operations and from a clearer understanding of what is to be operated upon. The effort to make design scientific is, from this perspective, an effort to extend the sense in which science is an art of controlled operations; the description of design methods in this context *is* scientific operationàlism. In other words, design is best understood in its most general form as the description or characterization of processes and operations of thought and decision-making such as one finds in the traditional sciences. The extension of scientific description to what are *popularly* known as the design professions is not a reduction of design to science; it is, in reality, a validation of science as design, a recovery of a domain of design thinking that has heretofore eluded proper operational understanding. In a sense, design reasoning is the only valid form of reasoning, and the task is to clarify the operation of design reasoning for those professional designers who work at the boundary zones of special applications. Graphic design, industrial design, engineering design, architectural design, and urban planning merely represent some of the special limitations in the scope of real problems that may be addressed by science.

Fourth, design may be regarded as the **dialectical interplay of science and art in action.** Design is the meeting ground for science, politics, and art. It is a place where doctrines clash, where alternative hypotheses about human nature and "the good" compete in shaping the made-world and the understood-world. All knowledge is relevant to the designer and, in turn, the scientist must be cognizant of the possible uses of scientific knowledge in the practical world. In short, design is a technology--a science of art--that may be used wisely or foolishly in the evolution and development of culture through a series of approximations towards an ideal state or condition. To bring science to design is to force the recognition of diverse values and beliefs in the scientific enterprise; to bring design to science is to force a recognition that knowledge is required for effective design thinking.

These strategies are mutually incompatible, if not also mutually incomprehensible. They are connected only by the ambiguous terms "design" and "science," whose meanings are subject to radically different interpretation as one moves from one line of discourse and exploration to another. Yet, the ambiguity can be productive. One of the reasons for participating in a conference on the relationships between design and science is to gather ideas and observations presented from one perspective and try to reapply them in one's own line of inquiry--with appropriate changes of meaning and application, of course. Thus, even if the philosophic assumptions behind different approaches to design and science are not made explicit, and the actual meanings of the terms are not clarified with precise understanding for each participant, one may yet benefit and leave the meeting believing that time was not wasted. Although mutually exclusive, each of the four approaches sketched earlier has its own rationale for participation. One may discover specific new applications of scientific knowledge or techniques; or further clarify the disciplines of design thinking; or persuade others to enter into one's own version of design studies directed towards this or that agenda; or advance the evolution of the culture of design and science towards a new level of awareness and synthesis of competing ideas and values. All of these are valid reasons for joining in the collaborative project of design studies, although the pluralism of the project ranges from weak and trivial, to strong and complex, to complex and intensely political, to culturally necessary but insufficient for the ultimate synthesis towards which conversation is believed to be moving.

Any effort to sort out and clarify the different philosophic assumptions present in the area of design studies is fundamentally *semantic* in nature. It is the clarification of concepts and methods and their various meanings as used in inquiry. The accomplishment of such a task is not futile, but it may be thankless. There is no neutral framework for characterizing the diverse meanings of concepts and methods. In fact, each participant is likely to possess a schematism for locating the positions of other participants. If semantic investigation is pursued in design studies--as it has been in other fields as part of their maturation and intellectual development--the goal should not be to establish acceptable meanings for the concepts and methods of design, excluding some meanings and certifying others. Rather, the goal should be to clarify the *alternative directions for seeking meaning*, the alternative lines of investigation that lead to different interpretations of terms such as "function," "ergonomics," "useful," "form," "value," "communication," and so forth. While this does not advance any particular agenda, it may become a platform for more productive debate. Clarifying alternative views on the fundamental concepts of design should not be underestimated for its value in supporting further research. Old lines of research sometimes run dry. New concepts or new interpretations of old concepts often help to create opportunities for new inquiry. There is reason to believe that a rich gene-pool of ideas is as valuable in intellectual discourse as a rich gene-pool in biological populations.

More significant, however, is the substantive question of why alternative meanings of fundamental concepts and methods are possible at all in design studies--or in any other field of inquiry. Granted that some of the various meanings and lines of inquiry in design studies have less rationale than others, because of internal contradictions, obvious inconsistencies, or the lack of sufficient objective evidence to support a particular line of thought. There remain many contrasting or even mutually exclusive views of design that are not easily dismissed, except perhaps on grounds of intellectual politics and conflicting agendas. What is there in the nature of design that allows alternative lines of investigation such as the four identified earlier?

This is more than a semantic question, although a semantic analysis of discourse in design studies may help to make the issue more visible, forcing attention on what is surely one of the most important yet ignored empirical facts of the human community: *systematic diversity of intellectual perspectives persists despite the most powerful and successful results of one or another line of thought at a particular moment in history*. This is a problem for substantive inquiry, although the precise form of the question may be shaped and developed in a wide variety of ways. Perhaps what ultimately counts as success in a particular line of investigation of design--in contrast to what is momentarily persuasive or seductive in shaping the popular fashions of design discourse--is a contribution, direct or indirect, towards understanding how the nature of design can support alternative, often incommensurate, lines of investigation.

3. Rhetoric and the Productive Sciences

Questions about philosophic assumptions and diversity in the study of design are not idle. The field of design studies, although no more than a few decades old, is rapidly entering a period of change in which new assumptions and new approaches are likely to proliferate. The philosophic assumptions that have guided the "design methods movement" have not been refuted, and the tangible results of the movement have not been proven invalid. Yet, there is dissatisfaction with this line of inquiry, even among some of those who have pursued it for nearly thirty years. Has the time been wasted? What problem has emerged to require a new direction? Can the study of design methods be revitalized through the introduction of a new perspective, perhaps based on different philosophic assumptions?

The study of design methods is an effort to discover instrumentalities of design thinking to replace intuition, trial-and-error, and other vague practices. It is an effort to give precision to professions that are notably imprecise because of the contingencies of particular circumstances and the influence of individual human perspective, whether manifested in preferences, values, habits, or ideas and ideals. However, the effort has taken two distinct forms. Some investigators have sought to eliminate, to the greatest degree possible, the idiosyncratic influence of the designer. This is an approach through cognitive science that leads to expert systems, artificial intelligence (AI), and, in general, operational models of cognitive processes. In contrast, other investigators have recognized that the perspective of the designer is an important feature of design thinking and have sought to characterize methods that incorporate point-of-view as a resource. Indeed, some of the methods described in this approach involve systematic techniques for *shifting* or *changing* the point-of-view of the designer to reveal new possibilities, opportunities, or neglected factors in the design problem or to incorporate the point-of-view of the potential user of a product or the perspective of the individual or group that manufactures a product. In the area of computerization, this approach has emerged as a concern for 'intelligence augmentation' or 'intelligence amplification' (IA), where computers are used to do what humans do poorly or slowly, freeing humans to do what they do well--invent ideas, select goals and purposes, make novel or meaningful connections among disparate data and events, and evaluate success or failure.

However, what is missing in both of these approaches is, paradoxically, an understanding of the broad strategies of design thinking that guide the uses of particular methods. The study of design methods has yielded many useful tools, but it has not yielded an understanding, except within the limited specification of rules of applicability, of how diverse techniques and methods take on direction and purpose in the hands of a designer. There has been an assumption that in characterizing the tools of designing one characterizes designing itself. But this is a dubious assumption--as if one could understand the nature and purposes of science by describing the instruments of science. True, one may infer many things about science from its instruments--as archeologists infer many things about a lost culture from its tools and implements. But the data are inadequate to the task of full understanding.

This realization seems to lie behind Christopher Alexander's cutting remark in the 1971 preface to the paperback edition of **Notes on the Synthesis of Form.** Commenting on the method of creating diagrams of forces to help resolve design problems, he writes: "[S]ince the book was published, a whole academic field has grown up around the idea of 'design methods'--and I have been hailed as one of the leading exponents of these so-called design methods. I am very sorry that this has happened, and want to state publicly, that I reject the whole idea of design methods as a subject of study, since I think it is absurd to separate the study of designing from the practice of design."

This is a distressing statement for those who study design methods, and it is all the more difficult to understand since Alexander does not repudiate the design methods described in his book. At worst, he seems to have taken a position similar to one of the types of conference participants described earlier--the type that holds to the philosophy of anti-philosophy, believing that reflection is irrelevant to the practical problems faced by designers. At best, however, Alexander has reminded those who study design that the central problem of design remains what it has been throughout the twentieth century: to establish arts of design thinking that can serve as architectonic, integrative disciplines. What could such arts and disciplines be? And, could their identification and exploration provide a new program for research in design?

Any effort to describe a new program for research in design is futile and quixotic if it does not grow out of issues and themes that are embedded in current inquiry. Yet, if such issues and themes were obvious, there would be no need to suggest a new program because it would already exist. The problem is one of perspective and, again, assumptions. One of the dominant, though seldom examined, assumptions of design studies is that the subject matter of design is in some sense fixed and determinate, rather than quintessentially ambiguous and indeterminate. This is evident in the continuing belief of many people that design is merely a form of applied fine art, applied natural science, or, most recently, applied social and behavioral science. If the subject matter of design is determinate--or undetermined and awaiting determination--then the principles and methods of designing must be found among the humanistic, natural, and social sciences whose task is to study and understand what exists in the world. The only uncertainties are those of application, depending on the contingencies of particular circumstances. This leads to the cognitive science approach: the effort to identify and characterize the operations of designing, which properly include decision-making and procedures for taking account of all available knowledge about what is to be operated upon.

However, there may be a reasonable sense in which the subject matter of design is no more and no less than what the designer conceives it to be. Design is often regarded as a kind of universal art, because design thinking may be applied to virtually any area of human experience. Obviously, it may be applied to all areas of production, ranging from graphic design and visual communication, to the construction of material objects by engineers and industrial designers, to the planning of processes and activities by industrial and management consultants or others involved in practical decision-making, to the planning and construction of architecture and urban spaces. In addition, however, design thinking may also be applied to the formation of policies and institutions which may guide and enable practical action, help to shape human character, and provide the framework for distributive and rectificatory justice. And, in its furthest reach, design thinking may be applied to the making of philosophic, scientific, social, or aesthetic theories which attempt to explain some aspect of the world. (Illustrations include Umberto Eco's opening pages of **A Theory of Semiotics**, where he speaks of a design for the theory of semiotics, or Kant's more-than-casual reference to the design of the various "houses" of philosophical schools in **The Critique of Pure Reason.**)

It is perhaps troubling and frightening to consider the wide range of non-trivial uses of the term design. No wonder some design scholars are quick to exclude from consideration most of these applications of the theme of design, favoring the smaller and more manageable areas such as graphic or industrial design, whether on pragmatic grounds or on the ideological ground that extended meanings are not "practical" enough in their application--as in the "empirical" position described earlier. Yet, the sense of universality in design thinking is difficult to avoid.

How is it that designers may range over a subject matter that is potentially universal in scope yet always invent or create a particular subject as the fruit of their work? The answer to this question is critical in understanding the relationship that exists between design and science. Yet, there is no single answer that stands out as obviously superior, except from the perspective of one or another set of philosophic assumptions. In fact, the question gains its force from the ambiguity of the term "subject matter" and the consequent meaning of "universal," and "particular."

In the process of applying design thinking, designers must always discover or invent a *particular* subject to resolve the problems of specific circumstances. They conceive and plan this or that graphic image, this or that piece of furniture, this or that computer program, or this or that building. The design is a particular response to particular circumstances.

In this context, it is useful to distinguish between *design as a kind of rhetoric of the artificial world*--an art of invention and planning--and *productive sciences*, devoted to understanding made-things. On the one hand, rhetoric is a kind of universal art comparable to design in scope. It depends on the ability of a human being to **invent, judge, make decisions about how to plan and develop ideas**, and, finally, to **evaluate**. In turn, these abilities are *enabled* by correlative disciplines of **communication**, **construction**, **strategic planning**, and **systemic integration**. On the other hand, productive sciences are distinguished by the kinds of products made by human beings and by the kinds of activities that are appropriate to each kind of making. Design as rhetoric addresses the universality of conception and planning. It provides a way of understanding how designers work in the world, inventing ideas and carrying those ideas forward in concrete activities of planning. Furthermore, it provides a way of discussing the sense in which design is a debate about social life, where issues of utility, beauty, pleasure, and justice or human good are explored from alternative perspectives. In contrast, productive sciences provide a way of understanding the particular constraints of making to which designers must accommodate their ideas. Productive sciences are the **poetics** of the artificial world, balanced against the designer's **rhetoric** of the artificial world.

INNOVATION AND DESIGN FOR DEVELOPING TECHNOLOGICAL CAPABILITIES IN GENERAL EDUCATION

D. BLANDOW
Pädagogische Hochschule Erfurt
Federal Republic of Germany

ABSTRACT. The paper explaines the scientific background of technological literacy (technological competancies, technological capabilities) as a part of general education. It is shown, how the pillars of general technology, technical artifacts (theory of technical systems) and innovation and designing act together in technological projects with the interests and experiences of the students. Especially the intellectual toolbox for change over from one level to another (active principles, functional principles, technical principles, manufacturing, etc.) will be shown.

1. Introduction

Technology has to be considered a very complex discipline based on the network of relationships among the individual, nature, society and knowledge. There is also a causal link between development of society and development of technology. The context in which technology operates involves all spheres of daily life of all people and indicates the effectiveness of societies as a whole. Therefore the relationships between people and technology are best characterized as development relations in this special dimension may be considered as irreversable.

Technology is:"...the know-how and creative process that may utilize tools, resources and systems to solve problems and to enhance control over the natural and man-made environment".

The purpose of this paper is to present, from our German perspective, the basic principles that frame a coherent theory relating technology and work, both in their existing condition as well as technology's developmental/innovation aspect.

We have, in Germany, and now I mean in both of the formerly halves, a well developed system of technological education - in both its important dimensions: General education and vocational education.

M. J. de Vries et al. (eds.), Design Methodology and Relationships with Science, 277–308.

In Germany, teachers of physics, mathematics, language and history work along side the technology teacher. The subjects taught by the latter is generally named with some combination of Technic, Work and Economics. The approach used for the 11th, 12th and 13th school years varies from land to land (Germany now consists of 16 lands). But the most important point is that the concept has changed over to the innovating product-development, see Figures 1 and 2.

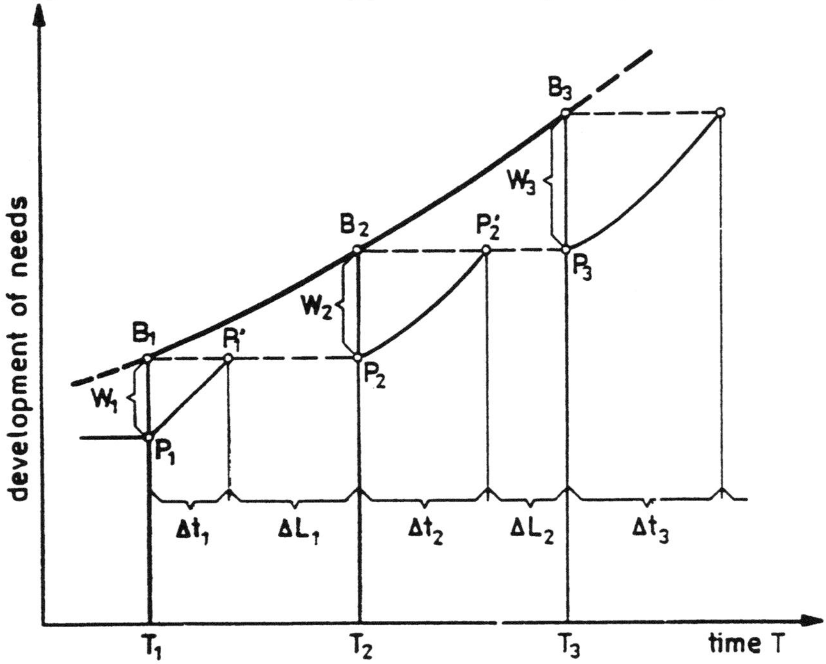

B_1	= needs fixation at time T_1
P_1	= level of product and process development at time T_1
W_1	= hence following contradiction to the development
$P_{1'}$	= reached level of development of the new products
D t_1	= needed development time for P_1
D L_1	= production time for P_1
W_2	= newly appearing contradiction between needs and P_1
B_2	= needs fixation at time T_2
P_2	= analogue to P_1, but at T_2
D t_2	= analogue to D t_1, but at P_2
D L_2	= analogue to D L_1, but at P_2

<u>complete result</u>

- always algorythm needs = = = > product and process development
- increase P_i - P_i' is a degree for the standard of development work
- the longer the production time the larger the contradiction to the objectively progressing needs development

Figure 1 Classical product-development.

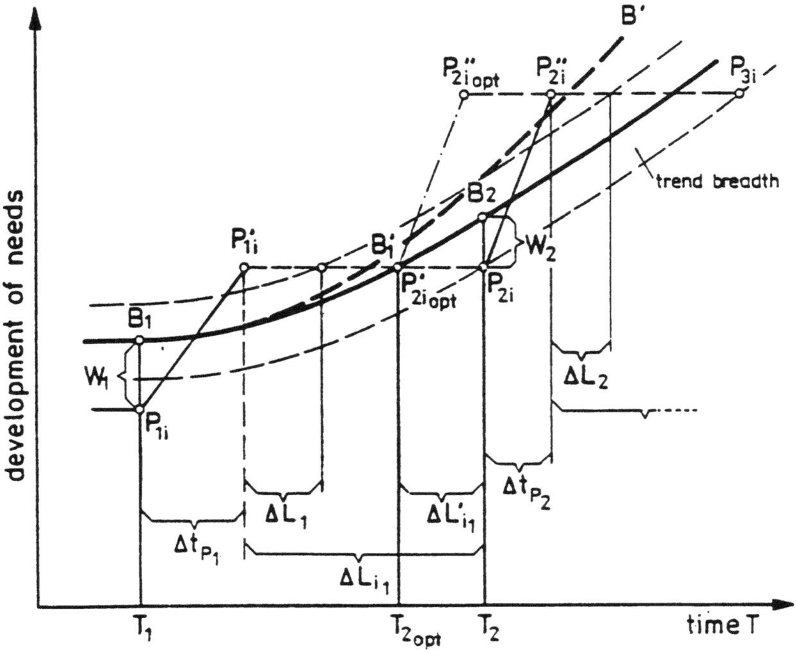

B'	= by new products influenced needs development
B_1	= needs fixation at time T_1
P_{1i}	= level of product and process development at time T_1
W_1	= hence following contradiction to the development
$P_1{}'_i$	= preliminary data collection of product - and process development
$P_2{}''_i$	= preliminary projected P_{2i}
Dt_{P1}	= development time for the new product $P_1{}'_i$
DL_{i1}	= production time for $P_1{}'_i$
DL_1	= market dominating time for $P_1{}'_i$
DL_{i1}	= begin of production stop
W_2	= newly appearing contradiction between $P_2{}'_i$ and the needs of the development $P_2{}'_i$
$P_{2\ opt}$	= optimally, projected development of $P_2{}'_i$ with market dominance
$T_{2\ opt}$	= optimal development date for P_{2i}

<u>complete result</u>

- always the longer DL, the clearer the product and process development influences process of the needs development, increase P_i - P_i' is a degree of the standard of the development
- optimal development moment is determined by the reaching of the objective needs development

Figure 2 Innovating product-development.

2. The Understanding And Overview Of The Structure Of Production Processes And The Principles Of Technology

Explanation of this point begins best with a short historical example.

Prehistory until about 1500	Transition/ prerequisites 1500-1800	Development 1800-1850	Consolidation 1850-1920	Transition to systems theory since 1920	
Practical oriented	Knowledge oriented	Process oriented		Methodology oriented	Strategy oriented
Technical knowledge/skills	Systematic descriptive technical knowledge	Classical technical disciplines		Engineering and technological disciplines	
Practical methodical knowledge ↓ technological knowledge ↓ Constructive technical knowledge→	Development of the natural scientific mechanics → Constructive technical knowledge (descriptive) ↓ Technological knowledge (descriptive) ↓ ↓ Specific general engineering (technology)→	Traditional disciplines of the mechanical cycle (machine building) Technological disciplines Mechanical technology ↓ ↓ Division into techniques ↓ ↓ Classical production engineering	Differentiation into → ↓ ↓ internal external ↓ ↓ structure industrial (basic branches components) Chemical, etc. technology ↓ ↓ Division into processes ↓ ↓ Classical process engineering	Formation of an independent theory of machine design from the classical engineering and technological disciplines	Formation of an independent theory of innovation processes based on the methodology of engineering and technological disciplines

Blandow/Dyrenfurth/Lutherdt, 1991

Figure 3. Unity of construction (design) and technology (engineering) in the structure of the sciences.

The figure documents a long history involving key questions about the relationships between science, technology and engineering continually arise. Similarly the tension between general versus specialization in technology-oriented education is a perennial thorn.

If this perennial reoccurrence is actually the case, one must ask, what causes it to emerge with such regularity? Two answers appear with similar frequency. One involves the acceptance/respect of one discipline's practitioners for those of another. Frankly, I've thought about offering an "interdisciplinary tolerance" course at my institution.

The second, and more important problem, is the lack of recognition of those in a discipline for their field's necessary interactions with and dependence on essential components of other disciplines. Therefore we must be clear about two insights:

1. There exist natural overlaps among disciplines

2. The further development of society depends on the interrelationships among disciplines and the differences of their individual perspectives and approaches.

To examine the relationships of the unity of disciplines versus their uniqueness, I use the concepts shown in Figure 4. From the author's point of view, the key to developing the capability for arriving at the proper answers is to find the right starting points in one's research activities. Especially for technological subjects, the keys to all technological solutions are found in the terms z and Dz.

The condition or situation, the changing of conditions or situations, the relations between input and output and feedback are the key points for understanding technical artifacts as well as technological processes. Note the pertinent question here is in fact the key question. It is not so important to know what is being produced. Instead, the salient issue is what changes and processes are being used -- i.e., how are we processing!

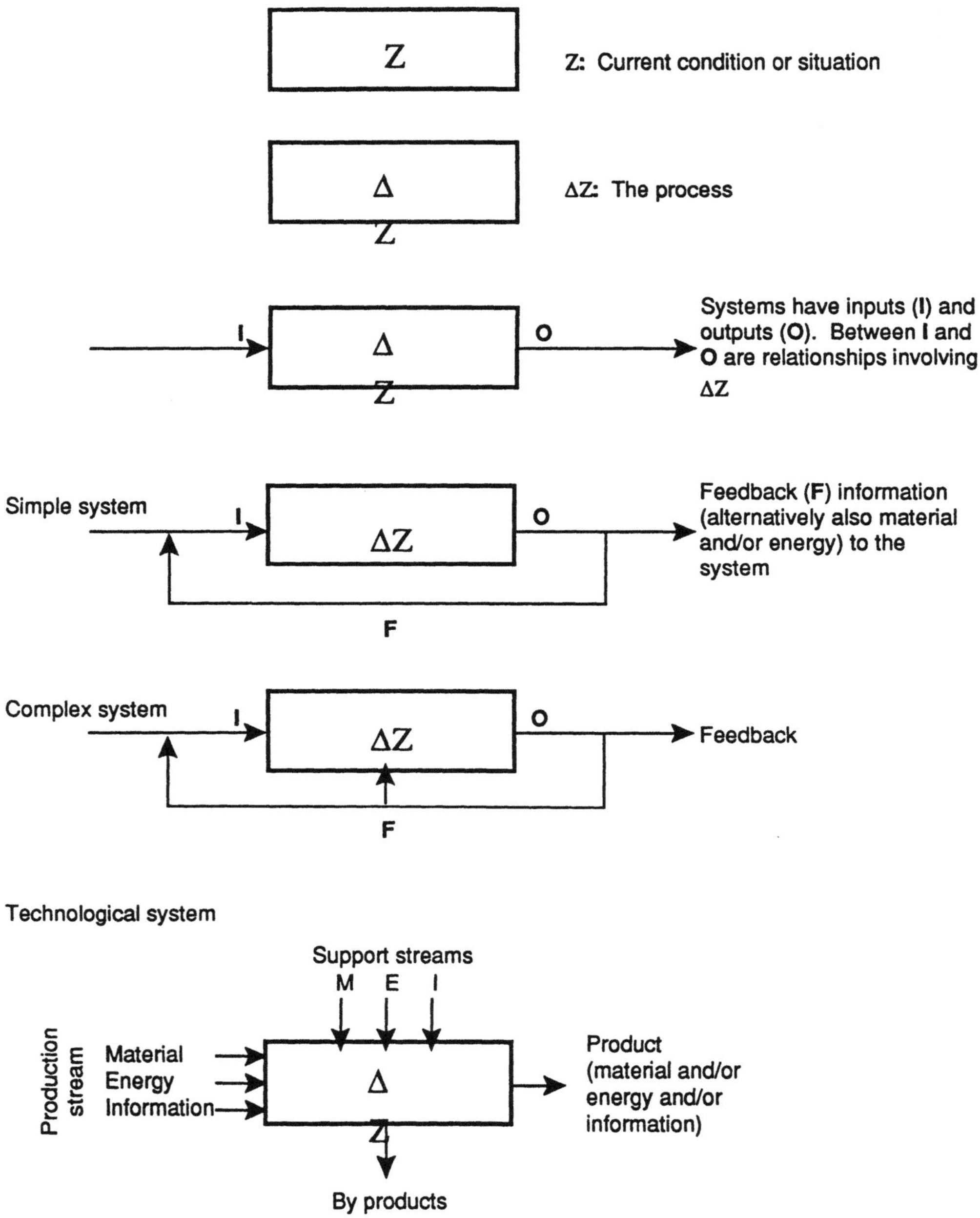

Blandow/Dyrenfurth, 1991

Figure 4. Key elements of production systems.

Naturally there are also general models which resulted from research to establish an overview. Figure 5 is one such model.

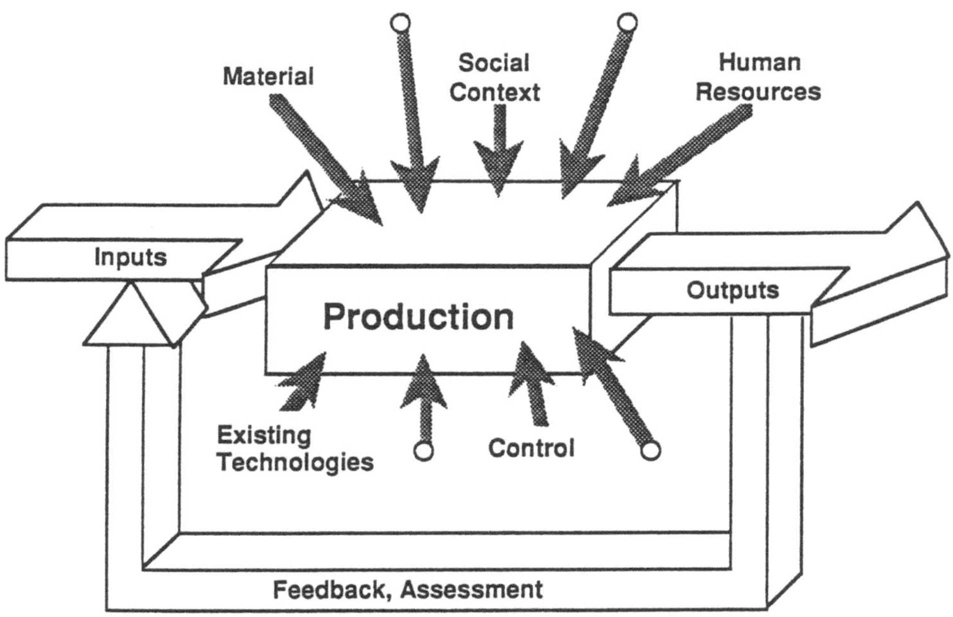

Blandow/Dyrenfurth, 1991

Figure 5. Model for the production process.

The German approach emerges from the dynamic tension between the reductionist methodology involving specification of elements (as in Figure 4) and the holistic systems view such as the example depicted in Figure 5. Of interest now is the question of how one reconciles these two poles of approach. Our goal was to establish characteristic planes between these poles in order to identify key constants that facilitate understanding and enhance generalizability. Examination leads to the same four key frames of reference and they are used at all levels of the overall model.

1. The change of condition/situation Z1, Z2, Z3.

2. The process, DZ.

3. The location and/or point of time characteristics.

4. The technical artifacts or means.

In reality, these four key frames of reference interact in various planes and levels (see Figure 6).

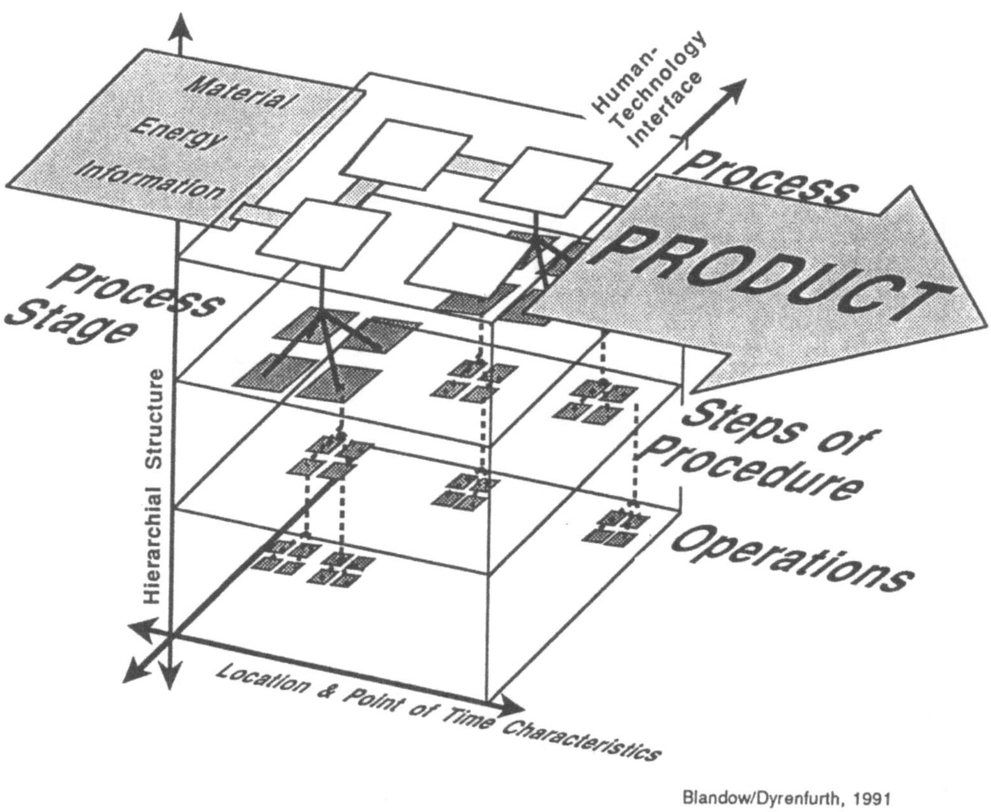

Figure 6. Hierarchical structure of production processes.

This structure, shown in Figure 6, may be used to describe, explain and/or analyze production in any of the realms typically used. It is most important to note that the model is equally applicable to the traditional industrial realms of paper, metal, wood, and plastics work; as well as to the food processing industry, e.g., cheese, sugar, meat and milk processing; or the process industries such as petro-chemical, waste-water, and bio-technical. Furthermore, although not as frequently applied because of history, the model clearly also fits agriculture. From this illustration, it becomes clear that we know much about each individual level. We know much detail about material, energy and information -- but the principles that each embodies are less well known. Even less well known than these are the principles governing vertical linkage between the levels. Another great weakness is our ability to formulate variables and make decisions. The same situation is observed in the hierarchical system of technical artifacts shown in Figure 7.

Level Six	Cybernetic Systems	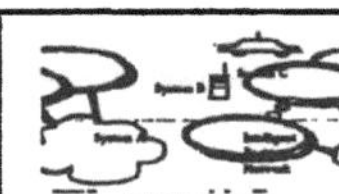 Infrastructure
Level Five	Integrated Networked	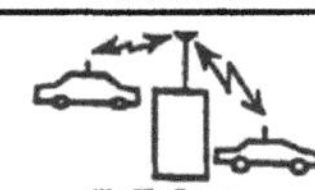 Traffic System
Level Four	Networked Systems	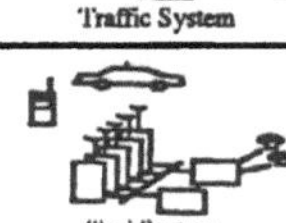 Taxi System
Level Three	Systems	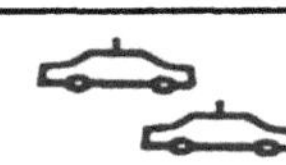 Separate Automobiles
Level Two	Standard Assemblies	 Clutch
Level One	Standard Components	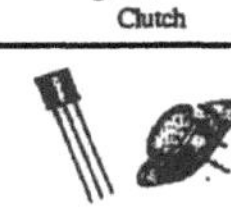 Transistors

Illustration: Tim Trogden

Blandow/Dyrenfurth, 1991

Figure 7. Hierarchy of technical artifacts.

Made visible however, is the concept that hierarchy can be used to establish understandable order -- just like we learned from our study of technical artifacts (see Figure 7). The relationships between these two models, the hierarchical model of production processes and the hierarchical model of technical artifacts, are critically important but they fall outside the scope of this paper's main discussion.

The first result of our work to make sense of technology was a useful structure to make sense out of the multiplicity of production by creating a matrix of the three pillars (materials, energy, information) against the nature of change (shape, structure, location, time) was shown in Figure 8.

Object of work	Nature of Change			
	Form/Shape	Altering	Location	Time
Material	Material shaping	Reconstituting	Materials handling	Aging, Wine, Patina
Energy	Energy processing	Energy conversion	Energy transfer	Half-life
Information	Information handling	Information processing	Communication	Obsolescence

Blandow/Dyrenfurth, 1991

Figure 8. Matrix between objects of work and nature of change.

If one includes the principles of technological organization together with the preceding matrix, one arrives at the principles of production. This combination, depicted in Figure 9, has not yet been found elsewhere in the literature. When found or developed, however, it would become a most useful tool to help understand the correlation between the several planes (as in Figure 6) and to understand the strategy of their combination for solving specific technologic/innovation challenges.

Sample goals of process operation	Implications of the Goals on:			
	Changing conditions or situations, Z_1, Z_2, Z_n	The process, ΔZ	Location and point of time characteristics	The technical artifacts/means
Minimization of resources	Capitalization on material characteristics [Material structure]	Process integration WRT time [Work hardening]	Production timing (continuous/intermittent) [continuous casting, Just-in-time organizing]	Energy supply at point of use [Solar telephone, Integral wheel-motors]
Increase of variability	Substitution, Alternative sequences [Recycled material aggregate]	Adaptation [Photochromatic glass]	Flexible reactions [Combine threshing drum pressure]	Standardized components [Microchip logic elements]
Increase of stability	Feedback systems [Sensor technology, Dash warning indicators]	Networking, optimization [Feedback driven control systems, assembly line buffers]	Quality assurance [Zero defect programs]	Parallel/redundant structure [Pilot/co-pilot]
Reduction of production cycle time	Activation characteristics [Catalysts, hardeners]	Increase of energy/work density at the work station [halogen lights, microcomputers]	Parallel processing, assembly lines [automobile manufacturing, Matrix computers]	Increase of frequency [Newspaper production]
Reduction of product planning and setup time	Standardized stock, modular construction components [Rolled steel, 4x8 panels, DIN paper]	Utilization of standardized (modular) process elements [Canned cycles in NC machines]	Modular machines, handling systems [Flexible manufacturing systems]	Automatic tool/jig fixture changers, robots [Computerized plant change over]
Increase in ecological responsibility	• • •	• • •	• • •	• • •
• • •	• • •	• • •	• • •	• • •

Blandow/Dyrenfurth, 1991

Figure 9. Implications and goals of process operations.

Consequences of point 2

Our scientific research yielded several insights that make things easier for teachers and researchers.

1. When one combines the constants of objects with the constants of processes of technology, one establishes one plane of the overall model (see Figure 10).

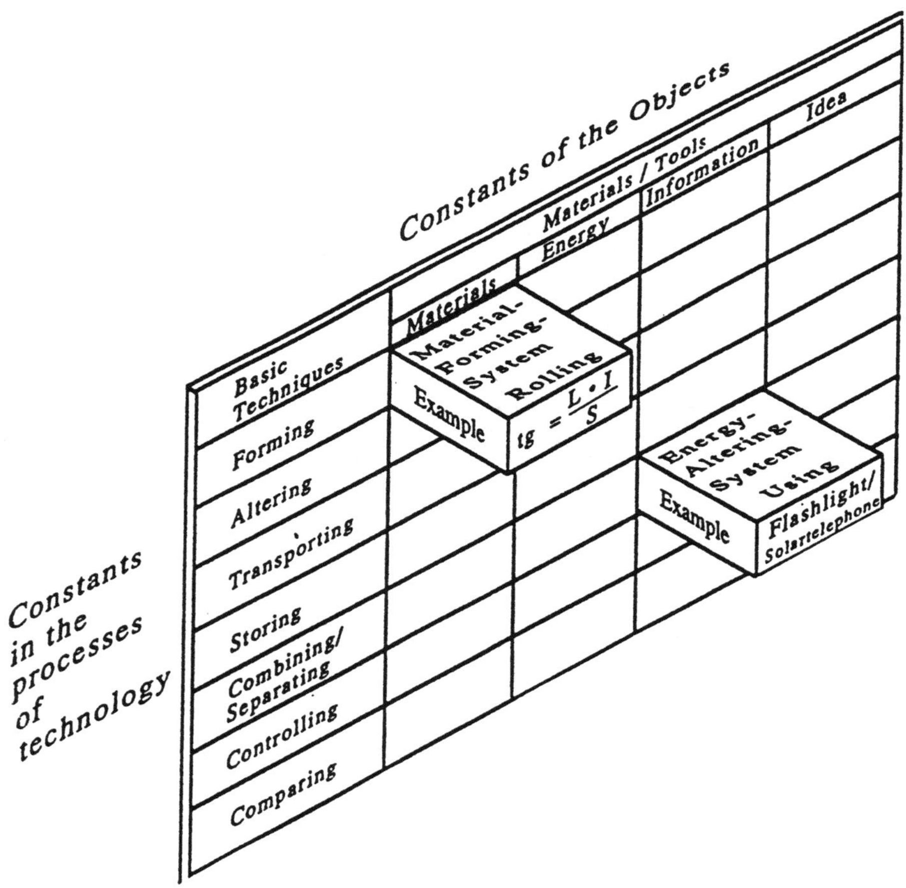

Figure 10. Model plane 1

2. The second plane is formed by combining the constants of the processes, with the constants of the objects. Then, by adding the constants of technology interaction sites one generates a three dimensional model that represents a matrix of technological activity. Selected provocative examples might include (also see Figure 11):

- Energy transport in space
- Energy storage in water
- Material transport in hospitals
- Material altering in space
- Information storage in water
- Information forming in factories

Note that these are not merely academic musings or wishful thinking. Immediately behind them emerge questions regarding the emerging possibilities of technological capability. For example, and one step more concrete than the preceding, are:

- Lasers as surgical tools
- Lasers as dynamic measuring devices
- Electrophoresis gene identifying
- Biotechnology use in laboratories

Furthermore, the emergence of such technological possibilities brings with it a responsibility to incorporate an inclination towards such forward looking tendencies into our various technician, technologist and technology educator curricula.

3. The large field of production principles (see Figure 9) is also used as an activity guide during our degree programs when we prepare students and researchers for practical work as engineers and teachers of technology.

4. Technology is interdisciplinary and all its applications represent compromises.

5. With the increase of relations between materials and energy (e.g., refer to column DZ in Figure 9) we can distinguish between laws of development and laws of structure (see Figure 12).

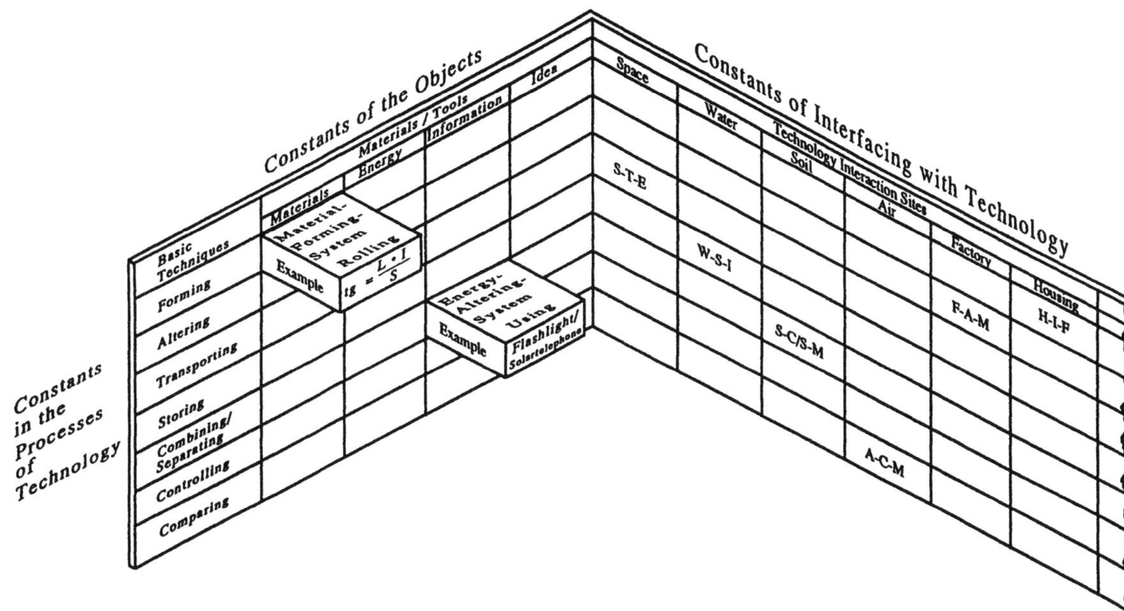

Figure 11. Model Planes 2 and 3: Constants of the technological processes vs objects vs interaction sites

Types of laws	Changing conditions or situations, Z_1, Z_2, Z_n	The process, ΔZ	Location and point of time characteristics	The technical artifacts/means
Structural laws	Coupling Feedback Parallelity	Structure of active principles • Form • Energy • Movement • Location • State	Degrees of freedom Flexibility Hierarchy Variability Adaptation	Flows are closed Σ Material = 0 Σ Energy = 0 Σ Information = 0
Development laws	Integration of function Continuous/ Intermittent Integration of present production with new developments	Increase of energy density, Integration of production and recycling	Shortening of the information flows Reduction in changeover time Increased variability of mass-produced items	Flexible workstations Increased output per unit mass Multiple use components

Blandow/Dyrenfurth 1991

Figure 12. Types of laws for understanding the structure and development of production processes.

6. For instructional purposes, eight groups of themes emerged as important:

a. Principles for determining the structure of technical and technological processes with respect to:
- ° Structure of processes
- ° Changing conditions
- ° Location and time characteristics
- ° Structure of technical artifacts

b. Principles of development of technical and technological conditions:
- ° Structure of processes
- ° Changing conditions
- ° Location and time characteristics
- ° Structure of technical artifacts

3. The Understanding Of The Strategies And Structure Of Innovation

The recognition and solution of technological problems, the purposeful development of technical artifacts, the strategies, methods and tools of assessment are the alternative to a technical education, which is based on various traditional disciplines. This new element will involve the perception of systems and principles from single fields, will necessarily lead to connection, will train creativity and competence for decision of the personality etc.

Whenever one generation succeeds the other, even within product generations, innovation will be based on nearly the same principle, the Multiplicity in Unity. The example of ICARUS points out that important concept. He was forced to learn to survive and to translate his wish for freedom into activities to keep busy. It was his vision of a palatable future, stimulated by the example of the freedom of birds flying over him, that triggered his innovation--his modeling of their wings. He builds wings and attempted to escape. The conclusion is known. This example's lessons for today's situation is shown in Figure 13.

Problematic situation (thought initiator) = > Overcoming thought barriers = > imagining a vision of a possible solution = > model development (resolution of contradictions) = > development of approach strategies = > development of time and activity plan = > execution of the plan = > evaluation of the results = > new situation/problematic situation

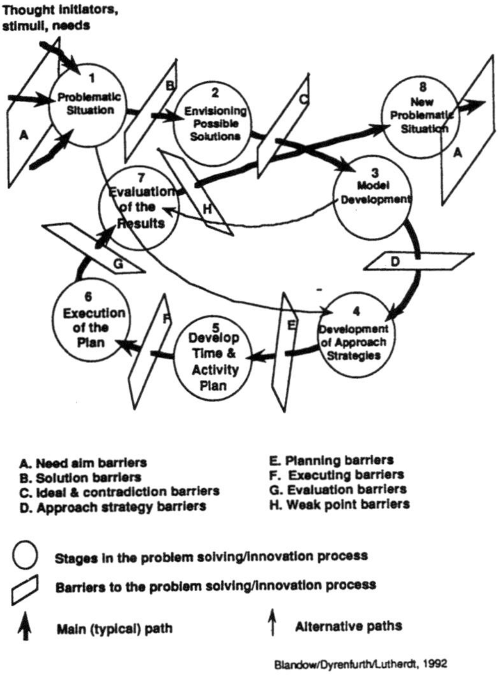

Figure 13. Stages of problem solving/innovation, barriers and paths.

The cycle is important today. The example from the fable shows us today, as it did in the past, the relationships between the tangible world, our mental concepts of it, the initiating value of our concepts/visions and the power of translating our ideas into practice via a systematic strategy. In further figures we depict the same concepts albeit with addition of time and economics as factors (see Figure 14).

But this example also shows us, which barriers we experience in our thought processes and how difficult it is to overcome them. The main point, however, is that we can now complete the modular concept from the activity point of view. With the seven key stages identified by analysis of over 100 documents, and our knowledge of the characteristics of the typical barriers, we are able to genuinely help people develop the thought processes necessary for effective technological problem solving/innovation.

All over the world, from Icarus to today, the problematic situation is the starting point for all activities. But today, internationally, up to 60 percent of the total development and lifetime of a product are used to overcome the 1st and 2nd barriers (the recognition of trends and the definition of function, contradictions and ideal systems). Therefore, consequences for technology education concerning the need--aim--motive and problem transformation are inevitable (see Figures 15, 16)

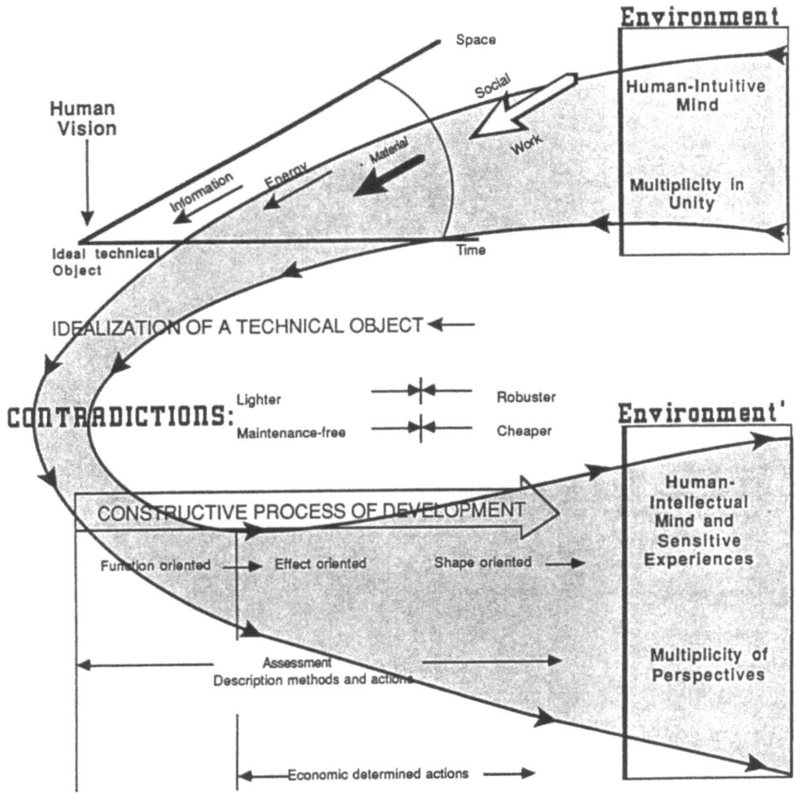

Blandow/Dyrenfurth, 1991

Figure 14. Concrete-abstract-concrete' -- The routes of technical thinking.

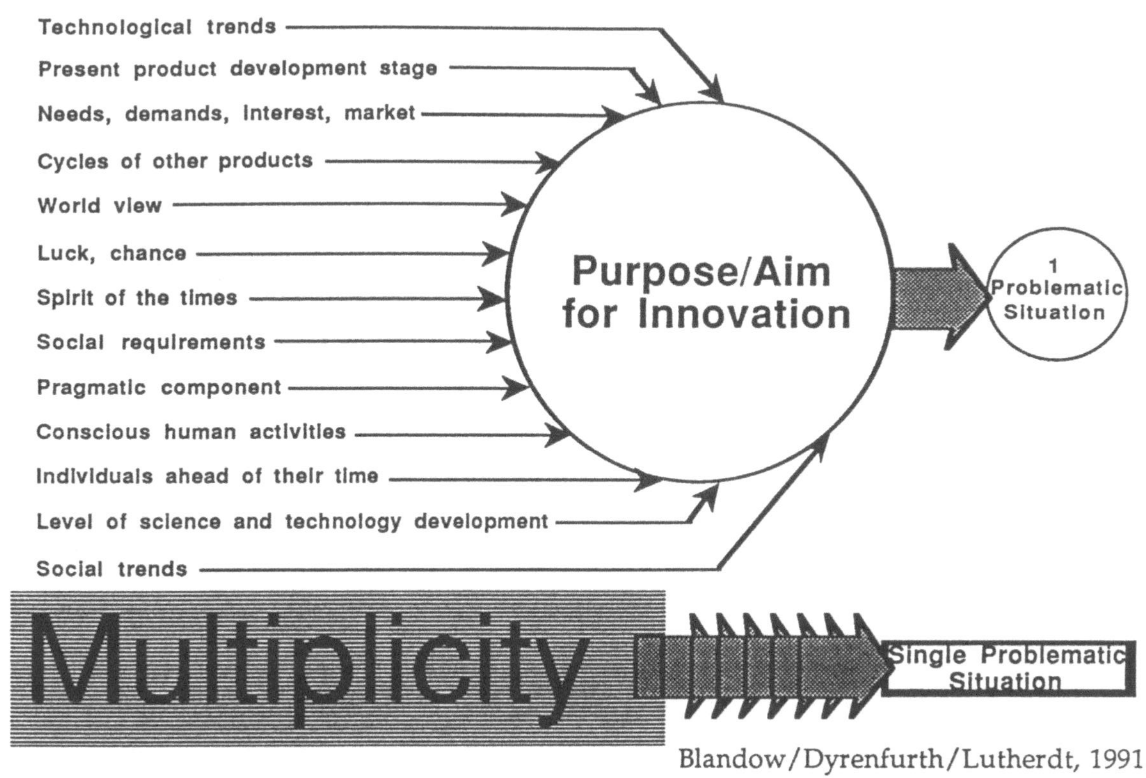

Figure 15. Strategies for overcoming need-aim barriers.

Consequences of point 3

1. The approach used to define a problem or task is most decisive because along this route of abstraction and decomposition creative solutions will be generated by observing technical/technological principles and the reality given:

- consideration of trends and formation of ideals,
- using contradiction as a heuristic means,
- planing, put into practice and evaluating experiments
- initiative spirit situations (brain-wave, association,...)

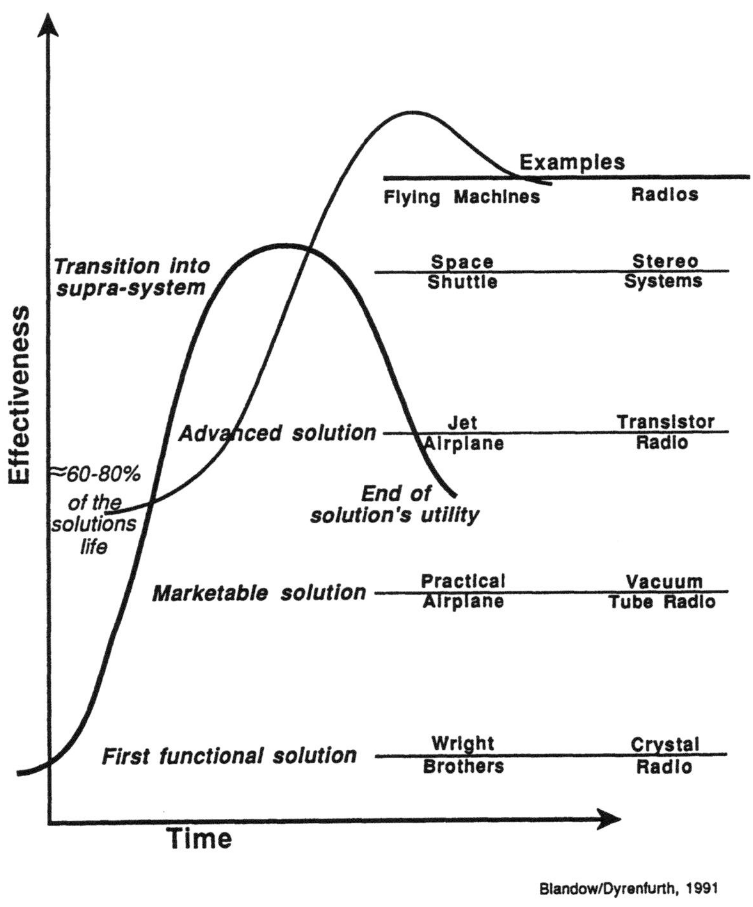

Figure 16. Life-stages of products.

The results of our findings are also summarized in Figure17. As was stated previously, the problem solving/innovation process necessitates the overcoming of transition barriers which requires the access to and proper utilization of, different information-masses. These information-masses are each unique in terms of hierarchy structure, nature of information and acessing method. Remember that not all steps in the problem sloving/innovation process need to be used in every situation. The logical consequence of this is also that not all barriers are encountered every time.

How to find contradictions and to define them will be demonstrated in the *Worksheet for the formulation of contradictions to trigger innovation*. The heart of the matter involves the identification of the requirements, their grouping by appropriate characteristic, and the framing of the key contradictions. Thinking in contradictions will help the people in charge, by their own creative characteristics, to find new orders for answering their purpose.

They will continue to restructure and rearrange their thoughts which will also be kept in tension and motion due to verbal operations such as formulating the contradictions, see Figure 18.

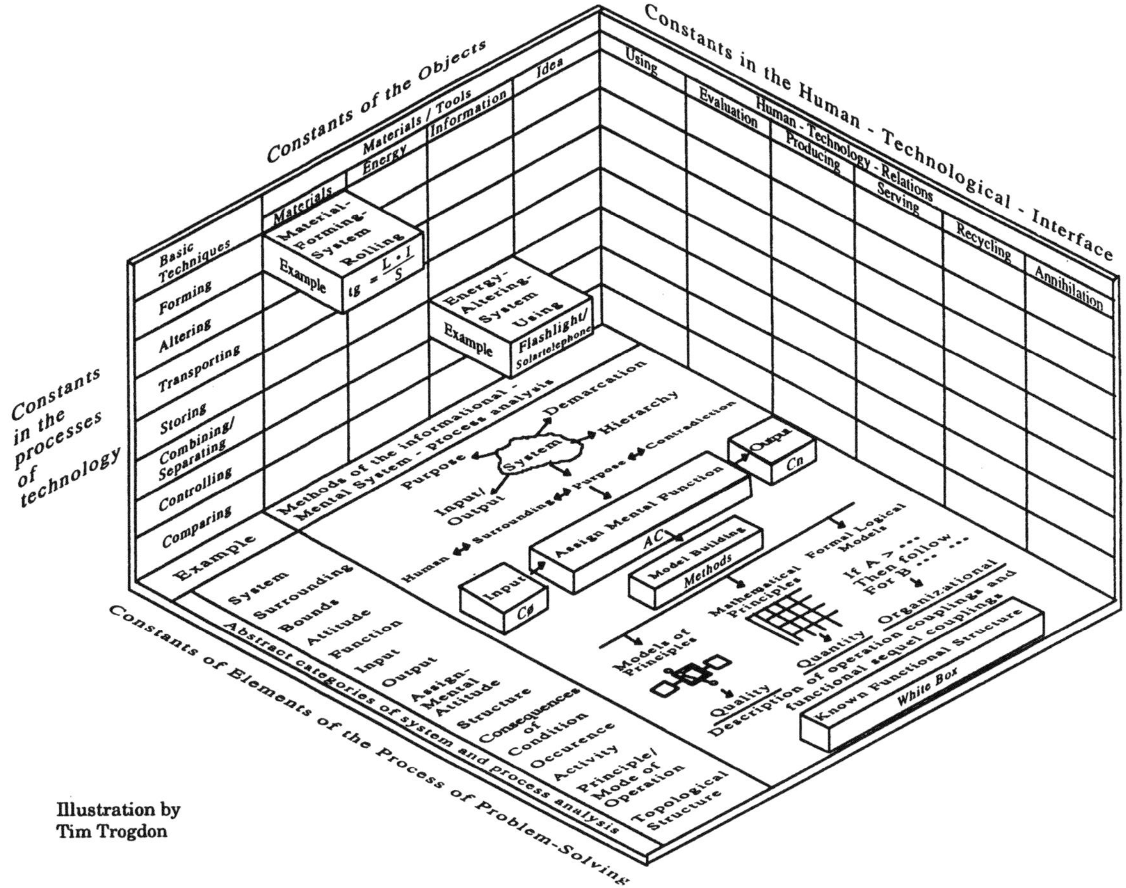

Illustration by
Tim Trogdon

Figure 17. The modular concept of technology and work

(31)

Contradiction	-->|<--		Examples
Stiff	-->|<--	Elasticity	Spring, tree branch, tire
Porous	-->|<--	Holder	Filter, unglazed clay pot, skin
Dry	-->|<--	Humidity	Moss, Pampers, bandaids
Loose	-->|<--	Rigidity	Polystyrene bead boards, ice cream
Open	-->|<--	Enclosure	First law of thermodynamics, window
Accelerating	-->|<--	Delay	Energy conservation systems (commuter trains, elevators)
Light	-->|<--	Darkness	Infrared imaging, radar
...			...
...			...

Blandow/Dyrenfurth/Kahmann, 1991

Figure 18. Worksheet for the formulation of contradictions to trigger innovation.

2. Recognition and solution of technical problems will happen in sequence of emerging of technical contradictions, their solution and further emergence. The first step taken in working on contradictions will be abstraction In a heuristic way, we will find on our road to determine the tasks, levels of abstractions (related to the above explained four elements of production).

° social-economic-technological contradictions for the completeness of the technical system to be developed
° technological contradictions between the process DZ and the conditions Z_1, Z_2, Z_3, Z_n
° technological- technical contradictions for any partial systems (partial function)
° contradiction based on both, technology and nature within a partial system.

3. We have thought as contents of an innovation methodology laws of:

° the development of social and individual needs,standards, resources,
° the development of the market and of demands
° laws of:
-the structure and development of technical systems
-the structure of technological assessment
-the constants in the human technological interface laws of:
-the social creativity and the structure of information systems used on different levels in the process of problem solving and so called heuristic principles, rules
-principles of recognizing problems (of contradictions, of ideals)
-principles of problem solving (association analogy, variation, combination,...)
-principles of materialization (dimensioning, designing,...)

4. To arrive at solutions involves the problem solving process. The simplest views of problem solving may be depicted by a sequence and/or path network of alternative solution steps and solution stages as shown in Figure 19.

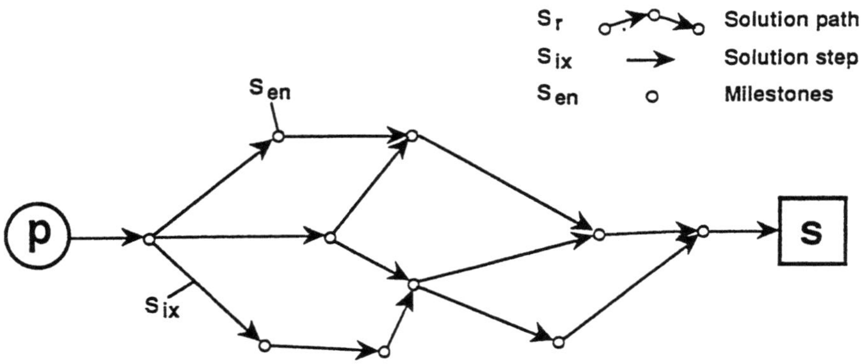

Fig. 19. Problem - solution path alternatives.

The way from p to s leads:

- along a solution road S_r	- in data processing
- through a solution network S_n	- set (strategies)
- with the solution steps S_i	- program (algorithms)
- and the solution elements S_e	- data (facts), milestones.

From this model we can come also to a hierarchically structured network of problem solving as well as to the modelling of the production process From the latter, we also come to the constants in the human-technological interface (see Figure 10).

5. The constants in the Human-environment relationship, such as using, evaluating, etc. can be folded against the other planes of the technology model to create new interaction fields. With such a concept of modular planes one has a useful instrument/methodology to organize the multiplicity of technological applications/examples and on the other hand, it enables teachers and teacher educators to generate thousands of ideas and examples for teaching activities, insights and the furthering of innovative thinking.

4. New Information-Masses As A Part Of Innovative Thinking

The Human-Technology relationships are curious (see Figure 14). As more and more insights are developed and put into practice, each generation finds it increasingly difficult to work with these accumulated practices because of their exponentially increasing number and complexity. The main evidence of this, the well known knowledge--time problem, makes itself particularly visible in the information explosion. To overcome this problem, capabilities such as the following are being addressed:

- ° Information interpretation
- ° Information production
- ° Information structuring
- ° Information combination
- ° Information elimination
- ° Information selection
- ° Information acquisition
- ° Information ...
- ° Information reduction

To understand the fundamental process of the irreversible Human--Technology relationship, one must first change paradigms and acknowledge information both as a product and as a key target for assessment. For our strategy-oriented concept (see Figures 13 and 14), it is important to note that each barrier and the strategy necessary to overcome it requires mastery of its own information-mass. In this paper we are providing only selected examples of our theory.

Consequences of point 4

To demonstrate the application of the theory, and to heighten your interest, we have provided three examples.

- ° Information-mass to overcome the Function--Ideal Model barrier.
- ° Information-mass to overcome the Function--Structure barrier (see Figure 20).
- ° Worksheets for overcoming the Environment-Need barrier (Figure 21).

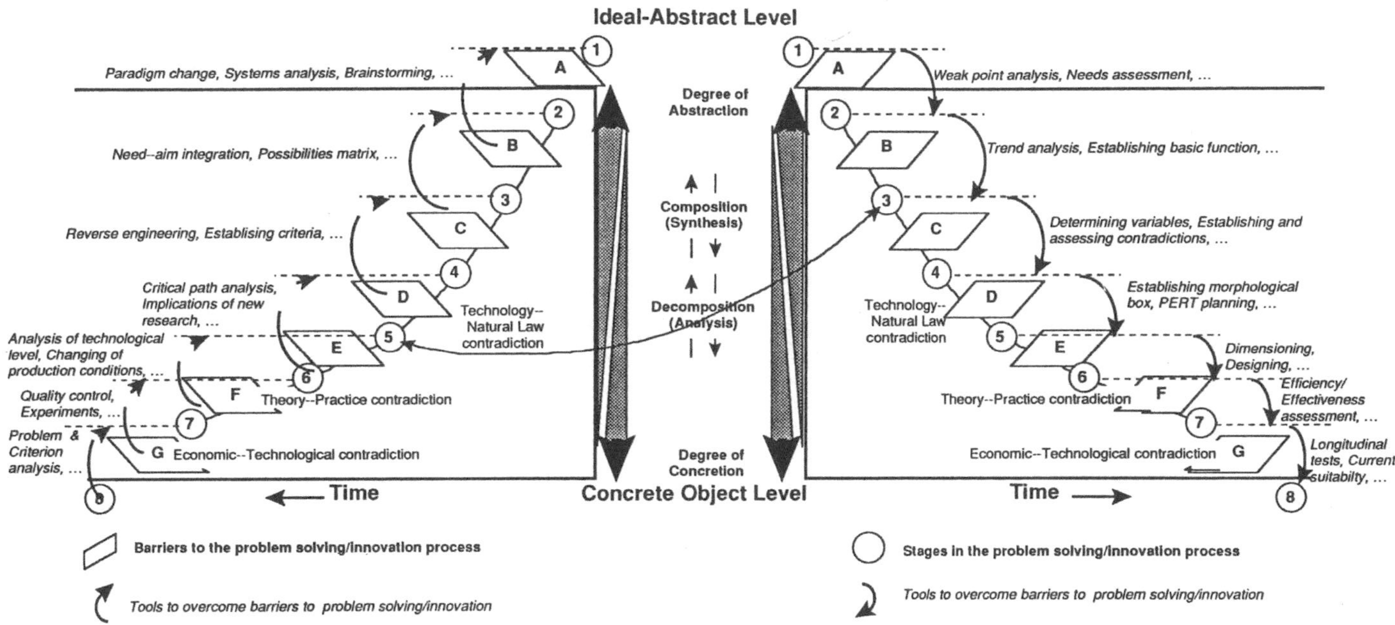

Figure 20. Overcome function-structure barriers.

Worksheet for Trends Analysis

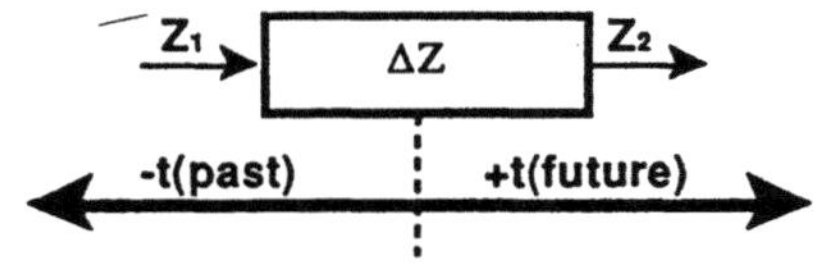

Structure	ΔZ, the ways and means of change (process) that changes the conditions/situations (**Z_1, Z_2, Z_n,...**) is charted in chronological sequence and the most likely next development is extrapolated on the basis of the trend curve.
Characteristic questions	How were the changes of conditions/situation (**Z_1, Z_2, Z_n,...**) accomplished? What technical means and/or processes (ΔZ) are likely developments along the trend line
Situation	**Known Trends**
1. Automobile radio station selection	1. Faster travel, digital tuning, integrated controls, international symbols, error readouts, self-correcting circuits, ...
2. Home windows	2. Greater durability, thermal-control/insulation, noise abating, stronger, easier cleaning, variable trim, light transmission control (amount and direction), ...
Potentially new examples	Automobile radio station selection: Steering wheel mounted controls; self-seeking according to type; emergency message superimposition; ... Home windows: Adjustable, i.e., thermostat-controlled heat absorption/rejection setting; variable light transmission and direction; image diffusion control;...
Your suggestions (any situation)	

Blandow/Dyrenfurth/Lutherdt, 1991

Fig. 21. Worksheet for trend analysis.

Worksheet for Identifying Contradictions

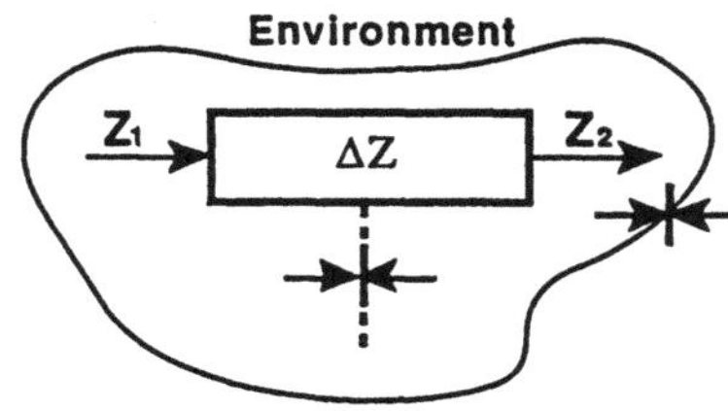

Structure	ΔZ, Effectiveness = f(Trends, needs, demands, ...) Factors E ⇑ = $f(e_1⇑, e_2⇑, \ldots, e_n⇑)$ Components $e_1⇑ = f(x_1⇑, x_2⇑, \ldots, x_5⇑)$ Components $e_2⇑ = f(y_1⇓, y_2⇑, \ldots, y_5⇓)$ E ⇑ = $f(x_5 \rightarrow \mid \leftarrow y_5)$ Identification of contradictions that operate within ΔZ and between the outcomes and the environment through analysis of factors and their composition
Characteristic questions	What contradictions operate within ΔZ and/or between its output and the environment? How can one counteract the individual effects, trends, demands, ...
Situation 1. Buildings 2. Ironing (clothes) 3. Bicycle lighting	**Known Examples and [Contradictions]** 1. Pneumatic structures [Area vs Mass], Moving form construction [Size vs Time], ... 2. Teflon-soled irons [Friction vs Pressure], Temperature controlled iron [Fabric protection vs Operator intelligence]... 3. Generator powered light [Light vs Effort], [Light vs maintenance], ...
Potentially new examples	1. Energy efficient buildings [Material mass vs Energy storage], Environmentally protective buildings [Internal oxygen generator vs Complexity] 2. Induction powered iron [Energy supply vs Mobility], Magnetically pressured iron [Downward force vs Operator fatigue] 3. Visibility to others [High visibility vs Power demand], Forward lighting [Energy source vs Operator effort]
Your suggestions (any situation)	

Blandow/Dyrenfurth/Lutherdt, 1991

Figure 22. Worksheet to identify contradictions.

To properly understand the problem solving/innovation process and to properly develop this capability in others, it is not sufficient to merely depict the key stages it involves--these are typically well known. One must also depict the information-masses that must be mastered and the barriers that must be overcome. Most important, however, is the learning of the intellectual tools and strategies (their organization) necessary to master each information-mass and surmount each barrier.

5. Summary And Conclusions

It was our goal to present an overview of the structure and the distinguishing features of technology. With the view of the developmental stages in traditional technical and technological disciplines; which evolved from the practical-oriented approaches, through knowledge-, process- and methodology-oriented approaches to the strategy-oriented approach; it should be emphasized that this involves a most critical change of paradigm from a subject- or discipline-oriented one to one that is much more focussed on goals to be accomplished. The specific problem to be surmounted is the predominant focus, not the individual disciplines that will make-up the solution. An integrative perspective is necessary, taking from each discipline what it has to offer and then synthesizing these contributions into a solution that addresses the problem in a new way.

From these point the key elements of modern structured technology were indicated. They were the process, the changing of conditions or situations, the location and point of time characteristics and the technical artifacts or means. These key elements are involved in all levels of hierarchically structured production processes. This also yielded the insight that such views of technology are useful in all of technology's arenas including those of agriculture, industry, chemical processes as well as in home economics for example. Also identified were the goals for the development of an organizational structure of processes. These served to guide the development of the elements of processes.

In the paper's second part the determining factor in the human-technology relationship--particularly in the field of production--is identified as the further development/advancement of capability -- not the mere satisfaction of need. From this point, new questions arise with respect to the handling of information-masses as well as the capabilities for choosing the appropriate storage and retrieval mechanisms. The development process involves seven key stages: Recognition of a problematic situation (thought initiator), overcoming thought barriers, envisioning possible solutions, model development (resolution of contradictions), development of approach strategies, development of time and activity plan, execution of the plan, evaluation of the results and recognition of the new situation/problematic situation.

Given the presented concept of a modular view of technology and work, we have synthesized two kinds of thinking. One is object-oriented thinking and the other is a kind of innovative thinking. By combining these two approaches with the modular concept, and then emphasizing the development of strategies, we hope that the result of our work can be used for a diverse set of problem situations--both industrial and educational as well!

References

Blandow, D. : (1985),UNESCO, Symposium on the Teaching of Technology, Paris, France

Blandow, D.: (1989) Probleme des Erkennens and Förderns wissenschaftlich - technischer Begabungen im Unterricht; Wissenschaftliche Schriftenreihe der Technischen Universität Karl - Marx - Stadt (Chemnitz), Heft 6, 1989, S. 19 ff

Blandow, D. & Wolffgramm, H. : (1975) Zur Spezifik der fachwissenschaftlichen Grundlagen der Ausbildung von Diplomfachlehrern für Polytechnik. Wissenschaftliche Zeitschrift der Pädagogischen Hochschule "Dr. Theodor Neubauer". 11(1), pp. 5-14 [Erfurt, Germany-East]

Buresch, J.: (1992) Specialists with all-round competence -- A reflection on advanced technology in vocational training. [International conference on technology education, Weimar, Germany]

Dyrenfurth, M. J.: (1990) Rethinking Technology Education in the Secondary School: Missouri's Approach to Technological Literacy; Prepared for Landesfachkonferenz Polytechnik -Arbeitslehre. [Thüringen, German Democratic Republic]

Eder, W.E.: (1992) Science in engeneering, one component of the science of engeneering design. [NATO ARW, Eindhoven, NL]

Erasmus Projects Group : (1992) Pädagogische Hochschule Erfurt, Fachbereich Technik-Technolog and the Christelijke Hogeschool Noord-Nederland, Leeuwarden, department for basic education.

Herrmann Holliger - Uebersax : (1987) Morphologisches Institut Zürich; Integrale Systeme oder Denkkatastrophen; Technische Rundschau, Heft 34.

Hill, B.: (1990) Aufbau von biologischen Funktionsspeichern, ein Beitrag zur Überwindung von Barrieren im Analogiebereich; [Pädagogische Hochschule Erfurt, Forschungsgruppe Polytechnik]

Klix, F.: (1987) Erwachendes Denken; Deutscher Verlag der Wissenschaften. [Berlin]

Lutherdt, M.: (1990) Zu den konstituierenden Elementen ausgewählter technischer Wissenschaften; Wissenschaftliche Zeitschrift, Pädagogische Hochschule Erfurt.

Nacken, W.: (1986) Lernen im Dialog: Jugend - Wirtschaft - Politik; in Sozialwissenschaften und Berufspraxis; Jahrgang 9 / Heft 4, 1986

J.H. Raat.: (1988) Onderwijs in Techniek, afscheidscollege, gegeven op 18 november aan de T.U. Eindhoven.

Rubinstein, S. L.: (1968) Das Denken und die Wege seiner Erforschung; Deutscher Verlag der Wissenschaften. [Berlin]

Schaefer, G.: (1989) Systems thinking in biology education; UNESCO, Paris 1989, Division of Science, Technical and Environmental Education; No. 33, Document Series

Schmidt, V.: (1989) Bewerten technischer Objekte - ein methodologischer Ansatz zur Erschließung der Komplexität der Technik; Thesen zur Dissertation B, Pädagogische Hochschule. [Erfurt]

Szücs, E.: (1992) Technology Education in General Culture. [International conference on technology education, Weimar, Germany]

Theurkauf, W.E.;Weiner, A.: (1992) Schlüsselqualification als eine orientierungsmöglichkeit für eine technische bilding. [Hildesheim]

DESIGN EDUCATION AND SCIENCE: PRACTICAL IMPLICATIONS

R. MCCORMICK
The Open University
United Kingdom

ABSTRACT. When students perform design activities a teacher is faced with a dilemma about how scientific knowledge should be provided and used. When should the students be provided with the necessary science to enable them to carry out the design task? This dilemma, not restricted to scientific knowledge, comes about because the knowledge required in design activities is potentially extensive and unpredictable in nature. The dilemma has implications at three levels: the whole school level, the course level, and the level of an individual project. The article examines these three levels, focusing upon that of the individual project, where the evidence of a number of areas of research is outlined and the implications considered. The article concludes that an effort is needed by science teachers to teach the use of scientific knowledge, and by design educators to recognize the context and domain sensitive nature of cognitive processes such as design.

1. Introduction

When students perform design activities a teacher is faced with a dilemma about how scientific knowledge should be provided and used. When should the students be provided with the necessary science to enable them to carry out the design task? This dilemma, is not restricted to scientific knowledge, comes about because the knowledge required in design activities is potentially extensive and unpredictable in nature. This is exacerbated by activities that offer students a wide choice in the design brief. For example, a student designing an aid for an elderly person with arthritis may choose a wide range of aids, some electrical or electronic and some mechanical, and even if a particular aid is defined (e.g. for opening bottles), a wide range of solutions are possible. Each of the solutions may require different kinds of scientific knowledge (e.g. levers, friction, material properties). Indeed different design briefs may themselves require different knowledge just to understand the context of the 'problem' or design brief.

M. J. de Vries et al. (eds.), Design Methodology and Relationships with Science, 309–319.

This dilemma is manifest at three levels. First, at the level of the whole curriculum, decisions are made about what science is taught at the various stages of a student's education. These decisions are part of the logic and pedagogy of science, and will not necessarily correspond to the needs of design education, wherever that is found in the curriculum. Second, at the level of a particular design education course, decisions about the sequence of certain projects or design tasks may be made on the basis of certain knowledge areas or the requirements of a certain class of tasks (e.g. environmental design as opposed to product design). Both of these levels are essentially curriculum planning issues and, although the problems are particularly acute in schools, they apply equally in higher education. The third level at which the dilemma is faced is at the level of particular design activities or projects. When students are faced with a particular design brief, or are exploring possible designs they may or may not have the scientific knowledge required to carry out the activity adequately. The teacher has to devise a strategy that will make use of what knowledge students have and aid them in developing any they need to acquire.

Although scientific knowledge can be both conceptual (e.g. the concept of force) and procedural (e.g. how to carry out an investigation), my focus in this article will be on conceptual knowledge. The problems of ensuring that pupils have sufficient procedural knowledge (to, for example, test whether a fabric has the required properties of tensile strength and fire resistance) are different because such knowledge is built up over a long period of time. However, it is also true that scientific conceptual and procedural knowledge are linked (Gott & Murphy, 1987), and so problems about developing procedural knowledge are linked to those of conceptual knowledge. The discussion of three levels of the dilemma, which I will deal with below, will be confined to conceptual knowledge. In addition I will focus upon school design education (mainly within the context of technology education), although the main thrust of my arguments apply equally to higher education.

2. The Whole Curriculum Level

The literature on curriculum change attests to the dominance of subjects as the basis for organizing the curriculum (Goodson, 1983 "Goodson, 1983"), and in recent years the policies of many governments have reinforced this by central efforts to define a 'national curriculum' (Skilbeck, 1990XE "Skilbeck, 1990"). This has meant that efforts to encourage cross-curricular activities have often been thwarted by the subject divisions that exist. These divisions are not simply intellectual constructions, but have social dimensions in the communities of teachers they are based upon. Thus, as I have explored elsewhere (McCormick, 1991 "McCormick, 1991"Hacker, Gordon, de Vries), the formation of a cohesive group of teachers to deliver technology education in schools is affected (even hindered) by the subject traditions that have been brought together to form the group (e.g. home economics and craft, design and technology teachers). The implication of the power of subjects is that any initiative that requires some form of activity that crosses subject boundaries is not straightforward.

If design education takes place within, say, the subject of 'technology' there will be a need to draw upon the work of science teachers. At the very least some form of co-ordination needs to be carried out so that the technology teacher will be aware of what students learn and when. This will allow the technology teacher to make assumptions about what science students have encountered, although she can make limited assumptions about what they understand. Thus if a design activity requires an understanding of electric circuits the teacher needs to know what students' starting points are, even if only to decide what further knowledge to provide or develop. A feature of some technology teaching which is based upon design activities, and draws heavily upon science, its that much repetition takes place. Students find themselves learning about energy, for example in the context of a design activity on alternative forms of energy generation, a topic also dealt with in science lessons. There may be nothing inherently wrong with this provided that the two approaches build upon one another and avoid wasteful repetition. More commonly a topic like energy will be dealt with in many parts of the curriculum, using different ideas of energy (in food, in moving objects) and different units (calories and joules). The lack of co-ordination leads to the possibility of confusion on the part of students, who are left with trying to make sense of the various interpretations. Part of the confusion may simply be that the levels of explanation in science lessons is different from that needed in the design activity. Thus if a student is designing a house then the crucial idea about heat loss will be the 'U-value' of the materials, whereas in science an elaborate molecular model may be employed to explain heat flow. The model in science is concerned with explanation, whereas the one in design is for use in design. (I will return to this difference when I consider the third level of the dilemma.) Of course technology teachers, for example, do not hold the monopoly of understanding about what constitutes appropriate science. Science teachers have many years of experience of dealing with students' conceptual confusions and in teaching in such a way so as to avoid them. A teacher supporting a student on a design project may add to the conceptual confusion (even cause it) because she is not so careful in the distinctions etc., that she makes (e.g. by confusing 'mass' and 'weight').

On the face of it remedying this situation is simple; the design teacher needs to find out what the science teacher teaches. Unfortunately schools are not such simple places. A teacher from another department, especially a junior one, asking a head of a science department when and what he teaches about energy, or whatever, may be met with a rebuff that amounts to "mind your own business". Even obtaining the curriculum plans of the science department may be difficult, although where there is a well defined national curriculum (as in some parts of the UK), this should be easier. Any prospect of a co-ordination of the curriculum plans of the science and technology (or design) departments to enable the science to be encountered at an appropriate time for certain design tasks (or visa versa) is unlikely. This is partly because of the isolationism just discussed, but also partly because this co-ordination is just too complex (the geography department could make a similar request to the science department), and would probably disrupt the development of scientific understanding of pupils as seen by the science department.

The isolationism is compounded by the fact that teachers in different departments also have different views on what knowledge might be necessary for a task. Paechter (1992) has investigated this and shown how a 'sub-culture' is set up around teachers of a particular subject, even when they are collaborating with those of other subjects. They therefore take a different view of the requirements of a design task, viewing it from the perspective of the norms of their own subject background. Thus the practical and social problems of ensuring the sensible use of science in a design task are compounded by a cultural-intellectual difference among teachers.

3. The Course Level

Even where a teacher can control the knowledge within a course problems exist. The teacher can provide the knowledge before a design task is attempted, so that the students are equipped to tackle it. Thus in the UK there was the Modular Technology course, where modules of content (e.g. mechanisms, electronics) were followed by project work that allowed students to use in a design activity the concepts developed. The alternative approach is to only teach content on a 'need to know' basis, i.e. when a student reaches a point in a design activity where knowledge is required then the teacher provides that either by individual or group instruction. Both approaches have a sound rationale. Providing the conceptual base first means that the students are able to sensibly use the knowledge, and create a more sophisticated design based upon more than common sense and everyday knowledge. Giving knowledge on a 'need to know' basis increases the chances that students will be able to understand the knowledge and use it successfully in a meaningful context. Both have disadvantages. Providing knowledge first leads to design education becoming science education, and possibly bad science education at that given the experience of some design teachers. Design education will therefore suffer from the problems of abstract and unattractive teaching that science is desperately trying to avoid, and leads to artificially constrained design activities where the solution has to use the knowledge provided in the earlier instruction (an electronic timer would be expected for cooking eggs following a module on electronics whether it was the most appropriate solution or not). Knowledge provided on a 'need to know' basis can be a nightmare for teachers if students are undertaking a variety of design tasks, or choose widely different solutions. Providing a variety of kinds of knowledge 'on demand' can be inefficient and taxing for the teacher. The use of resource material that will allow the student to learn what is necessary independently is possible for older students and where the conceptual density is not too great. The advent of multi-media databases will increase the attractiveness of this approach, but there are issues of accessibility, that I will return to later. Such an approach also runs the risk of students only being able to learn relatively superficial knowledge while they are in the midst of their design activity. Indeed this points to a more profound problem of how students approach learning in individual design tasks, an issue I will now turn to.

4. Design Activity Level

At this level the dilemma of how to provide the scientific knowledge needed in a design activity is informed by our understanding of how that activity takes place. It is common in design education to assume that the capability to design, can be expressed by for example (DES/WO, 1990), being able to:

identify and explore a problem;
generate design ideas;
plan and make the design;
evaluate the design.

This is part of a wider belief in a general problem-solving ability, that sees this ability as independent of the context of the of the problem. In other words it is more important to teaching students to solve problems and this can be done with little reference to content. (See Down, (1983) for a review of this stance.) This approach therefore assumes that conceptual knowledge and procedural knowledge are independent; it does not deny the importance of each but implies that they can be learnt separately. The Assessment of Performance (APU) testing programme in design and technology in the UK recognized the importance of both kinds of knowledge (Kimbell, et al, 1991: 23). But, because of the difficulty of testing students who may have experienced quite different content (and the fact, indicated earlier, that design activity can draw upon a large range of concepts), the APU team effectively separated conceptual and procedural knowledge by assessing only the latter (identifying and clarifying, investigating, planning, generating and developing, appraising - Kimbell, et al, 1991: 30). They in fact switched the concern for assessing conceptual knowledge to mapping it, i.e. finding out what knowledge was used by students. This left them unable to judge the inter-relationship of the two types of knowledge, although they did report profiles (Kimbell, et al, 1991: Section 13) that indicated the relative performance of four areas of concepts (materials, energy, aesthetics, people).

Their approach was in contrast to the Science APU testing team who found a close inter-relationship between conceptual and procedural knowledge (Gott & Murphy, 1987). Thus they found that students' performance in constructing an investigation was affected by the level of their conceptual understanding of the situation being investigated. This finding is in line with recent work on problem solving and general thinking skills (Glaser, 1984; Simpson, 1988). There are several areas of relevant research, namely, novice and expert problem solving, situated cognition, and the application of knowledge. Each of these areas of work has shown the importance of the domain of knowledge and the context of the problem. I will deal with each of these areas of research in turn, drawing upon a review of the literature by my colleague Sara Hennessy (1992)[1]. (See also Layton, 1991) In addition there is the large body of work on constructivist approaches to conceptual learning that has some bearing upon the use of science in design activities.

4.1. NOVICE AND EXPERT PROBLEM SOLVING LITERATURE

This literature is based upon finding out how experts solve problems and comparing them with how novices perform. The important conclusion of this work is that experts perform on the basis of a considerable amount of prior knowledge relevant to the problem, which they acquire in the context of real tasks that make them goal directed. Novices do not share these goals, and do not have the extensive knowledge. In addition novices tend to focus upon the superficial elements of the problem and spend too much time 'doing' rather than 'thinking'. What is more the interaction between the domain-specific knowledge and the procedural knowledge is such that the latter cannot be used effectively without the former. Consequently it is not sensible for novices to learn procedural knowledge outside the context of specific knowledge domains.

4.2. SITUATED COGNITION

This research rejects general cognitive abilities as existing free from the context of specific problems (e.g. Rogoff, 1990: 5-6). Cognitive processes are seen to differ according to the domain of thinking and the specifics of the task context (as was the case in the novice and expert literature). But further than this thinking is seen as a cultural activity. Like the novice and expert literature it sees effective action as being goal directed, but with a social and cultural definition of the goals and the means of handling problems. In reviewing the traditional cognitive psychology research on problem solving that attempted to find evidence of general problems solving abilities which would transfer across contexts, Lave (1988: 23-39) concluded that the evidence of transfer was weak, with strained

---

[1] This review was in preparation for a research programme at the Open University, and a bibliography and review paper will subsequently be made available.

explanations of results. Also the problems posed were of the kind that had correct answers, defined by the experimenters. Real problem solving, however, is characterised by the resolution of dilemmas as part of an ongoing activity where conflicts in values leads to the need to choose between viable alternatives, without right or wrong answers. (A view that would find a great deal of sympathy in the design education world.) Trying to represent such activity as a sequence of 'recognizing a problem, representing it, implementing a resolution, and evaluating the results' is to ignore the multitude of ways problems are tackled and the fact that often some of these activities take place simultaneously (Lave, 1988: 142). The fact that thinking is seen as a cultural activity means that a focus on individual learners is inadequate because learning will be interpersonal. The role of other learners and the teacher is to guide the thinking of an individual. This leads to an apprenticeship model of learning where: a student takes an active role in making use of social guidance; tacit and routine activity are important (rather than instruction); students achieve a shared understanding with those who guide them (Rogoff, 1990: 8).

4.3. APPLICATION OF KNOWLEDGE

In fact a more appropriate title for this area is 'practical knowledge' or 'knowledge in action'. Layton (1991) gives an overview of the issues in this area, based partly on a review of the literature on the history of technology. This history shows how technology has come to be recognized as distinct from science, and that applying scientific knowledge to technology is an art in itself, and certainly not unproblematic. Quoting Staudenmaier (1985), Layton (1991) argues that scientific knowledge needs to be restructured to form technological knowledge, which is geared to the demands of design. He also notes the importance of experiential learning, and of tacit knowledge, in a field of practical action such as technology.

The view that scientific knowledge is not structured for use in design has a parallel in the use of science in everyday activity, whether it be in the home or in the context of special activities, such as the general public tackling an environmental issue. Layton (1991) and Jenkins (1992) both show how such everyday situations are messy, value laden and require context-specific knowledge. Again the knowledge needs to be reworked and translated for everyday use[2]. Jenkins (1992) goes further and questions whether science can be taught as universal knowledge applicable in all situations, thus reinforcing the context-specific nature of knowledge that has been a theme of all the areas of

[2] A recent example of this was given by Brian Wynne (Times Higher Educational Supplement, 10 April 1992, p. 10) where he described the failure to predict that sheep in the Lake District would be affected by the fall-out of Chernobyl. The scientific knowledge upon which the prediction was made considered only the physical mobility of the radiocaesium (the pollutant), whereas it was the chemical mobility that mattered. The scientific research had posed questions and provided answers in a way relevant to what scientists thought was important and this choice involved assumptions that needed to be exposed. In other words the validity of the scientific knowledge was at stake in this practical context.

research. In part the arguments of Layton and Jenkins are an expression of the differences between the views of scientists and technologists (and hence designers) that have their parallel within schools between teachers. This is a point I made when discussing the difference between the views of science and technology teachers, the former who want to explain and the latter who want to use knowledge to design.

What we need, however, is to consider the reality of students using scientific knowledge within a design project. Sadly little research has been undertaken on this topic. Job (1991), in discussing differences between science and technology, gives a sketch of a student who designed a bar-code reader so that a visually impaired person could identify tins of food in the home. This involved some sophisticated electronics to read the bar code and convert it to a voice synthesised output. Yet this student's performance on basic electrical circuits in science was poor. The key to the student's success was the fact that she only needed a qualitative understanding of most concepts, and a grasp of the linkage between a number of functional blocks (not an understanding of the actual functioning of individual blocks). As Job (1991: 104) concludes "This way of using scientific principles poses major challenges for the science curriculum. It is an uncomfortable fact that, certainly for school students, 'wrong' science is as good as, and often better than, 'correct' science in this respect. Aristotelian physics, the caloric theory of heat, fluid theories of electricity can all deliver the technological goods".

This is indeed a challenge, given the extensive work on students' conceptual understanding by constructivists. Their work assumes that students bring to science lessons mis-conceptions of, say heat flow (for example that a liquid cooling in a can only loses heat through the vapour rising from the top), and that some deliberate strategy will have to be employed to change them; they are very resistant to change. Given this difficulty, the use of such concepts in design activities, especially when 'less than correct' science will do, is a tall order.

4.4. THE PRACTICAL IMPLICATIONS OF THE RESEARCH

If design education is based upon teaching a generalized process that applies across any domain of knowledge or context then from the above discussion there must be some doubt as to its success. This is in spite of the attractiveness of this approach as a basis for teaching. Idealized capabilities, such as put forward in the 1992 UK National Curriculum for Technology (design and technology profile component)[3], may be of doubtful value in encouraging the 'designerly ways of knowing' advocated in design education (Cross, 1982). Equally teaching science as universal knowledge, unrelated to how it will be used, may be doomed, assuming the use of such knowledge is valued. We need more work on how

---

[3] Currently it is under revision and not only may the Attainment Targets change, but knowledge, especially from science and mathematics, may be emphasised.

science is "deconstructed" (Layton, 1991: 62-3)[4], for use by students in situations such as is found in design activities. A more intimate relationship is needed between science teaching and activities like design that require the use of scientific knowledge, in the face of the lack of co-ordination noted in the discussion on the whole curriculum level.

If learning to design is "guided participation", to use an idea Rogoff (1990: 8) thought was the key to her idea of apprenticeship, then both the role of the teacher and of other students may need to be re-examined. In schools the idea of apprenticeship, common in craft teaching in the past, is unfashionable and not associated with cognitive aspects of learning. Further the stress on individual work in design, driven in part by the assessment system, reduces the possibility of the inter-personal context being important. In this regard the team work of industry, and its reliance on expert roles for providing knowledge, may be a good model to follow.

5. Conclusion

The dilemma posed at the beginning of this article has not been resolved. But questions have been raised about the practicality of students being able to pick up scientific knowledge on a 'need to know' basis, on the assumption that the process of design is all important. Knowing what you need to know, a common view in design (e.g. Kimbell, et al, 1991), already implies some conceptual understanding of the domain. Evidently more research is needed to see how the results of the work, in for example situated cognition, apply in the area of design[5]. The implications are, as I noted above, that a more intimate relationship is required between science teaching and design education, which in turn requires changes in the relationships among teachers at the whole curriculum level. At present such relationships are poor in the UK. In the United States of America there are more curriculum projects and the like that are attempting to combine mathematics, science and technology teaching (e.g. Bachmeyer et al, 1990), but these are weak on design and do not appear to be informed by the literature reviewed here.

Multi-media databases, although having some limitations noted earlier, are being developed with models of student use that are sensitive to the importance of context, while trying to encourage general investigative procedures. (See McCormick (1992) for a review of such developments.) These may allow some kind of apprenticeship model, however, at present they are not being developed specifically for design activities, rather they are specific to knowledge domains.

---

[4] This is opposite to the process of 'construction' that science teachers currently focus on in aiding students' conceptual development (hence the name constructivists).

[5] Some of this work has its origins in mathematics education (e.g. Lave, 1988).

Given our present state of understanding of the use of scientific knowledge in design activity it is likely that we will have to look for simpler solutions which recognize the need to consciously address the use of such knowledge.

References

Bachmeyer, S. Durbahn, M. Gauger, R. Olson, C. (1990) 'Science, Math, and Technology teachers learning together', The Technology Teacher 50(2), 25-9.

Cross, N.(1982) 'Designerly ways of knowing', Design Studies 3(4), 221-7.

Department of Education and Science/ Welsh Office (DES/WO) (1990) Technology in the National Curriculum, Her Majesties' Stationary Office, London.

Down, B.K. (1983) 'Problem-solving, CDT and child-centredness', Studies in Design Education, Craft and Technology 16(1), 38-43.

Glaser, R. (1984) 'Education and thinking: the role of knowledge', American Psychologist 39(2), 93-104.

Goodson, I.F. (1983) School Subjects and Curriculum Change, Croom Helm, London.

Gott, R. and Murphy, P. (1987) Assessing Investigations in Science at Ages 13 and 15, Association of Science Education, London.

Hennessy, S. (1992) Discussion Paper No.1 for Problem Solving in Technology Project Team, (Second draft), The Open University, Milton Keynes (unpublished mimeograph).

Jenkins, E. (1992) 'Knowledge as action: science as technology?' in R. McCormick, P. Murphy, M.E. Harrison (eds.) Teaching and Learning Technology, Addison-Wesley, London.

Job, G.C. (1991) 'The relationship between science and technology in the school entitlement curriculum', in M. Hacker, A. Gordon, M. de Vries (eds.) Integrating Advanced Technology into Technology Education, Springer-Verlag, Berlin, pp. 93-105.

Kimbell, R. et al (1991) The Assessment of Performance in Design and Technology, Schools Examination and Assessment Council, London.

Lave, J. (1988) Cognition in Practice: mind, mathematics and culture in everyday life, Cambridge University Press, New York.

Layton, D. (1991) 'Science education and praxis: the relationship of school science to practical action', Studies in Science Education, 19, 43-79.

McCormick, R. (1991) 'The evolution of current practice in technology education', in M. Hacker, A. Gordon, M. de Vries (eds.) Integrating Advanced Technology into Technology Education, Springer-Verlag, Berlin, pp. 41-55.

McCormick, R. (1992) Integrating Advanced Educational Technology into Technology Education, Paper presented to NATO Advanced Study Institute, 17-28 August, University of Salford, UK.

Paechter, C. (1992) 'Subject subcultures and the negotiation of open work', in R. McCormick, P. Murphy, M.E. Harrison (eds.) Teaching and Learning Technology, Addison-Wesley, London.
Rogoff, B. (1990) Apprenticeship in Thinking: cognitive development in social context, Oxford University Press, New York.
Simpson, M. (1988) 'Improving learning in schools - what do we know? A cognitive science perspective', Scottish Educational Review, 20(1), 22-31.
Skilbeck, M. (1990) Curriculum Reform: an overview of trends, OECD, Paris.
Staudenmaier, J.M. (1985) Technology's Storytellers: reweaving the human fabric, MIT Press, Cambridge MA.

LIST OF AUTHORS AND RAPPORTEURS

NATO-countries

Prof.dr. S. Kasse
Babbage Institute for Knowledge and Information Technology
J. Plateaustraat 22
B-9000 Gent
Belgium

Prof. W.E. Eder
Royal Military College of Canada
Dept. of Mechanical Engineering
Kingston, Ontario K7K 5L0
Canada

Prof. R. Levy
Université de Montreal
Dept. Médicine Sociale et Préventive
Pavillon Marguerite d' Youville, C.P. 6128, succersale 'A'
Montreal, Quebec H3C 3J7
Canada

Prof.dr.ing. D. Blandow
Päedagogische Hochschule Erfurt
Dept. Technik/Technologie
50 Brühler Herrenberg 28
Erfurt 5063
Federal Republic of Germany

Dr. J. Schlattmann
Universität Paderborn
Lab. für Konstruktionslehre
Pohlweg 47/49
4790 Paderborn
Federal Republic of Germany

Dr. F. Le Guet Tully
Observatoire de la Cote d' Azur
CERGA-URA 1360
BP 229
06304 Nice Cedex 4
France

Dr. P.A. Kroes
Eindhoven University of Technology
Fac. of Philosophy and Social Sciences
P.O. Box 513
5600 MB Eindhoven
Netherlands

Dr. S.L. Marzano
Philips Consumer Electronics
P.O. Box 80002
5600 JB Eindhoven
Netherlands

Ir. N. Roozenburg
Delft University of Technology
Fac. of Industrial Design Engineering
Jaffalaan 9
2628 BX Delft
Netherlands

Prof.dr. A. Sarlemijn
Eindhoven University of Technology
Fac. of Philosophy and Social Sciences
P.O. Box 513
5600 MB Eindhoven
Netherlands

Dr. M.J. de Vries
Eindhoven University of Technology
Fac. of Philosophy and Social Sciences
P.O. Box 513
5600 MB Eindhoven
Netherlands

Dr. H.P. Hildre
University of Trondheim
The Norwegian Institute of Technology
Division of Machine Design
R. Birkelandsv. 2B
7034 Trondheim
Norway

Prof.dr. N. Bayazit
Istanbul Technical University
Dept. of Architecture
ITU Mimarlik Fak. Teknik Universite, Taksim
Istanbul 80191
Turkey

Prof.dr. N. Cross
Open University
Fac. of Technology/Design Discipline
Walton Hall
Milton Keynes, MK7 6AA
UK

Mr. J. Heinen
Brunel University of West London
Dept. of Manuf. & Engin. Systems
Uxbridge, Middlesex UB8 3PH
UK

Dr. R. McCormick
Open University
School of Education
Walton Hall
Milton Keynes, MK7 6AA
UK

Prof. R. Buchanan
Carnegie Mellon University
Dept. of Design
110 Margaret Morrison Hall
Pittsburg, PA 15213-3890
USA

Dr. M.D. Eckersley
University of Minnesota
Department of Design
240 McNeal Hall, 1985 Buford Av.
St Pauls, Minesota 55108
USA

Dr. D.P. Grant
California Polytechnic State University
Architecture Dept.
One Grand Avenue
San Luis Obispo, CA 93407
USA

Prof. W.R. Spillers
New Jersey Institute of Technology
Dept. of Civil & Environmental Engineering
Newark, NJ 07102
USA

Non-NATO countries

Prof.dr. W.W. Gasparski
Polish Academy of Sciences
Institute of Philosophy and Sociology
ul. Nowy Swiat 72
00-330 Warszawa
Poland

SUBJECT INDEX

MIX
Papier aus verantwortungsvollen Quellen
Paper from responsible sources
FSC® C105338

If you have any concerns about our products,
you can contact us on
ProductSafety@springernature.com

In case Publisher is established outside the EU,
the EU authorized representative is:
Springer Nature Customer Service Center GmbH
Europaplatz 3, 69115 Heidelberg, Germany

Printed by Libri Plureos GmbH
in Hamburg, Germany